Theoretical Physics

Theoretical Physics

V.K. Mathur

Theoretical Physics

ISBN 978-93-5111-968-5

Published in 2016 in India by

RANDOM PUBLICATIONS

4376-A/4B, Gali Murari Lal, Ansari Road
New Delhi-110 002
Phone : +9111-43580356, 011-23289044, 011-43142548
e-mail: sales@randompublications.com,
info@randompublications.com, randomexports@gmail.com

Reprinted 2026

Type Setting by : Friends Media, Delhi-110089

Preface

Theoretical Physics is that branch of physics, which deals with natural phenomena, and their relations with Physical properties of matter. It holds a paramount importance in life and has helped define a number of laws for the benevolence of mankind in areas of business construction, mechanism, traveling, defence, communication and so on.

Today's theoretical physicists are often working on the boundaries of known mathematics, sometimes inventing new mathematics as they need it, like Newton did with calculus.

. Newton was both a theorist and an experimentalist. He spent many many long hours, to the point of neglecting his health, observing the way Nature behaved so that he might describe it better. The so-called "Newton's Laws of Motion" are not abstract laws that Nature is somehow forced to obey, but the observed behaviour of Nature that is described in the language of mathematics. In Newton's time, theory and experiment went together.

. Today the functions of theory and observation are divided into two distinct communities in physics. Both experiments and theories are much more complex than back in Newton's time. Theorists are exploring areas of Nature in mathematics that technology so far does not allow us to observe in experiments. Many of the theoretical physicists who are alive today may not live to see how the real Nature compares with her mathematical description in their work. Today's theorists have to learn to live with ambiguity and uncertainty in their mission to describe Nature using math.

The book will also serve for a foundation course for allied subjects such as astrophysics, geophysics, meteorology, laser physics and plasma physics. Rather, it is a book with a variety of tools for improving both teaching and learning of physics.

– ***Author***

Contents

1

Atomic Spectra

INTRODUCTION

It was noticed as early as 1855 by Bunsen and Kirchoff that when atoms of a particular element, say hydrogen, are energized by heating or by electrical discharge, they emit (give off, or produce) light of a characteristic colour. In the case of hydrogen, the light is pale magenta; in the case of neon, it is orange; and so on. Further, when this light is collimated (that is, gathered into a narrow beam by passage through a slit) and then passed through a prism, it is found that only certain wavelengths of light emerge from the prism, with wide regions of darkness at wavelengths between the specifically emitted values. The collection of emitted lines is called the emission spectrum of the element. (It is also called a line spectrum, because the emitted beams take on the shape of the slit through which they are passed.)

Figure shows the line spectrum of helium. Most light sources (for example, the sun and stars, flames, and so on) produce a continuous spectrum of wavelengths when passed through a prism, just like the rainbow produced by passage of sunlight through mist or rain. That is, all wavelengths in the visible region of the spectrum are produced, with none missing. The discrete line spectra of atoms were therefore considered unusual, and could not be readily explained.

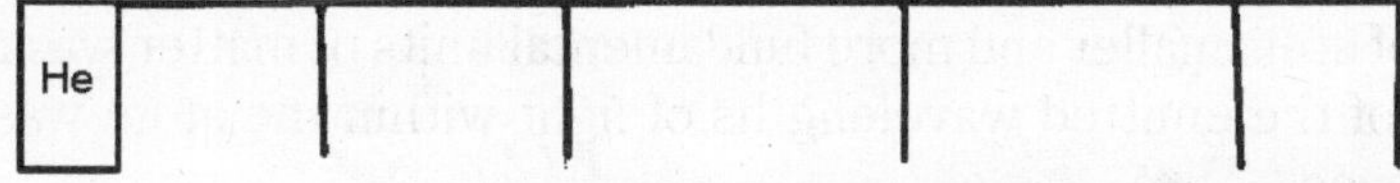

Fig. Lline Spectrum of Helium

The discrete emission spectra of certain atoms meshed nicely with an observation that had been made in 1814 by Fraunhofer. Using a high quality prism, he discovered that the spectrum of light from the sun, although essentially continuous, contains a number of dark (black) lines representing wavelengths where no light is seen. The positions of these dark lines matched exactly the positions of lines in the emission spectra of a number of elements, notably hydrogen, helium, and sodium.

The eventual interpretation of the Fraunhofer lines is that they result from absorption of light at the dark wavelengths by elements present in the outer portion of the gaseous cloud of the sun. The Fraunhofer lines constitute the absorption spectrum of these elements. The absorption and emission spectra of an element are complementary. The absorption spectrum shows dark lines against a bright background. The dark lines are at wavelengths where light is absorbed, and thus not seen.

The emission spectrum shows bright lines against a dark background. The positions of the lines in the two types of spectra coincide exactly. Line spectra were studied for many years without adequate explanation. A number of scientists focussed on the spectrum of hydrogen, which appeared the simplest. The series of lines falling in the visible region, called the Balmer series.

The series is so named for Jacob Balmer, who found that the reciprocal wavelengths of the lines obey equation.

$$1/\lambda = R_H(1/2^2 - 1/n^2)$$

R_H is a constant with value 1.09678×10^{-2} nm^{-1}. Note that reciprocal wavelength is closely related to frequency, hence energy, via equations. Balmer was unable to explain the meaning of the integers occuring in the equation. Nonetheless, their occurence, and the very simple form of equation, are remarkable.

By recording the emission spectrum of hydrogen on light sensitive film, scientists were able to "see" not only the visible emission lines, but also emission lines in the ultraviolet and infrared regions of the electromagnetic spectrum. Thus between 1906 and 1924, four additional series of lines were discovered. All of the observed spectral series of the hydrogen atom were found to obey a generalized version called the Rydberg equation below:

$$1/\lambda = R_H(1/n_1^2 - 1/n_2^2)$$

The unified fit of the hydrogen spectrum provided by equation suggested a regular internal structure to the atom that is changed in some way by interaction with light. Nothing was known about this internal structure in 1885, when Balmer proposed his equation. In fact, the knowledge that atoms are composed of still smaller and more fundamental units of matter was not known. The origin of the emitted wavelengths of light within the atom was therefore not understood.

The first definite knowledge of the substructure of the atom emerged from the experiments of JJ Thomson in 1897. Using a device known as a cathode ray tube, Thomson demonstrated that the "cathode ray" consists of a beam of negatively charged particles that emerge from the metal cathode of the tube under an applied voltage. He called these particles electrons. By observing the manner in which the beam deflected when subjected to magnetic and electric fields of known strengths, Thomson was able to calculate a value of about 10^8

for the charge-to-mass ratio of the electron. The modern value is given in equation:

$$\text{charge/mass} = e/m = 1.76 \times 10^8 \text{ C/g}$$

Subsequent experiments by Robert Millikan between 1908 and 1917 provided the value 1.6×10^{-19} coulombs for the charge, leading to a mass of 9.11×10^{-28} g. Subsequent to Thomson's work, the atom was viewed as a uniform distribution of electrons embedded in a uniform distribution of positive charge, required to be present to ensure the electrical neutrality of the atom. However, in 1911 Rutherford published the so-called nuclear model of the atom, based on the results of experiments in which he and his students bombarded thin gold foil with alpha particles, which are the nuclei of helium atoms.

The results of these experiments were interpreted by Rutherford as indicating that the positive charge of the atom was concentrated in a tiny and extremely dense region at the centre of the atom, which he called the nucleus. He proposed that virtually all of the mass of the atom resides in the nucleus. The electrons were then proposed to orbit the nucleus, much as the planets orbit the sun.

ATOMIC SIZE

The molar volume for an element, in units of mL/mole can be calculated readily by dividing the atomic weight in g/mole by the density of a condensed phase (solid or liquid) of the element in g/mL. When molar volume is plotted as a function of atomic number, Figure results.

Examination of the plot shows two clear trends. First, molar volume tends to decrease from left to right across a period. Second, molar volume increases top to bottom down a family. If we make the reasonable assumption that there is a relationship between the molar volume and the size of an atom of the element, then the second trend is expected, but the first is counter-intuitive. Most of us would probably predict that atoms should get larger as the number of electrons increases across a period.

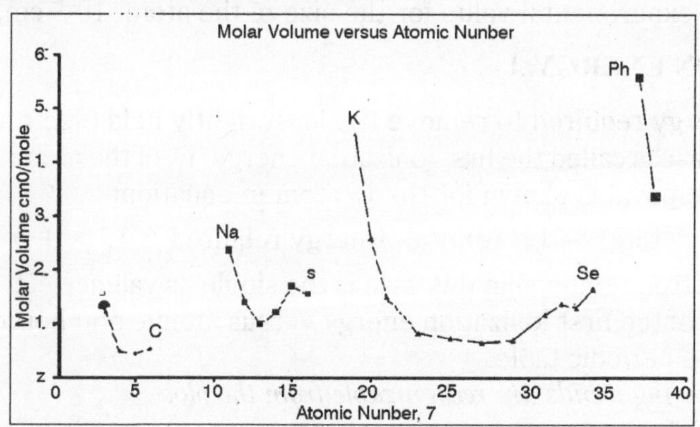

This is quite readily understood in terms of the trend in Z_{core} discussed above. Across a row, the added electrons enter the same main shell while experiencing an ever increasing pull from the nucleus.

The result is a steady shrinkage of the electron cloud. The elements of a family (column) of the periodic table have the same number of electrons in the shell of largest n. For example, the elements of group 2 all have 2 electrons in the outermost main shell. But they apparently do not. The results of x-ray diffraction experiments have given us fairly reliable values for the radii of atoms of many elements. Some representative values are presented in Table. It is clear from the data that atomic radii do indeed decrease left to right across a row of the periodic table.

Table. Atomic Radii, pm (Values are Van der Waals radii)

H	132						
Li180	Be---	B---	C168	N155	O150	F155	Ne 160
Na230	Mg170	Al---	Si210	P185	S180	Cl180	Ar190
K 280	Ca---				Se190	Br190	Kr 200
							Xe 220

Thus the core charge is the same for all members of a family. The vertical trend in atomic size therefore reflects the increasing distance from the nucleus that accompanies an increase in the value of the principle quantum number, n. Before ending this discussion of periodic trends in atomic size, it is useful to make a rough calculation of the size of an atom from the experimentally measured atomic volume.

Mercury, a liquid metal, has a density of 13.6 g/mL and atomic weight 200.6 g/mole. The molar volume of mercury is thus,

$$(200.6 \text{ g/mole})/(13.6 \text{ g/mL}) = 14.8 \text{ mL/mole}$$

We divide by Avogadro's number to convert this to an atomic volume: (14.8 mL/mole)/(6.02×10^{23} atoms/mole) = 24.5×10^{-24} cm^3

To approximate the atomic radius, we take the cube root of the atomic volume: $(24.5\times10^{-24})^{1/3} = 3\times10^{-8}$ cm. This very simple calculation provides a very approximate experimental value for the size of the atom: 10^{-8} cm, or 100 pm.

IONIZATION ENERGY, I

The energy required to remove the least tightly held electron of an atom in the gas phase is called the first ionization energy, I_1, of the atom. The process of electron removal is shown for the Li atom in equation:

$$Li(g) \rightarrow Li^+(g) + e\text{- Energy required} = I_1 > 0$$

The electron removed in this case is the single 2s valence electron. Figure shows a plot of the first ionization energy versus atomic number for the first 3 periods of the periodic table.

The following trends are recognizable from the plot:

- Ionization energy tends to increase left to right across a period. For example, look at the section of the curve corresponding to period 2 (Z = 3 through 10). However there are downward "jogs" at boron and at oxygen.
- Ionization energy smoothly decreases down a family, with no jogs.

The force exerted on the outer electron by the core charge that it feels is given by Coulombs Law:

$$F = -Z_{core}e^2/r^2$$

Here Z_{core} is the core charge felt by the electron, e is the charge of the outermost electron, and r is the average distance of this electron from the nucleus. It is clear from equation that the force, and therefore the required input energy, increases as Z_{core} increases. We have seen above that Z_{core} increases steadily from left to right across a period. Thus F and I_1 are expected to increase also.

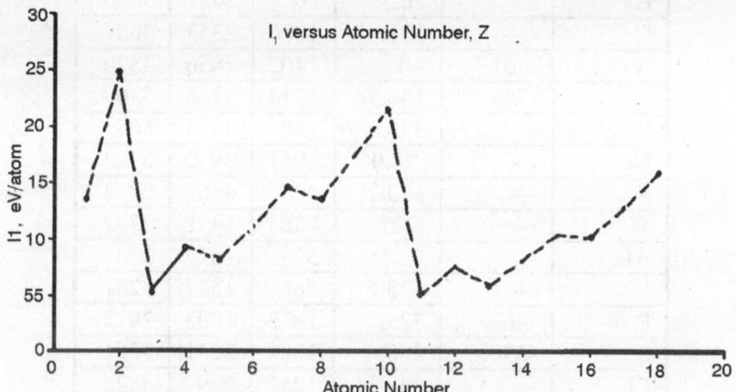

Fig. Ionization Potential versus Atomic Number

The overall increasing trend is readily understood. Similarly, the force decreases as r increases. Since we have seen that r increases with increasing n down a family, the decrease in I_1 down the family is also understandable in terms of 2-6-3. For the time being we will not concern ourselves with the "jogs" in the left-to-right trend in I_1. We will provide an explanation for them at the point that they become important. It is of course possible to remove more than one electron from an atom. Sequential removal of the first 3 electrons from an atom, E, is represented in equations:

$$E(g) \rightarrow E^+(g) + e \text{ Energy required} = I_1$$

$$E+(g) \rightarrow E^{2+}(g) + e \text{ Energy required} = I_2$$

$$E^{2+}(g) \rightarrow E^{3+}(g) + e \text{ Energy required} = I_3$$

The energies required to remove the second and third electrons are called the second ionization energy, I_2 and the third ionization energy, I_3, respectively. No matter what particular atom E represents, it is invariably true that $I_3 > I_2 >$

$I_1 > 0$. Of course, it requires the input of energy to remove the outermost electron of the atom because the electron must be pulled away from the effective positive nuclear charge that it feels.

The second electron is therefore harder to remove because it must be pulled away from the positive ion, E^+. For the same reason, the third electron is even harder to remove than the second. The first 3 ionization energies for the elements through Ar are given in Table.

Table. Ionization Energies and Electron Affinities of the Elements, kJ/mole

Element	$EA_2(I_{-1})$	$EA_1(I_0)$	I_1	I_2	I_3
H	---	---	1312		
He	---	---	2372	5250	
Li	---	59.6	520	7298	11815
Be	---	-241	899	1757	14848
B	---	26.7	801	2427	3660
C	---	121.85	1086	2353	4621
N	-801	<0	1402	2856	4578
O	-780	140.98	1314	3388	5300
F	---	328.0	1681	3374	6051
Ne	---	-28.9	2081	3952	6122
Na	---	52.87	496	4562	6912
Mg	---	-231	738	1451	7732
Al	---	42.55	578	1817	2745
Si	---	133.6	786	1577	3231
P	-463	72.0	1012	1903	2912
S	-590	200.4	1000	2251	3361
Cl	---	349	1251	2297	3822
Ar	---	-34.7	1520	2666	3931

When the first electron is removed, the remaining electrons are pulled in more tightly to the nucleus because there is now a net positive charge of 1 unit on the atom.

Example: The ionization energy of hydrogen is 1312 kJ/mole. This is roughly the amount of energy that you would spend in curling a 220-lb (100 kg) barbell 1000 times. How many times would you have to curl the barbell in order to produce 1 mole of Mg^{2+} ions from Mg atoms? Solution. The sum of I_1 and I_2 for Mg is 2189 kJ/mole. This is 1.67 times larger than I for hydrogen. 1700 curls will do it.

Study of the table provides an explanation for a statement presented. Elements in Groups 1, 2, and 13 of the periodic table form cations with positive charge equal to the last digit of the group number. Another way to say this is that in forming compounds, these elements lose all of the electrons in the outer main shell, but do not lose any of the electrons in lower shells.

Successive values of the ionization energies for these elements show why. For Mg, I_2 is larger than I_1, for reasons discussed above, but by a factor of only

2. Loss of two electrons depletes the outer main shell of the magnesium atom.

If further electrons are to be removed, they must be taken from the next lower (n = 2) shell, which is substantially closer to the nucleus than the n = 3 shell. Consistent with this, I_3 is more than 5 times larger than I_2! The cost in energy to remove the third electron is too high. Mg therefore stops at the Mg^{2+} cation. The sodium atom has only one electron in the outer shell. Its removal requires the energy, I_1, which is only 496 kJ/mole. However, removal of a second electron from Na+ requires disruption of the n = 2 shell. Roughly 10 times more energy is required to accomplish this than is required to remove the first. Again, the cost is too high, and sodium stops at Na+.

ELECTRON AFFINITY, EA

The electron affinity is closely related to ionization energy. It provides a measure of the tendency of an atom to gain, rather than lose an electron. The first electron affinity, EA_1, is defined as the energy required to remove an electron from the anion, E-, forming the neutral atom E:

$$E^-(g) \rightarrow E(g) + e \quad \text{Energy required} = EA_1$$

This equation is written analogously to equations; that is, it shows loss of an electron by a species, forming a related species having one more positive (or one less negative) charge. Thus the first electron affinity is sometimes called the zeroth ionization energy, I_0 (because a species of zero charge is formed). The second electron affinity (also I_{-1}) is defined analogously:

$$E^{2-}(g) \rightarrow E^-(g) + e \quad \text{Energy required} = EA_2 \text{ or } I_{-1}$$

As explained above, successive removal of electrons becomes progressively more and more difficult because the electron must be pulled away from a centre of increasing positive charge. This is no less true of species that are initially negatively charged. Thus I_{-1} and I_0 fit in the expected way into the series of ionization energies:

$$I_3 > I_2 > I_1 > I_0 > I_{-1}$$

In contrast to ionization energies, however, electron affinities are not always positive. That is, for some species, E-, the electron comes off spontaneously. No energy need be put in to cause it to happen. Instead, energy is produced as the electron comes off. EA_1 is positive for some elements towards the right of the periodic table, but is in fact negative for many elements. EA_2 is negative for all elements.

Just as the trend in values of ionization energies serves to rationalize the observed positive charges of cations from groups 1,2, and 13, electron affinities rationalize the complementary rule: that elements from groups 15-17 tend to form anions with a number of negative charges equal to (8–group number).

Another way to say this is that these elements can add electrons to the point of filling the current main shell, despite the fact that EA is assuredly negative for any electrons past the first one added; but they will not add more electrons than this, because these would have to occupy a new main shell, further from the nucleus, where the added electrons feel a greatly diminished attraction from the nucleus. Supplement. Experimental Evidence for the Existence of Shells—Photoelectron Spectroscopy. Spectroscopy is the study of the interaction of light and matter. Our understanding of atoms and molecules is largely attributable to spectroscopy.

In this section, we briefly discuss Photoelectron Spectroscopy, PES, in which light is used to eject electrons from atoms (or molecules).

By subtracting the kinetic energies of the ejected electrons from the energy of light used to eject them, the ionization energies of the electrons can be calculated:

Energy of photon = ionization energy of electron + kinetic energy of electronhn = I + $KE_{electron}$

By varying the frequency of light used, electrons from various shells of the atom can be removed. This gives a measure of the energies levels of the shells. This idea is illustrated schematically in Figure.

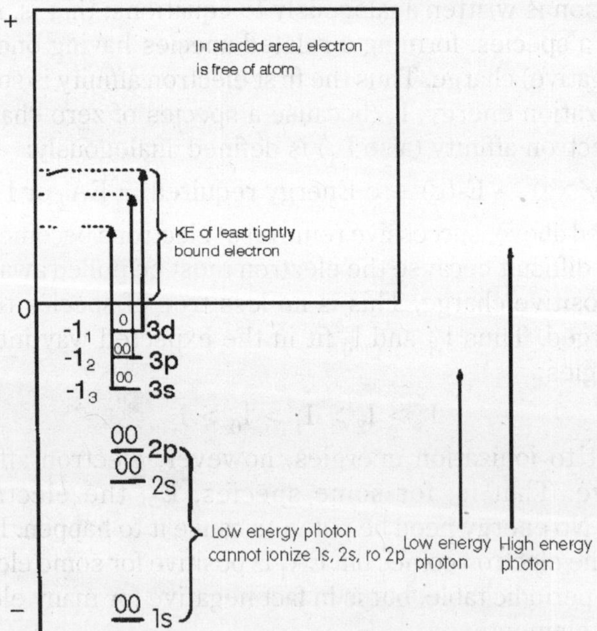

Fig. Photoelectron Spectroscopy

Table shows ionization energies obtained by photoelectron spectroscopy for the first 10 elements of the periodic table, H through Ne. Let's see what we can make of these data.

First, we see that H has only one ionization energy; we would expect this, because it has only one electron. Similarly, He has only one ionization energy, even though it has two electrons, but the intensity of the photoelectron signal is twice that for hydrogen.

This indicates that there are twice as many electrons susceptible to ejection by a photon of light. A single ionization energy is understandable if the two electrons are just alike; that is, if they occupy the same subshell. This is consistent with our understanding of helium, for which we write the electron configuration $1s^2$.

We also see that the single ionization energy for He is larger than that for hydrogen. This is perfectly consistent with its greater core charge, Z_{core}. For Li, there are two distinct ionization energies with intensities in the ratio 1 to 2. The first ionization energy is of relatively small magnitude, corresponding to an electron that is easily removed.

The second ionization energy is quite substantial in magnitude, corresponding to two electrons that are much more difficult to remove. This is in accord with the quantum picture, in which there is a single electron in the outer n=2 shell, and two electrons in the inner filled n=1 shell. The PES for Be gives similar information.

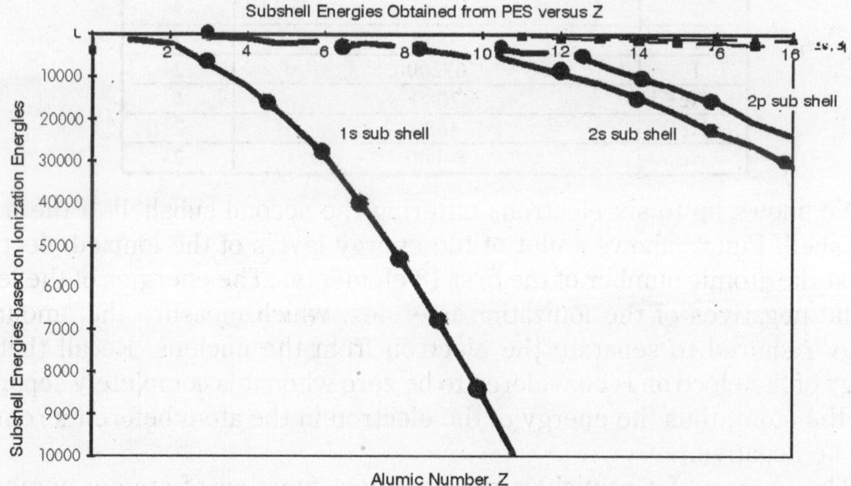

However, for boron, three types of electrons are observed, in relative numbers 1, 2, and 2. There is a single electron that is relatively easily removed. There are then two electrons that are also relatively easily removed (about the same as the single electron of hydrogen), though more difficult than the first. Finally, there is a pair of electrons that is very tightly held by the nucleus of the boron atom.

These results are consistent with a picture in which the the three easily ionized electrons are in two different subshells of the n = 2 main shell. The remaining two electrons are then deeply buried in the core shell with n = 1. Proceeding from B to.

Table. Results of Photoelectron Spectroscopy for H through Ne

Element	*Ionization Energy of Electron, kJ/mole*	*Intensity*
H	1310	1
He	2370	2
Li	520	1
Li	6260	2
Be	900	2
Be	11500	2
B	800	1
B	1360	2
B	19300	2
C	1090	2
C	1720	2
C	28600	2
N	1400	3
N	2450	2
N	39600	2
O	1310	4
O	3040	2
O	52600	2
F	1680	5
F	3880	2
F	67200	2
Ne	2080	6
Ne	4680	2
Ne	84000	2

Ne shows up to six electrons entering the second subshell of the n = 2 main shell. Figure shows a plot of the energy levels of the ionized electrons against the atomic number of the first 18 elements. (The energies of the levels are the negatives of the ionization energies, which measure the amount of energy required to separate the electron from the nucleus. Recall that the energy of the electron is considered to be zero when it is completely separated from the atom; thus the energy of the electron in the atom before its removal must be negative.)

The energy of a particular subshell as a function of atomic number is indicated by connecting the appropriate points with a curve.

Several things are evident from the plot:

- The shell, subshell structure of the atom is obvious;
- The energy of a particular subshell (*e.g.*, 1s) decreases very dramatically as Z increases;
- The gap in energy between main shells increases dramatically as Z increases;
- The gap in energy between the subshells of a particular main shell increases substantially as Z increases across a row of the periodic table.

- The ionization energies of electrons in the valence shell fall within a fairly narrow range (500-5000 kJ/mole, judging from Figure). This range is a measure of the energies involved in chemical reactions.

SINGLE ELECTRON IN AN ATOM

Let us begin with atoms that contain only a single electron. Hydrogen is of course the only electrically neutral species of this kind, but by removing electrons from heavier elements we can obtain one-electron ions such as He^+ and Li^{2+}, etc.

The most widely studied set of quantum numbers is that for a single electron in an atom: because it is not only useful in chemistry, being the basic notion behind the periodic table, valence and a host of other properties, but also because it is a solvable and realistic problem, and, as such, finds widespread. In non-relativistic quantum mechanics the Hamiltonian of this system consists of the kinetic energy of the electron and the potential energy due to the Coulomb force between the nucleus and the electron.

The kinetic energy can be separated into a piece which is due to angular momentum, J, of the electron around the nucleus, and the remainder. Since the potential is spherically symmetric, the full Hamiltonian commutes with J^2. J^2 itself commutes with any one of the components of the angular momentum vector, conventionally taken to be J_z. These are the only mutually commuting operators in this problem; hence, there are three quantum numbers.

These are conventionally known as:

- The principal quantum number ($n = 1, 2, 3, 4...$) denotes the eigenvalue of H with the J^2 part removed. This number therefore has a dependence only on the distance between the electron and the nucleus (*i.e.*, the radial coordinate, r). The average distance increases with n, and hence quantum states with different principal quantum numbers are said to belong to different shells.
- The azimuthal quantum number ($l = 0, 1... n-1$) (also known as the angular quantum number or orbital quantum number) gives the orbital angular momentum through the relation $L^2 = \hbar^2 l(l + 1)$. In chemistry, this quantum number is very important, since it specifies the shape of an atomic orbital and strongly influences chemical bonds and bond angles. In some contexts, l=0 is called an s orbital, l=1, a p orbital, l=2, a d orbital and l=3, an f orbital.
- The magnetic quantum number ($m_l = -l, -l+1... 0... l-1, l$) is the eigenvalue, $L_z = m_l\hbar$. This is the projection of the orbital angular momentum along a specified axis.

Results from spectroscopy indicated that up to two electrons can occupy a single orbital. However two electrons can never have the same exact quantum state nor the same set of quantum numbers according to Hund's Rules, which addresses the Pauli exclusion principle. A fourth quantum number with two possible values was added as an *ad hoc* assumption to resolve the conflict; this supposition could later be explained in detail by relativistic quantum mechanics and from the results of the renown Stern-Gerlach experiment.

- The spin projection quantum number (m_s = −1/2 or +1/2), the intrinsic angular momentum of the electron. This is the projection of the spin $s=1/2$ along the specified axis.

Example: The quantum numbers used to refer to the outermost valence electron of the Fluorine (F) atom, which is located in the 2p atomic orbital, are; $n = 2, l = 1, m_l = 1$, or 0, or −1, $m_s = -1/2$ or 1/2. Note that molecular orbitals require totally different quantum numbers, because the Hamiltonian and its symmetries are quite different.

QUANTUM NUMBERS WITH SPIN-ORBIT INTERACTION

When one takes the spin-orbit interaction into consideration, l, m and s no longer commute with the Hamiltonian, and their value therefore changes over time. Thus another set of quantum numbers should be used.

This set includes:

- The total angular momentum quantum number (j = 1/2,3/2... n−1/2) gives the total angular momentum through the relation $J^2 = \hbar^2 j(j + 1)$.
- The projection of the total angular momentum along a specified axis (m_j = -j,-j+1... j), which is analogous to m, and satisfies $m_j = m_l + m_s$.
- Parity. This is the eigenvalue under reflection, and is positive (*i.e.* +1) for states which came from even l and negative (*i.e.* -1) for states which came from odd l. The former is also known as even parity and the latter as odd parity

For example, consider the following eight states, defined by their quantum numbers:

- $l = 1, m_l = 1, m_s = +1/2$
- $l = 1, m_l = 1, m_s = -1/2$
- $l = 1, m_l = 0, m_s = +1/2$
- $l = 1, m_l = 0, m_s = -1/2$
- $l = 1, m_l = -1, m_s = +1/2$
- $l = 1, m_l = -1, m_s = -1/2$
- $l = 0, m_l = 0, m_s = +1/2$
- $l = 0, m_l = 0, m_s = -1/2$.

The quantum states in the system can be described as linear combination of these eight states. However, in the presence of spin-orbit interaction, if one wants to describe the same system by eight states which are eigenvectors of the Hamiltonian, we should consider the following eight states:

- $j = 3/2, m_j = 3/2$, odd parity (coming from state (1))
- $j = 3/2, m_j = 1/2$, odd parity (coming from states (2) and (3))
- $j = 3/2, m_j = -1/2$, odd parity (coming from states (4) and (5))
- $j = 3/2, m_j = -3/2$, odd parity (coming from state (6))
- $j = 1/2, m_j = 1/2$, odd parity (coming from state (2) and (3))
- $j = 1/2, m_j = -1/2$, odd parity (coming from states (4r) and (5))
- $j = 1/2, m_j = 1/2$, even parity (coming from state (7))
- $j = 1/2, m_j = -1/2$, even parity (coming from state (8)).

ELEMENTARY PARTICLES

For a more complete description of the quantum states of elementary particles. Elementary particles contain many quantum numbers which are usually said to be intrinsic to them. However, it should be understood that the elementary particles are quantum states of the standard model of particle physics, and hence the quantum numbers of these particles bear the same relation to the Hamiltonian of this model as the quantum numbers of the Bohr atom does to its Hamiltonian.

In other words, each quantum number denotes a symmetry of the problem. It is more useful in field theory to distinguish between spacetime and internal symmetries. Typical quantum numbers related to spacetime symmetries are spin the parity, C-parity and T-parity. Typical internal symmetries are lepton number and baryon number or the electric charge. It is worth mentioning here a minor but often confusing point. Most conserved quantum numbers are additive.

Thus, in an elementary particle reaction, the sum of the quantum numbers should be the same before and after the reaction. However, some, usually called a *parity*, are multiplicative; ie, their product is conserved. All multiplicative quantum numbers belong to a symmetry in which applying the symmetry transformation twice is equivalent to doing nothing. These are all examples of an abstract group called Z_2.

MULTIELECTRONIC ATOMS

The motion of a multi-electronic atom in an external electromagnetic field is reconsidered.

STRUCTURE OF ATOM

Just as in the atom of hydrogen, in more composite atoms electrons can rotate around of a kern on allowed orbits only. The orbits in composite atoms cluster in system of shells. Each shell contains a particular number of orbits, an electron can be on any og them. A *K*-shell is the closest to a kern and contains two orbits. The second is a *L*-shell has eight orbits. The third is a *M*-shell has eight orbits, the fourth *N*-shell has 18 orbits etc.

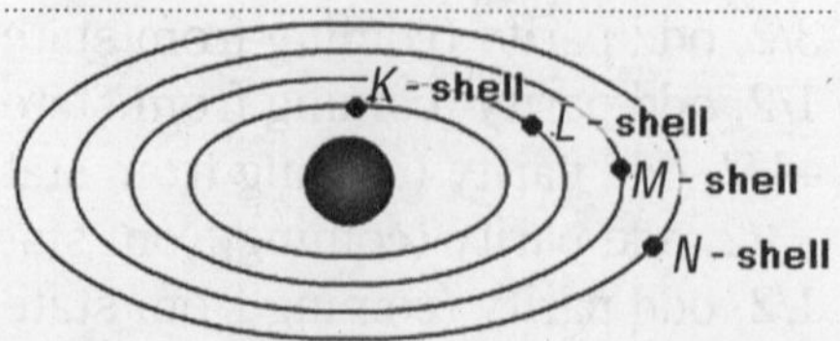

X-RAY SPECTRUMS OF ATOMS

The X-rays results from losses by atom of an inner-shell electron: The transition of electrons is accompanied by an emission of quantums of high energy with a small wave length, which are termed as Roentgen rays.

The X-rays emerges when atom losses inner-shell electron: The transition of electrons is accompanied by an emission of quantums of high energy with a small wave length called Roentgen rays. The atom can lose inner-shell electrons as a result of an electron bombardment with electrons speeded up to energy of several thousand electron-volt. The wave length of the X-rays is comparable to interatomic distance in solid bodies. The examination of the X-ray diffraction on crystal lattices of solid bodies enables, knowing a wave length of Roentgen rays, to determine interatomic distances in them.

The examination of Roentgen rays also allows to determine nuclear charge, because it depends on a prime relation of an electron-binding energy in atom, which is expressed through energy of the X-rays. This relation is called Moseley's law:

$$v \quad = A\,(Z - B)$$

v and -frequency of a line in a X-ray

characteristic spectrum of a device

A and B - constants Z -serial number of a device

PARAMETERS OF ATOM

When the electron travels from an orbit of greater radius to an orbit of smaller radius, the energy is oozed. But the transition from outer shells on interior is possible only when there are free orbits on the leter. The outer-shell electrons are less bound to a kern than interior ones: they are on greater distance from a kern, and the electrons, which are on inner shells, screen the nuclear charge.

Therefore just a small amount of energy is sufficient for excitation of outer-shell electrons. The quantums of energy radiated at transition of such electrons in a non-excited state have a quantity of about several electron-volt. Greater energy is necessary to detach inner-shell electrons from the atom. The greater the nuclear charge the greater energy is needed. Therefore, when there is a transition of electrons from outer shells to a K-shell in heavy atoms, the emission of quantums with energy of hundreds and even thousand electron-volt is possible. Such quantums of high energy are called Roentgen rays.

PROPERTIES OF LIGHT

In one phenomena, which are bound to distribution of light, its undular properties (interference, diffraction) are shown. In other phenomena its quantum properties (photoeffect, radiation and uptake of a light, process of interaction of a light with substance) are shown. The light has simultaneously undular and quantum properties, which are differently shown in the different phenomena. However, when the wave length of a light is diminished, its undular properties are attenuated and the quantum properties are shown.

LIGHT PRESSURE

Some properties, for example light pressure, can be surveyed simultaneously in the electromagnetic theory and in the quantum theory. According to the electromagnetic theory of light, the pressure p, which renders a luminous flux with power W, impinging normally on a simple surface, is equal:

$$p = \frac{W}{c}(1+R)$$

For an ideal black body ($R = 0$): $p = \frac{W}{c}$

For an absolute reflecting bodyh ($R = 1$) $p = \frac{2W}{c}$

If N of quantums of a luminous flux with frequency ν, which bears (carries) energy $W = N \cdot h\nu$, impinges on a simple surface for 1 second, then the blanket impulse at a reflectivity R will look like:

$$p = \frac{N \bullet h\nu}{c}(1+R) = \frac{W}{c}(1+R)$$

Thus, electromagnetic theory and quantum theory give identical results. However it is essential different representations will be utilized in the electromagnetic and quantum theories of light. The concept of an electromagnetic wave of application to a light guesses existence of a wave in particular field (area) of space, which is capable to contain even some lengths of waves of light.

The light quantums are considered as a particle, which are localized in very small volume of space. The representation about a light wave guesses a continuous distribution of energy in a spatial wave, the representation about quantums guesses localization of energy in a small part of space. Both these incompatible representations reflect unity of a wave-corpuscle nature of light, which is shown at distribution of light and its interaction with substance.

COMPTON'S EFFECT

The experience of Arthur Holly Compton (1892-1962) on a dispersion of Roentgen rays by substances, which consist of atoms of light devices, have shown, that the dispelled Roentgen rays have a major wave length λ', than impinging λ. Thus the residual $\Delta\lambda = \lambda' - \lambda$ depends on properties of scattering substance and wave length of an impinging light:

$$\Delta\lambda = \lambda' - \lambda = 2\lambda_K \sin^2\frac{\theta}{2}$$

The quantity λ_K is a stationary value for all substances also is termed as Compton's wave length, $\lambda_K = 2{,}43 \times 10^{12}$ m. This phenomenon, which is termed as Compton's recoil, is impossible to explain by a wave theory of light. In quantum theory the magnification of wave length of radiation at a dispersion is explained by interaction of photons of impinging radiation with electrons of atoms and molecules.

The part of energy of an impinging photon, which is transmitted to electrons of scattering substance, depends on a scattering angle of photon.

At $\theta = \pi$, that is provided that after a dispersion a photon will fly in the party, which is opposite to a tentative direction, the losses by a photon of energy will be maximal, that will give in maximal magnification of a wave length of radiation.

PROPERTIES OF SUBSTANCE

DE BROGLIE'S WAVES

The specified waves have received a title of associated waves. Thus, the associated waves are travelling waves for freely propellent electrons or standing waves for electrons, which are bound in atoms.

De Broglie's waves are not waves of propellent substance, they have not analog in classical physics. De Broglie's guess about existence of undular properties of particles carries universal character: undular properties should have the electron, positive proton, neutron, atom, molecule, any moving object. However objects, which have major masses and move with usual velocities, will have a so small wave length To in comparison with the sizes of objects, that the phenomena of an interference and diffraction for them can completely be neglected.

The undular properties of a light are clearly shown in cases, when the wave length has compared to the sizes of bodies, with which the light interreacts. The lengths of waves of electrons in usual requirements have the order of the nuclear sizes, hence, the effects are characteristic for them which are observed usually for Roentgen rays. For such small lengths of waves it is possible to observe a diffraction on nuclear crystal lattices.

DE BROGLIE'S THEORY

The French physicist Louis Victor duc de Broglie has applied wave-corpuscle treatment and to particles of substance: as the light detects corpuscular properties, also particles of substance should detect undular properties. De Broglie has utilized already wave-corpuscle properties, accepted for a light, and their quantum rule and has expressed a wave length through the performances of a particle. For a light the corpuscular point of view determines energy and impulse accordingly by expressions $E = mc^2$ and $p = mc$.

The wave length of a particle of substance is determined under the formula:

$$\lambda = \frac{h}{p} = \frac{h}{m\upsilon}$$

At adding in a rule of quantization of a moment of momentum of an electron the impulse of an electron, will turn out a relation $n\lambda = 2\pi r$. From it follows, that on length of a stationary orbit of an electron the integer of lengths of waves should be stacked.

Thus, the requirement of existence of allowed orbits, which was injected by Bohr frequency relation, acquires legible physical sense: the allowed orbits are orbits supposing formation on them of standing waves of an electron. For a rectilinear motion of particles de Broglie has offered expression, which is made by analogy with the equation for a spread flat light wave in the complex shape:

$$\psi = \psi_0 e\ i(wt - \frac{x}{\lambda})$$

EXAMINATION OF PROPERTIES

The existence of undular properties of substance was revealed at observation of patterns of a diffraction of beams of particles. Legible patterns of an interference and the diffractions give bundles of electrons, which have almost identical impulses. These patterns are very similar to optical patterns of an interference and diffraction.

At change of quantity of a momentum of electrons and measuring of a wave length, it is possible to receive a relation between these quantities. It has appeared equal to a known relation,

$$\lambda = \frac{h}{p}$$

This de Broglie's relation is valid for any kind of particles. The undular properties of particles were detected at observation of a pattern of the diffraction of the bundle of electrons on chips. At measuring a scattering angle it is possible to spot a wave length of electrons. Thus, at known quantity of potential, which was utilised for acceleration of a bundle of electrons, their velocity is possible to calculate, and then to receive independent value of a wave length of an electron. The value of a wave length of an electron, which was spotted in such a way, completely coincides with quantity of a wave length, which is found from a diffraction pattern. The undular properties of electrons also were detected in experience with a cathode rays at major velocity, which was passed through a thin metal foil on a photographic plate. After development of a photoplate on it the diffraction pattern from rings of different intensity was visible. However this diffraction pattern was stipulated not by Roentgen rays, which could arise at a concussion of high-velocity electrons with metal. It confirms presence of a stream of charged particles.

MECHANICS

Quantum Mechanics is the study of matter and energy on the subatomic scale. It was came into being because classical physics could not explain certain experimental results, such as why the photoelectric effect occurred only when the light shone is above a certain frequency. Quantum Mechanics and Relativity became the foundation of modern physics.

MAIN POINTS

PRINCIPLE OF QUANTATIZATION

The fundamental principle of Quantum Mechanics, and its namesake, is that energy came in discrete particles called "quanta". This idea, the Quantatization of Energy, was put forth by Planck in his Quantum Hypothesis (1900), an explanation of blackbody radiation. It was a departure from Classical Physics, which asserted that energy could be infinitely small.

PHOTOELECTRIC EFFECT

This quantatization of energy approach was used by Einstein in his 1905 explanation of the photoelectric effect, the emission of electrons by a metal when light is shone on it. No electrons are emitted when the light is below a threshold frequency, no matter how intense the light. Einstein postulated that light, while exhibiting wave properties as shone by the Double-slit Experiment, also came in the from of particles called "photons", discrete packets of energy.

The energy of a photon is proportional to the frequency of the light ($E=hv$); thus, light with higher frequency had more energy and could excite the electrons, while light whose frequency were below the threshold could not, no matter how many photons there were.

BOHR'S MODEL AND SPECTRAL LINES

Bohr also used the Quantatization of energy in his 1913 model of the atom. He stated that electrons revolved around the nucleus much like the planets revolve around the sun. They could be found in predefined, stationary orbits, radiating no energy as they revolve. Only when they gain a quantum of energy equivalent to the difference between the energy of the orbits do they become excited and enter a higher energy orbit. This is consistent with Einstein's explanation of the photoelectric effect, and explains why spectral lines only occur at certain places.

WAVE-PARTICLE DUALITY OF MATTER

In 1923 De Broglie expanded the Principle of Wave-particle Duality of Light put forth by Einstein to include all matter, called Principle of Wave-particle Duality of Matter. He asserted that all matter, not just light, exhibit both wavelike and particle-like behaviour,

QUANTUM MATRIX AND WAVE MECHANICS

By now there was enough theories to put together a quantum picture of the atom. However, it was very difficult to sort through all the math. Heisenherg and Shroedinger simultaneously and independently worked to put forth a complete quantum mechanics.

Heisenberg employed use of matrices, and his model was called the matrix mechanics. His approach emphasized the quantum—discrete properties while Schrödinger concentrated on the wave properties. He tried to find a function that would encompass all information about a particle or system at any given time. This is called the wave function. While debated ensued about which was correct, Schoedinger proved shortly after that the matrix mechanics and the wave function were mathematically equivalent.

UNCERTAINTY PRINCIPLE

Heisenberg then came up with the Uncertainty Principle, which stated that the position and velocity (momentum, which is mass*velocity) cannot be known precisely simultaneously. The more accurately one is known, the less accurately the other can be measured. The product of the uncertainty in position, and the uncertainty in momentum, is always greater than or equal to h/(2pi). While this is negligible on the macroscopic level, subatomically, it causes great uncertainty. As a result, the world exists in statistical probabilities.

For example, a electron cannot be found precisely at one location. At any given chance, it has a certain probability of being found anywhere in the universe, with higher probabilities in regions around the atomic nucleus and less likelihood elsewhere. This can be calculated with the wave function.

QUANTUM ATOMIC MODEL

The quantum model of the atom calls for arbitrarily defined orbitals. An orbital is a region around the nucleus where an electron has 90 per cent chance of being found. Electrons have spin, a term used for the intrinsic angular momenta subatomic particles have. They have either a spin of +1/2 or -1/2. Two electrons in the same orbital of the same energy level must have opposite spin (Pauli Exclsion Principle).

SCHRÖDINGER'S CAT

Since the wave function describes the world in terms of statistical probabilities and not certainties, a cat placed in an isolated system with equal chances of being found alive and dead at anytime *is* equally dead and alive at the same time. When an observer is introduced to the isolated system, the wave function collapses and the cat is either dead or alive.

MODELS OF THE ATOM

The first attempt to construct a physical model of an atom was made by William Thomson (later elevated to Lord Kelvin) in 1867. The most striking property of the atom was its *permanence*. It was difficult to imagine any small solid entity that could not be broken, given the right force, temperature or chemical reaction. In contemplating what kinds of physical systems exhibited permanence, Thomson was inspired by a paper Helmholtz had written in 1858 on *vortices*.

This work had been translated into English by a Scotsman, Peter Tait, who showed Thomson some ingenious experiments with smoke rings to illustrate Helmholtz' ideas. The main point was that in an *ideal* fluid, a vortex line is always composed of the same particles, it remains *unbroken*, so it is ring-like. Vortices can also form interesting combinationsA good demonstration is provided by creating two vortex rings one right after the other going in the same direction.

They can trap each other, each going through the other in succession. This is probably what Tait showed Thomson, and it gave Thomson the idea that atoms might somehow be vortices in the ether. Of course, in a non ideal fluid like air, the vortices dissipate after a while, so Helholtz' mathematical theorem about their permanence is only approximate. But Thomson was excited because the ether *was* thought an ideal fluid, so vortices in the ether might last forever! This was very aesthetically appealing to everybody- "Kirchhoff, a man of cold temperament, can be roused to enthusiasm when speaking of it." In fact, the investigations of vortices, trying to match their properties with those of atoms, led to a much better understanding of the hydrodynamics of vortices-the constancy of the circulation around a vortex, for example, is known as Kelvin's law. In 1882 another Thomson, J. J., won a prize for an essay on vortex atoms, and how they might interact chemically. After that, though, interest began to wane-Kelvin himself began to doubt that his model really had much to do with atoms, and when the electron was discovered by J. J. in 1897, and was clearly a component of all atoms, different kinds of non-vortex atomic models evolved. It is fascinating to note that the most exciting theory of fundamental particles at the present time, *string* theory, has a definite resemblance to Thomson's vortex atoms. One of the basic entities is the closed string, a little loop, which has fields flowing around it reminiscent of the swirl of ethereal fluid in Thomson's atom. And it's a very beautiful theory-Kirchhoff would have been enthusiastic!

FLOATING MAGNETS

In 1878, Alfred Mayer, at the University of Maryland, dreamed up a neat demonstration of how he imagined atoms might be arranged in molecules. He took a few equally magnetized needles and stuck them through corks so that they would float with their north poles all at the same height above the water, all repelling each other equally.

He then held the south pole of a more powerful magnet some distance above the water, to attract the needles towards this central point. The idea was to see what equilibrium patterns the needles would form for different numbers of needles. He found something remarkable-the needles liked to arrange themselves in shells. Three to five magnets just formed a triangle, square and pentagon in succession. but for six magnets, one went to the centre and the others formed a pentagon. For more magnets, an outer shell began to form.

Kelvin's immediate response to Mayer's publication was that this should give some clues about the vortex atom. Apparently it didn't, but twenty-five years later it guided his thinking on a new model.

PLUM PUDDING

Kelvin, in 1903, proposed that the atom have the newly discovered electrons embedded somehow in a sphere of uniform positive charge, this sphere being the full size of the atom. (Of course, the sphere itself must be held together by unknown non-electrical forces-which is still true of the positive charge in our modern model of the atom.)

This picture was taken up by J. J. Thomson too, and was dubbed the plum pudding model, after traditional English Christmas fare, a large round pudding (rich with suet) with raisins embedded in it. In 1906, J. J. concluded from an analysis of the scattering of X-rays by gases and of absorption of beta-rays by solids, both of which he assumed were effected by electrons, that the number of electrons in an atom was approximately equal to the atomic number.

This led to a picture of electron arrangements in an atom reminiscent of Mayer's magnets. Perhaps by analyzing possible modes of vibration of electrons in these configurations, the spectra could be calculated. The simplest case to consider was clearly hydrogen, now assumed (correctly) to contain just one electron.

How Does an Atom's Colour Depend on its Size?

By "colour" we mean here the spectral colours emitted when the atom is excited. In Thomson's plum pudding model, there is a clear relationship between the *size* of the pudding and the *frequency* at which the electron will oscillate, and hence presumably radiate, when excited.

The two are related because the assumption is that the total positive charge-which is uniformly spread throughout the sphere is just equal to the electron's negative charge. At rest in its lowest state, the electron just sits in the middle of this sphere of charge. When bumped somehow, it will oscillate about that point. If the electron is at distance x from the centre, it will feel a restoring force towards the centre equal to the attraction from that part of the positive charge it is "outside" of-that is, the charge within a sphere of radius x about the centre.

Therefore, the *larger* the whole atomthe pudding-the more thinly spread the positive charge is, and the *smaller* the amount of charge within the small sphere of radius x that is attracting the electron back towards the centre. So, the bigger the atom is, the slower the electron's oscillation is, and the lower frequency the radiation emitted.

It is straightforward to give a quantitative estimate of the size of the atom based on the observation that when excited it emits radiation in the visible range. Let us assume that the positively charged sphere has radius r_0 (this is then the size of the atom, which we know is about 10^{-10} meters).

If the electron is displaced from the centre of the atom in the x-direction an amount x, it is attracted back by all the charge that is now closer to the centre than itself, that is, an amount of charge equal to $ex^3/r_0{}^3$. (Recall e is the total amount of charge on the sphere, and $x^3/r_0{}^3$ is the fraction of the sphere closer to the centre than x.)

This charge acts as if it were a point charge at the origin, so the inverse-square law gives a $1/x^2$ factor, and the equation of motion for the electron is therefore:

$$m\frac{d^2x}{dt^2} = -\frac{1}{4\pi\varepsilon_0}\cdot\frac{e^2x}{r_0^3}$$

Provided it stays within the sphere, the electron will execute simple harmonic motion with a frequency

$$\omega^2 = \frac{1}{4\pi\varepsilon_0}\cdot\frac{e^2}{mr_0^3}.$$

Notice that, as we discussed above, as the size of the atom increases the frequency goes down. And we know the frequency, at least approximately it corresponds to visible light. Therefore, this model will predict a size of the atom, which we can compare with the size from other predictions, such as

Brownian motion (plus the assumption that in a liquid, the atoms are fairly close packed-they take up most of the room available). If we take visible light, say with a frequency 4.10^{15} radians per second, we find r_0 must be about 2.10^{-10} meters, a little on the large side, but encouragingly close to the right answer for a first attempt. Sad to report, though, no real progress was made beyond this in predicting spectra using Thomson's pudding.

Many attempts were made to find stable arrangements of electrons in atoms, not just hydrogen, using models like Mayer's magnets, and also having the electrons going around in circles.

It was hoped that if certain numbers of magnets formed a very stable arrangement, that might model a chemically nonreactive atom, etc.-but nobody succeeded in making any real predictions along these lines, the models could not be connected with the properties of real atoms. Evidently, then, the theorists were stuck-and the experimental challenge was to find some way to look *inside* an atom, and see how the electrons were arranged.

THE BOHR ATOM

In 1911, the 26-year-old Niels Bohr earned a Ph. D. at the University of Copenhagen; his dissertation was titled "Studies on the Electron Theory of Metals". He was awarded a postdoctoral fellowship funded by the Carlsberg Brewery Foundation, which enabled him to go to Cambridge in September to study with J. J. Thomson. Bohr was a great admirer of Thomson's many achievements, both experimental and theoretical. In his thesis work, he had closely studied some of the problems covered in Thomson's book *Conduction of Electricity through Gases.*

He had uncovered some apparent errors in Thomson's work, and looked forward to discussing these points with the great man. Unfortunately, by the time Bohr arrived, the Cavendish Laboratory had grown to the point where Thomson as director had more than he could manage.

He had no spare time to think about electrons, and was not happy to hear from Bohr that some of his earlier work might be incorrect. In fact, Thomson went out of his way to avoid theoretical discussions with Bohr. He did assign Bohr an experiment on positive rays, but Bohr was not enthusiastic. Bohr kept himself busy writing a paper on electrons in metals, reading Dickens to improve his English, and playing soccer. In December, Rutherford came down from Manchester for the annual Cavendish dinner. Bohr later said that he was deeply impressed by Rutherford's charm, his force of personality, and his patience to listen to every young man who might have an ideacertainly a refreshing change after J. J.! A little later, Bohr met with Rutherford again when he visited one of his father's friends in Manchester, someone who also knew Rutherford.

Although Rutherford was usually skeptical of theorists, he liked Bohr. For one thing, Rutherford was a soccer fan, and Bohr's brother Harald (only nineteen months younger than Bohr) was famoushe had played in the silver medal winning Danish soccer team at the 1908 Olympics in London. After talking it over with Harald, who visited Cambridge in January, Bohr moved to Manchester in March, and took a six-week lab course, given by Geiger, Marsden and others. Really, though, his interests were theoretical, and he talked a lot with Charles Galton Darwin"grandson of the real Darwin", as Bohr put it in a letter to Harald.

Darwin had just completed a theoretical analysis of the loss of energy of an a -particle going through matterthat is, an a that doesn't get close enough to a nucleus to be scattered. Such a 's gradually lose energy by churning through the electrons, and the rate of loss depends on how many electrons they encounter. In particular, Bohr concluded, after reviewing and improving on Darwin's work, it seemed clear that the hydrogen atom almost certainly had a *single* electron outside the nucleus.

WHAT DETERMINES ATOMIC SIZE?

A big problem with the nuclear hydrogen atom was: what determined its size? Classical mechanics gives a simple dynamical equation for circular orbits:

$$\frac{mv^2}{r} = \frac{1}{4\pi\varepsilon_0}.\frac{e^2}{r^2}.$$

Now this equation is satisfied by *any* circular orbit centered at the nucleus, however large or small. (Note, by the way, that multiplying both sides by $r/2$ gives that the magnitude of the kinetic energy in the circular orbit is just half the magnitude of the negative potential energy. We need this below.) There is no hint here that the atom in its "natural" ground state should have any particular radius. But it does! This means we're missing *something*.

But what? Bohr (and others) thought that Planck's constant must somehow play a role in determining the size of the orbit. After all, it *did* play a role in restricting allowed orbital changes in the oscillators in black body radiationand these oscillators, although not very clearly understood, were of the same general size as atoms.

So evidently the standard picture of how an oscillating charge radiated couldn't be right at the atomic level. Bohr concluded that in an atom in its natural rest state, the electron must be in a special orbit, he called it a "stationary state" to which the usual rules of electromagnetic radiation didn't apply. In this orbit, which determined the size of the atom, the electron, mysteriously, didn't radiate.

Just how to bring Planck's constant into a discussion of the hydrogen atom was not so clear, though. For the black body oscillators, it related the frequency f of the oscillator with the allowed energy change E by $E = hf$. The obvious parallel approach for the hydrogen atom was to identify the frequency f with the circular frequency of the electron in its orbit.

However, in contrast to the simple harmonic oscillator this hydrogen atom frequency *varied* with the size of the orbit. Still, it was the only frequency around, and, dimensionally, multiplying it by h gave an energy. What energy could that be identified with? Again, the choice was limitedthe electron had a kinetic energy E, the potential energy was $-2E$ and the total energy $-E$. If a hydrogen nucleus captured a passing electron into its ground state, and emitted one quantum of electromagnetic radiation, that quantum would have energy E, the same as the electron kinetic energy in the natural stationary state (called the *ground state*). Bohr suggested in a note to Rutherford in the summer of 1912 that requiring this energy be some constant (assumed to be of order of magnitude one) multiplied by hf would fix the size of the atom. Actually his argument was a bit more complicated, he considered the several electron atom, and took the electrons to form rings. However, the basic point is the samea condition like this constrains the atomic size, it would be fixed uniquely if we knew the constant. If we assume the constant is 1, for example, we have

$$\frac{1}{2}mv^2 = \text{hv}/2\pi\text{r}$$

Putting this together with the dynamic equation above determines the atomic radius r. It is easy to check that it predicts a radius of 4p e $_0h^2$/p $^2me^2$, which is just four times the "right answer" defined as the Bohr radius. The correct Bohr radius comes out if we choose the constant to be one-half, $E = \frac{1}{2}hf$, which Bohr used later. Hence the approximate size of the atom follows from *dimensional* arguments alone once one assumes that Planck's constant plays a role! Of course, the nucleus is irrelevant in determining the atomic sizeit just provides a fixed centre of electrostatic attraction. The relevant electronic parameters are the mass m and the strength of attraction e^2/4p e $_0$. Together with h, these parameters determine a length.

It should be mentioned that this assumption explained more than the size of the hydrogen atom. It was believed at the time that in the higher atoms, the electrons formed rings, thought to lie one outside the other, and various stability arguments indicated that there couldn't be more than seven electrons in a ring. The length scale above, $4\pi\varepsilon_0 h^2/2\pi^2 me^2$, would *decrease* for larger atoms, with e^2 replaced by Ze^2 essentially, for nuclear charge Z. Thus as the number of rings increased, the size of the rings would decrease, explaining the observed approximate periodicity in atomic volume with atomic number.

Also, in 1911 Richard Whiddington in Cambridge had found that to cause a substance having atomic number A to emit characteristic x-rays by bombarding it with electrons, it was necessary to use electrons of speed approximately A x 10^6 meters per second. Any substance on being bombarded with sufficiently fast electrons emits a continuum of x-ray frequencies up to a maximum frequency f given by hf = kinetic energy of electron, ***plus*** some sharply defined linesx-rays at a particular frequency, which does not change as the electron speed is further increased. The frequency corresponding to these lines was found to increase with atomic number. Applying his length scale argument to the innermost ring of an atom, Bohr found that an electron in that ring would have a speed proportional to the nuclear charge, and hence, at least approximately, to the atomic number. Furthermore, the predicted speed in orbit was of the same order as that of Whiddington's electrons.

NICOLSON: A CLEVER IDEA ABOUT A WRONG MODEL

Meanwhile, in Cambridge one J. W. Nicolson was struggling to incorporate Planck's ideas in a model of the atom, in an attempt to understand some strange sets of spectral lines observed in nebulae and in the sun's corona. He conceived a rather exotic (and quite wrong!) model, in which a ring of electrons, like a necklace, orbited the nucleus. (Actually, many people, including Bohr himself, investigated models like this.

The reason was that the classical radiation from a ring of electrons is a lot less that that from a single orbiting electron, the fields tend to cancel each other.) Oscillations of electrons in this ring gave the spectra. Nicolson predicted the frequencies emitted by a straightforward classical analysis of these oscillation frequencies, in the spirit of earlier work on the plum pudding model.

He did bring in Planck's constant, though. He knew that dimensionally it was a unit of angular momentum, and he suggested that the atom could only lose angular momentum in discrete amountspresumably constant multiples of h. Nicolson felt that, given the dimensionality of Planck's constant, quantization of angular momentum was more plausible than quantization of energy. Of course, for the simple harmonic oscillator they amounted to the same thing, but not for any other system.

BOHR RETURNS TO DENMARK

Bohr left Manchester in July 1912 and was married on the first of August. In the fall, he began work at the University of Copenhagen. At the same time, he began setting down on paper some of his Manchester ideas about atoms. He read Nicolson's work. As he wrote to Rutherford at the end of January 1913, he and Nicolson were really looking at different thingsNicolson was considering atoms in a very hot environment (like the sun's corona, or an electrical discharge tube) and the spectra gave information about how energy was emitted as the atom settled into its ground state.

Bohr himself was only interested in the state in which the system possessed the smallest amount of energy. He went on: " The question of calculation of the frequencies corresponding to the visible part of the spectrum". At that time, Bohr thought of spectra as pretty but peripheral, having as little to do with basic physics as the colours of a butterfly had to do with basic biology.

BOHR CHANGES HIS MIND ABOUT SPECTRA

In February 1913, Bohr was surprised to find out in a casual conversation with the spectroscopist H. R. Hansen that some patterns had been discerned in the apparent chaos of spectral lines. In particular, Hansen (a colleague and former classmate of Bohr) showed him Balmer's formula for hydrogen. They had very likely seen this in class together, but, given Bohr's opinion of the value of spectra, he probably hadn't paid much attention.

Balmer's formula is:

$$\frac{1}{\lambda} = R_H\left(\frac{1}{4} - \frac{1}{n^2}\right)$$

for the sequence of wavelengths of light emitted, with $n = 3, 4, 5, 6$ being in the visible, the lines used by Balmer in finding the formula. Hansen would doubtless have informed Bohr that the 1/4 could be replaced by $1/m^2$, with m another integer. The constant appearing on the right hand side is called the *Rydberg constant*, $R_H = 109{,}737\ \text{cm}^{-1}$. (This is the modern value-Balmer got it right to one part in 10,000, about the limit of spectral measurements at the time.) Bohr said later: "As soon as I saw Balmer's formula, the whole thing was immediately clear to me." What he saw was that the set of allowed *frequencies* (proportional to inverse wavelengths) emitted by the hydrogen atom could all be expressed as *differences*.

This immediately suggested to him a generalization of his idea of a "stationary state" lowest energy level, in which the electron did not radiate. There must be a *whole sequence* of these stationary states, with radiation only taking place as the atom jumps from one to another of lower energy, emitting a single quantum of frequency f such that

$$hf = E_n - E_m,$$

the difference between the energies of the two states. Evidently, from the Balmer formula and its extension to general integers m, n, these allowed non-radiating orbits, the stationary states, could be labeled 1, 2, 3..., n... and had energies –1, –1/4, –1/9..., $-1/n^2$... in units of hcR_H (using $\lambda f = c$ and the Balmer equation above). The energies are of course negative, because these are bound states, and we count energy zero from where the two particles are infinitely far apart. Bohr was very familiar with the dynamics of simple circular orbits in an inverse square field. He knew that if the energy of the orbit was $-hcR_H/n^2$, that meant the kinetic energy of the electron, $\frac{1}{2}mv^2 = hcR_H/n^2$, and the potential energy would be.

$$-(1/4\text{pe}_0)e^2/r = -2hcR_H/n^2.$$

It immediately follows that the *radius* of the nth orbit is proportional to n^2, and the *speed* in that orbit is proportional to $1/n$. *I*t then follows that the angular momentum of the nth orbit is just proportional to n.

Evidently, then the angular momentum in the nth orbit was nKh, where h is Planck's constant and K is some multiplying factor, the same for all the orbits, still to be determined. In fact, the value of K follows from the results above. R_H, m, h, and c are all known quantities (R_H being measured experimentally by observing the lines in the Balmer series) so the above formulas immediately give the electron's speed and distance from the nucleus in the nth orbit, and hence its angular momentum. Therefore, by putting in these experimentally determined quantities, we can find K.

BOHR FINDS THE RYDBERG CONSTANT WITHOUT DOING AN EXPERIMENT

The Balmer formula gave Bohr the essential clue that led to the realization that the angular momentum was quantized: $L = Kh, 2Kh, 3Kh...$ where h is Planck's constant, as usual, and K is some constant numerical factor, presumably of order 1. Bohr gave a very clever argument to find K without doing any experiment. First, think about how the size of K affects the *physical* properties of the hydrogen atom. How would the atom be different for $K = 10$ compared with $K = 1$? For $K = 1$, the allowed orbits would be those having angular momentum $h, 2h, 3h, 4h$....

For $K = 10$, the only allowed orbits would be those having angular momentum $10h, 20h$.... Evidently, for $K = 10$ there will be a lot fewer spectral lines, and the average *spacing* between them will be *greater* that for $K = 1$. Next, Bohr imagined a really immense hydrogen atom, an electron going around a proton in a circle of one meter radius, say.

This would have to be done in the depths of space, but really this is just a thought experiment in the spirit of Einstein. The point is that for this very large atom, the electron is moving rather slowly over a distance scale we are familiar with.

We know from many experiments that charges moving at these slow speeds over ordinary (human size) distances emit radiation according to Maxwell's equations. Or, more simply, if it's going round the circle at frequency f revolutions per second, it will be emitting radiation at that frequency f because its electric field, as seen from some fixed point a meter or so away, say, will be rotating f times per second. On the other hand, the angular momentum quantization condition must be true for all circular orbits of the electron around the proton, even for this very large atom. Furthermore, the radiation emitted must still be given by the difference in energies of neighbouring orbits,

$$hf = E_{n+1} - E_n.$$

But $E_{n+1} - E_n$, the energy *spacing* between neighbouring orbits, *depends on K*. Therefore, Bohr concluded *K is fixed* by requiring that the frequency of radiation emitted by a really large atom be correctly given by ordinary common sense-that is, the frequency of the radiation must be the same as the orbital frequency of the electron, the number of cycles a second. In other words, for a large orbit we must have,

$$E_{n+1} - E_n = hf = h.v/2\pi r$$

Here v is the speed of the electron in the orbit, and the orbit radius is r.

The strategy is then as follows: We assume that the only allowed orbits are those having angular momentum integral multiples of Kh, where K is some constant, so the nth orbit has angular momentum nKh. We can then use the equation of motion to determine the radius r_n, the electron speed v_n and the energy E_n for the nth orbit. Naturally, these all depend on K. *Now concentrate on orbits close to one meter in radius*. They will radiate at the frequency given by the equations of motion for an electron circling a proton one meter away.

But this must *match up* with the frequency given by the energy difference between neighbouring orbits divided by h. Now, if K is very small, this energy difference is small, and they won't match. If K is very large they won't match either. We must find the value of K for which these frequencies do match. That is what we do below in detail.

We establish below that there can only be agreement between the classical radiation frequency for a man-sized atom and Bohr's prediction if K = 1/2p. Therefore, we must assume the angular momentum is always quantized in chunks of size h/2p.

It follows that the allowed energy levels are:

$$E_n = -\frac{1}{4\pi\varepsilon_0}\cdot\frac{e^2}{2r_n} = \left(\frac{1}{4\pi\varepsilon_0}\right)^2\cdot\frac{me^4}{2K^2h^2}\cdot\frac{1}{n^2}$$

$$= -\left(\frac{1}{4\pi\varepsilon_0}\right)^2\cdot\frac{2\pi^2me^4}{h^2}\cdot\frac{1}{n^2}.$$

Putting this together with E_n-E_m = *hf we get the Balmer formula*:

$$f = \frac{E_n - E_2}{h} = \left(\frac{1}{4\pi\varepsilon_0}\right)^2\cdot\frac{2\pi^2me^4}{h^3}\cdot\left(\frac{1}{4}-\frac{1}{n^2}\right).$$

The new point is that there is no adjustable parameter! The Rydberg constant that appeared before is here given in terms of h, m and e.

The rather abstract argument that the quantum predictions must match the known classical results for large slow systems actually fixes the Rydberg constant.

That is to say,

$$R_H = \left(\frac{1}{4\pi\varepsilon_0}\right)^2\cdot\frac{2\pi^2me^4}{ch^3}$$

This formula was found to be correct within the limits of experimental error in measuring the quantities on the right. This matching for large systems is called the Correspondence Principle: in that limit, quantum predictions must correspond to known classical results.

Derivation of the Angular Momentum Quantization from the Correspondence Principle

Let us assume this large orbit is the nth (where *n* is of order 10^5!), so it has angular momentum

$$mvr = L = nKh.$$

Using,

$$\frac{mv^2}{r} = \frac{1}{4\pi\varepsilon_0}.\frac{e^2}{r^2},$$

we find,

$$m\sqrt{\frac{1}{4\pi\varepsilon_0}.\frac{e^2}{mr}}.r = nKh$$

from which we find the radii of the allowed orbits are given by,

$$r_n = \frac{4\pi\varepsilon_0}{me^2}.n^2K^2h^2.$$

Therefore the allowed energies are:

$$E_n = -\frac{1}{4\pi\varepsilon_0}.\frac{e^2}{2r_n} = -\left(\frac{1}{4\pi\varepsilon_0}\right).\frac{me^4}{2K^2h^2}.\frac{1}{n^2}$$

Thus for *n very large*:

$$E_{n+1} - E_n = -\left(\frac{1}{4\pi\varepsilon_0}\right)^2.\frac{me^4}{2K^2h^2}.\left(\frac{1}{(n+1)^2} - \frac{1}{n^2}\right)$$

$$@\left(\frac{1}{4\pi\varepsilon_0}\right)^2.\frac{me^4}{2K^2h^2}.\frac{2}{n^3}$$

Now this must be equal to $hf = hv/2\pi r$ for the appropriate *v*, *r* for this large orbit.

Using $mv_n r_n = nKh$, $v_n = nKh/mr_n$.

Thus $hv/2\pi r = nKh^2/2\pi mr_n^2$.

Putting now,

$$E_{n+1} - E_n = hf = h.v/2p\ r = nKh^2/2p\ mr_n^2$$

And using,

$$R_n = -\left(\frac{4\pi\varepsilon_0}{me^2}\right).n^2 2K^2h^2$$

gives,

$$E_{n+1} - E_n \cong \left(\frac{1}{4\pi\varepsilon_0}\right)^2 \cdot \frac{me^4}{2K^2h^2} \cdot \frac{2}{n^3}$$

$$= \frac{nKh^2}{2\pi m r_n^2} = \frac{nKh^2}{2\pi} \cdot \left(\frac{1}{4\pi\varepsilon_0}\right)^2 \cdot \frac{me^4}{n^4K^4h^4}$$

Now, the second and fourth terms in the above equation must be equal in the limit of large *n*. Canceling out common factors of the two, we find that the condition for equality is:

$$K = 1/2\pi.$$

This was the argument Bohr used to establish that angular momentum for his model is quantized in units $h/2\pi$.

PROJECT OF ATOMIC SPECTRA

The first person to realise that white light was made up of the colours of the rainbow was Isaac Newton, who in 1666 passed sunlight through a narrow slit, then a prism, to project the coloured spectrum on to a wall. This effect had been noticed previously, of course, not least in the sky, but previous attempts to explain it, by Descartes and others, had suggested that the white light became coloured when it was refracted, the colour depending on the angle of refraction.

Newton clarified the situation by using a second prism to reconstitute the white light, making much more plausible the idea that the white light was composed of the separate colours. He then took a monochromatic component from the spectrum generated by one prism and passed it through a second prism, establishing that no further colours were generated. That is, light of a single colour did not change colour on refraction.

He concluded that white light was made up of all the colours of the rainbow, and that on passing through a prism, these different colours were refracted through slightly different angles, thus separating them into the observed spectrum.

In 1752, the Scottish physicist Thomas Melvill discovered that putting different substances in flames, and passing the light through a prism, gave differently patterned spectra. Ordinary table salt, for example, generated a "bright yellow". Furthermore, not all the colours of the rainbow appeared - there were dark gaps in the spectrum, in fact for some materials there were just a few patches of light. By the 1820's, Herschel had recognized that spectra provided an excellent way to detect and identify small quantities of an element in a powder put into a flame.

Meanwhile, the white light of the sun was coming in for more detailed scrutiny. In 1802, William Wollaston in England had discovered (perhaps by using a thinner slit or a better prism) that in fact the solar spectrum itself had tiny gaps - there were many thin dark lines in the rainbow of colours. These were investigated much more systematically by Joseph von Fraunhofer, beginning in 1814. He increased the dispersion by using more than one prism. He found an "almost countless number" of lines. He labeled the strongest dark lines A, B, C, D, etc.

FOUCAULT CONNECTS MELVILL'S BRIGHT LINES AND FRAUNHOFER'S DARK LINES

In 1849, Foucault (of speed of light and pendulum fame) examined the spectrum of light from a voltaic arc between carbon poles. He saw a bright double yellow line at exactly the same wavelength as Fraunhofer's dark D line in the solar spectrum. Investigating further, Foucault passed the sun's light through the arc, then through a prism. He observed that the D lines in the spectrum were even darker than usual. After testing with other sources, he concluded that the arc, which emitted light at the D line frequency, would also absorb light from another source at that frequency.

This discovery did not surprise Sir George Stokes in Cambridge. He pointed out that any mechanical system with a natural frequency of oscillation will emit at that frequency if disturbed, but will also absorb most readily at that frequency from incoming disturbances, the phenomenon of resonance. *Question*: In a total eclipse of the sun, the only sunlight reaching earth comes from the hot gases of the sun's atmosphere, light from the sun's main disc being blocked by our moon. The light from these hot gases was analysed during an eclipse in 1870.

How do you think the spectrum observed related to that of full sunlight? The spectrum of *hydrogen*, which turned out to be crucial in providing the first insight into atomic structure over half a century later, was first observed by Anders Angstrom in Uppsala, Sweden, in 1853. His communication was translated into English in 1855. Angstrom, the son of a country minister, was a reserved person, not interested in the social life that centered around the court. Consequently, it was many years before his achievements were recognized, at home or abroad (most of his results were published in Swedish).

Meanwhile, in Freeport, Pennsylvania, in 1855, David Alter described the spectrum of hydrogen and other gases. In the 1840's, Alter had started the first commercial production of bromine from brines. He also found a way to extract oil from coal, but that proved uneconomic after the discovery of oil in Pennsylvania. His work was not widely recognized, either.

BUNSEN AND KIRCHHOFF

The first really systematic investigation of spectra was that of Bunsen and Kirchhoff, in Heidelberg, between 1855 and 1863. They used several techniques.

For one thing, they introduced various salts intowhat else?the flame of a Bunsen burner. This was a very effective way of viewing spectra, because the Bunsen burner flame itself gave out practically no light. They also used the cooler flame of alcohol burning mixed with water to generate a vapour to study absorption spectra. Finally, they studied the spectra of electric arcs between electrodes of different materials. Using iron electrodes gave a spectrum that coincided with dark lines in the sun's spectrum. Copper electrodes did not. They concluded that the sun's atmosphere contained iron, but not much copper, and that, they said, seemed very plausible since there is so much iron in the earth, and in meteors.

(*Cautionary note to philosophers*: In 1835, the French philosopher Auguste Comte (the founder of positivism) wrote: "...Our knowledge concerning the gaseous envelopes [of stars] is necessarily limited to their existence, size and refractive power, we shall not at all be able to determine their chemical composition or even their density. The collaboration of Kirchhoff and Bunsen was a major research effort, even by modern standards. They determined thousands of spectral lines, each to an accuracy of one part in ten thousand. They spectroscopically discovered new elements: rubidium and cesium. Their method was used to find fifteen more new elements before the end of the century. In 1869, Joseph Lockyer studied the spectra of solar prominences (in eclipses). He found the spectra to be slightly Doppler shifted, so was able to deduce the speeds of the gases whirling around the sunspots. He also found a spectrum never seen before, and conjectured that it came from a new element he named Helium.

In fact, helium was later discovered on earth in 1895, by Ramsay. At that time, it had just become evident that there was an inert component, argon, in the earth's atmosphere. Earlier, an inert gas had been observed to emanate from uranium salts when they were heated. Ramsay assumed this would be the same gas, but decided to check. On heating uranium salts and performing a spectral analysis of the emitted gas, much to his surprise he found it to be helium.

THE BALMER SERIES

It is clear from the above that a tremendous amount of scientific progress was made using spectral lines, yet no-one had the slightest idea why atoms emitted at the frequencies they did. It was appreciated that spectra implied that atoms had structure.

In 1852, Stokes had stated that probably the vibrations that produced light were vibrations among the constituent parts of molecules (a term which also included atoms at that time) and in 1875 Maxwell, in enumerating properties atoms must have, included the capability of internal motion or vibration. This worried Maxwell, though. As he said, the spectroscopic evidence forces the conclusion that the atom is quite complex, with many internal degrees of freedom. Yet apparently all these modes of vibration, or almost all of them, are not excited by heat, since if they were this extra capacity of the atom to absorb energy would be reflected in its specific heat. Obviously, if any pattern could be discerned in the spectral lines for an atom, that might be a clue as to the internal structure of the atom. One might be able to build a model. A great deal of effort went into analyzing the spectral data from the 1860's on. The big breakthrough was made by Johann Balmer, a math and Latin teacher at a girls' school in Basel, Switzerland. Balmer had done no physics before, and made his great discovery when he was almost sixty.

He decided that the most likely atom to show simple spectral patterns was the lightest atom, hydrogen. Angstrom had measured the four visible spectral lines to have wavelengths 6562.10, 4860.74, 4340.1 and 4101.2 in Angstrom units (10^{-10} meters). Balmer concentrated on just these four numbers, and found they were given by the formula:

$$\lambda = b\left(\frac{n^2}{n^2 - 4}\right)$$

where b = 3645.6 Angstroms, and n = 3, 4, 5, 6. Balmer suggested that there would be other lines - in the infrared - corresponding to n = 7, 8, etc., and in fact some of them had already been observed, unbeknownst to Balmer. He further conjectured that the 4 could be replaced by 9, 16, 25, ... and this also turned out to be true - but these lines, further into the infrared, were not detected until the early twentieth century, along with the ultraviolet lines generated by replacing the 4 by 1.

It is instructive to write Balmer's general formula in terms of the inverse wavelength. This is called the wave number - the number of waves that fit in one unit of length.

$$\frac{1}{\lambda} = R_H = \left(\frac{1}{n^2} - \frac{1}{m^2}\right)$$

where n, m are integers, and R_H is the Rydberg constant, 109,737 cm^{-1}.

This constant is named after the Swedish physicist Rydberg who (in 1888) presented a generalization of Balmer's formula, in which the integer n was replaced by n + constant, the constant being less than unity. Rydberg suggested that all atomic spectra formed families with this pattern. (He also said he was unaware of Balmer's work.) It turns out that there *are* families of spectra following Rydberg's pattern, notably in the alkali metals, sodium, potassium, etc., but not with the precision the hydrogen atom lines fit the Balmer formula, and low values of n give lines that deviate considerably.

(Modern footnote: atoms having spectral lines following Rydberg's formula are called Rydberg atoms. These Rydberg atoms have one electron orbiting at a much greater distance from the nucleus than the others. Consequently,

Rydberg atoms can only survive in a gas at very low pressure, otherwise that outermost electron gets knocked off. Prof. Tom Gallagher in our Department is a world expert on these atoms, which have proven a rich source of information on the quantum mechanics of atomic structure.)

One pattern that was noticed in the spectra of many atoms is Ritz' Combination Principle: if for a given atom there are spectral lines at two wave numbers, there is sometimes another spectral line at the precise sum of those two wave numbers.It is easy to see from Balmer's formula that this is true for some pairs of lines in the hydrogen spectrum. It also turns out to be true for atoms where the spectral lines have no other discernible pattern.

THE PERIODIC TABLE

The periodic table was formulated long before the advent of quantum theory, but the ordering of elements was understood only on an experimental basis. Quantum Theory provides an explanation for the appearance of the periodic table in terms.

Of the order of filling of orbitals. In fact, this order of filling may be read directly from the table.

We recognize correlations between its appearance and the order of filling given above:

- A new row (period) of the table begins each time occupation of a new main shell begins. The period number is the value of n for the new main shell.
- The block of elements in the two left-most columns of the table are those in which the ns subshell is being filled. This is the s-block.
- The block of elements in the 6 right-most columns are filling the np subshell. These constitute the p block.
- The block consisting of 3 rows of elements, each 10 blocks long, are filling the (n–1) d subshell. This is called the d-block.
- The remaining block of 2 rows, 14 elements long, are filling the (n–2) f subshell. This is called the f block.

We now show by example how to "read" the electron configuration of an element from the periodic table.

Example: Determine the electron configuration of thallium, Tl, from its position in the periodic table.

Solution: We read the configuration by scanning the rows (periods) one by one, top to bottom, until we reach thallium.

- *Period:* $1s^2$ (there are only 2 elements in this row, H and He, corresponding to filling the n = 1 shell).
- *Period:* $2s^2\ 2p^6$ (there are 8 elements in this row, corresponding to filling the 2s and 2p subshells of the n = 2 main shell).
- *Period:* $3s^2\ 3p^6$ (again, scanning left to right we find 8 elements).

- *Period:* $4s^2\ 3d^{10}\ 4p^6$ (Here we find 18 elements as we scan left to right. We encounter for the first time a row of d block elements in this period. The first 2 elements in the period give us $4s^2$, the d-block row gives us $3d^{10}$, the p-block row gives us $4p^6$.)
- *Period:* $5s^2\ 4d^{10}\ 5p^6$ (Similar to the 4th period).Period 6: $6s^2\ 4f^{14}\ 5d^{10}$, $6p^1$ (Scanning left to right, we encounter 2 s block, 14 f block, and 10 d block elements. Thallium is the first p block element in this row, hence $6p^1$.)

Now string these together in order to obtain the complete configuration. To obtain the configuration in shorthand notation, we need read only from the previous noble gas.

Thus,

$$[Xe]\ 6s^2\ 4f^{14}\ 5d^{10}\ 6p^1$$

You are certainly free to memorize the order of filling if you want to. However, with practice it is much quicker and easier to read the configuration from the periodic table. There remains one matter with which we must deal before moving on. The electron configuration for nitrogen is given below.

$$N\ 1s^2\ 2s^2\ 2p^3$$

Because there are three 2p orbitals, it is not clear from the configuration how the three electrons are to be distributed in them. There are several possiblities, a few of which are shown here. It is found experimentally that the last of these arrangements, with one electron in each of the 2p orbitals and all electron spins parallel, is preferred. Generally, when several electrons occupy a set of equal-energy (degenerate) orbitals, the most stable arrangement is the one with the maximum number of unpaired electrons.

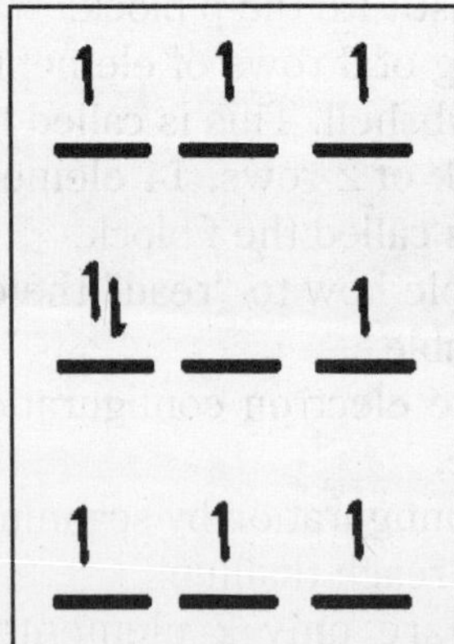

Fig. Electron Arrangements for Nitrogen

This statement is known as Hund's Rule. In accordance with Hund's Rule, the valence orbital occupations of carbon, oxygen, and iron are shown here. Electron Shielding and Effective Nuclear Charge We now return to an idea that we earlier stated as fact, without attempt at explanation. That is, that in an atom with several electrons, the energy ordering of subshells within a main shell is s < p (< d (< f)). The core electrons (in 1s, 2s, and 2p) are shown as a sphere that is inside the distance of maximum probability for the n = 3 shell. The sodium atom has a single electron in the n = 3 main shell, which must occupy one of the three subshells in Figure. Experimentally, it is found to occupy the 3s subshell, which we conclude must be more stable (lower energy) than the 3p and 3d subshells. Why? We can make a few observations about these plots that will hopefully tell us.

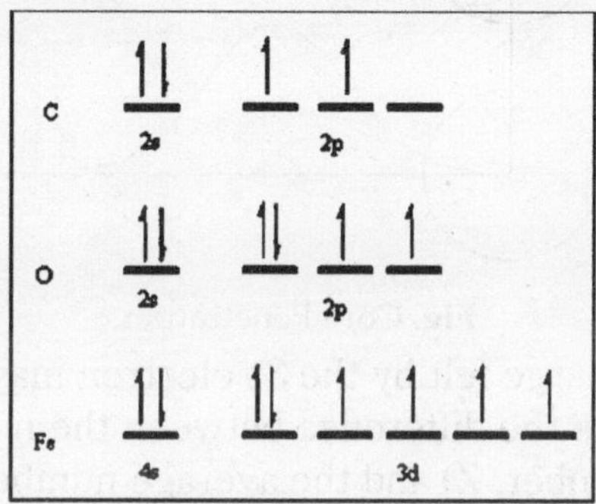

Fig. Configurations for Carbon, Oxygen, and Iron.

- The maxima in the 3s, 3p, and 3d radial plots all occur at approximately the same radius. However, the probability of finding the electron at distances closer than this most probable distance is greatest for the 3s subshell and least for the 3d subshell. We make this judgement based on the magnitude of the probability at distances less than the curve maximum. The electron in the n = 3 shell spends part of its time inside the electrons in lower lying subshells (1s, 2s, and 2p) because the radial curves in Figure above are non- zero inside the sphere representing the core electrons. We say that the n = 3 electron penetrates the core electron cloud.

The parentheses remind us that these subshells do not occur in shells with n < 3 or 4. We will try to develop an understanding of why this is true in terms of the radial plots for the 3s, 3p, and 3d subshells of the sodium atom shown qualitatively in Figure.

- Generally, the greater the amount of penetration, the more on average the electron feels the positive attraction of the nucleus, and the lower its potential energy. The amount of penetration is greatest for the 3s subshell and least for the 3d. The electron therefore occupies 3s.

The penetration of the core electron cloud by the 3s electron is of course incomplete, so the 3s electron does not feel the full effect of the 11 positive charges in the nucleus. The core electrons shield the 3s electron to some extent from the nucleus by coming between them.

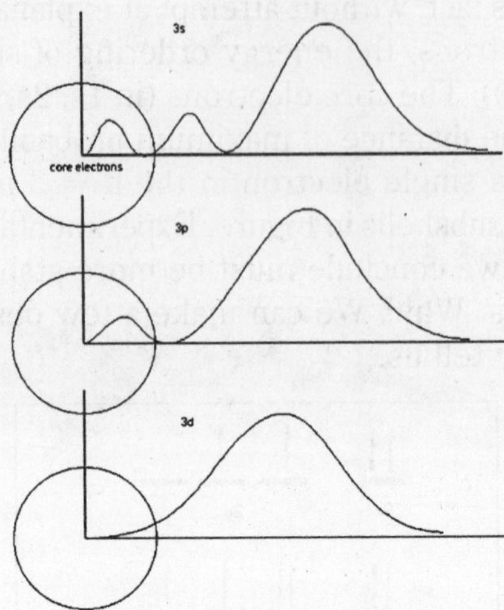

Fig. Core Penetration

The net positive charge felt by the 3s electron may be called the effective nuclear charge, Z_{eff}. It is the difference between the number of protons in the nucleus (the atomic number, Z) and the average number of electrons between the 3s electron and the nucleus, which we symbolize S:

$$: Z_{eff} = Z–S$$

S, the so-called shielding factor, is related to but always less than the number of shielding electrons. It is less than the number of shielding electrons because the shielded electron penetrates the core to some extent. Calculation of the value of S for a particular atom is somewhat complex; consequently, we will not attempt it.

Instead, we will discuss a quantity called the core charge, Z_{core}, which is an easily-calculated quantity that roughly parallels Z_{eff} in value.

The core charge of an atom is the difference between the number of protons in the nucleus and the number of core electrons. Core electrons are those in completely filled shells below the valence shell.

The core charge therefore measures the net positive charge pulling on the valence electrons. The core charge is very simple to calculate.

For example, the core charge for lithium is its atomic number (the number of protons) minus the number of electrons in filled shells (the atomic number of the preceding noble gas): Z_{core}(Li) = 3–2 = 1.

Similarly, the core charges for carbon and fluorine are 6-2 = 4 and 9-2 = 7, respectively.

Core charges for the period 3 elements are worked out below:

Note that the core charge for an element is equal to the number of valence electrons (which is the last digit of the group number). Two trends in Z_{core} are of importance to us. First, Z_{core} increases as we proceed left to right across a row of the periodic table, because the number of protons in the nucleus increases while the number of core electrons stays the same.

Element	*Z*	*Number of core electrons*	Z_{core}
Na	11	10	1
Mg	12	10	2
Al	13	10	3
Si	14	10	4
P	15	10	5
S	16	10	6
Cl	17	10	7
Ar	18	10	8

Second, Z_{core} remains constant down a family of the periodic table, because the elements in a family all have the same number of valence electrons. These simple trends are the basis for the experimentally observed variations in atomic size, ionization energy.

PERIODIC CHART OR TABLE OF THE ELEMENTS

THE PERIODIC CHART OR TABLE OF THE ELEMENTS

The Periodic Chart of the Elements is just a way to arrange the elements to show a large amount of information and organization.

As you read across the chart from right to left, a line of elements is a Period. As you read down the chart from top to bottom, a line of elements is a Group or Family. We number the elements, beginning with hydrogen, number one, in integers up to the largest number. The integer number in the box with the element symbol is the atomic number of the element and also the number of protons in each atom of the element.

PROPERTIES OF MATTER

The Periodic Chart is based on the properties of matter. A property is a quality or trait or characteristic. We can describe, identify, separate, and classify by properties. How would you describe a person? A young man impressed with a young lady might describe her, "She has long dark hair that she keeps in a pony-tail, brown eyes, a long neck, and a very light complexion. She is about 180 centimeters tall and has pierced ears."

He has used some of her properties to describe her. You might be able to pick her out of a small group of people based on his description if it is not too inaccurate, too vague, or too biased. Similarly, you can collect a number of properties to describe an element or compound. The properties of the element or compound, though, are true for any amount of the material anywhere. South American gold is indistinguishable from South African gold by its properties.

There are two types of property of matter. Physical properties describe the material as it is. Chemical properties describe how a material reacts, with what it reacts, the amount of heat it produces as it reacts, or any other measurable trait that has to do with the combining power of the material. Properties might describe a comparative trait (denser than gold) or a measured trait (17.7 g/cc), a relative trait (17.7 specific gravity), or an entire table of measurements in a table or graph form (the density of the material through a range of temperatures).

Physical properties include such things as: colour, brittleness, malleability, ductility, electrical conductivity, density, magnetism, hardness, atomic number, specific heat, heat of vaporization, heat of fusion, crystalline configuration, melting temperature, boiling temperature, heat conductivity, vapor pressure, or tendency to dissolve in various liquids. These are only a few of the possible measurable physical properties.

Chemical properties include: whether a material will react with another material, the rate of reaction with that material, the amount of heat produced by the reaction with the material, at what temperature it will react, in what proportion it reacts, and the valence of elements.

We can separate or purify materials based on the properties. We can separate wheat from chaff by throwing the mix into the wind.

The less dense chaff is moved more by the wind than the denser wheat. We can separate a mixture of sand and iron filings by magnetism. The iron filings will stick to a magnet dragged through the mixture. We can separate ethyl alcohol (good old drinking alcohol) from water by boiling point. This process is called distillation. A mixture of water and insoluble material with alcohol mixed in it will release the alcohol as vapor at the boiling point of alcohol (78 °C). We can separate by solubility. A mixture of table salt and sand can be separated by adding water. The salt dissolves and the sand does not.

PERIODIC PROPERTIES

The periodic chart came about from the idea that we could arrange the elements, originally by atomic weight, in a scheme that would show similarity among groups. The original idea came from noticing how other elements combined with oxygen. Oxygen combines in some way with all the elements except the inert gases.

Each atom of oxygen combines with two atoms of any element in Group 1, the elements in the row below lithium. Each atom of oxygen combines one-to-one with any element in Group 2, the elements in the row below beryllium. From here as we investigate the groups from left to right across the Periodic Chart, the story is not quite so clear, but the pattern is there. Group 3 is the group below boron. All of these elements combine with oxygen at the ratio of one-and-a- half to one oxygen. Group 4, beginning with carbon, combines two to one with oxygen. The group of transition elements (numbers 21-30 and 39-48 and 71-80 and 103 up) have never been adequately placed into the original scheme relating to oxygen. The transition elements vary in the ways they can attach to oxygen, but in a manner that is not so readily apparent by the simple scheme. Gallium, element number thirty-one, is the crowning glory of the Periodic Chart as first proposed by Mendeleev. Dmitri Ivanovich Mendeleev first proposed the idea that the elements could be arranged in a periodic fashion. He left a space for gallium below aluminum, naming it eka- aluminum, and predicting the properties of gallium fairly closely. The element was found some years later just as Mendeleev had predicted. Mendeleev also accurately predicted the properties of other elements.

Most Periodic Charts have two rows of fourteen elements below the main body of the chart. These two rows, the Lanthanides and Actinides really should be in the chart from numbers 57 - 70 and from 89 - 102. To show this, there would have to be a gulf of fourteen element spaces between numbers 20 - 21 and numbers 38 - 39.

This would make the chart almost twice as long as it is now. The Lanthanides belong to Period 6, and the Actinides belong to Period 7. In basic Chemistry courses you will rarely find much use for any of the Lanthanides or Actinides, with the possible exception of Element #92, Uranium. No element greater than #92 is found in nature. They are all man-made elements, if you would like to call them that.

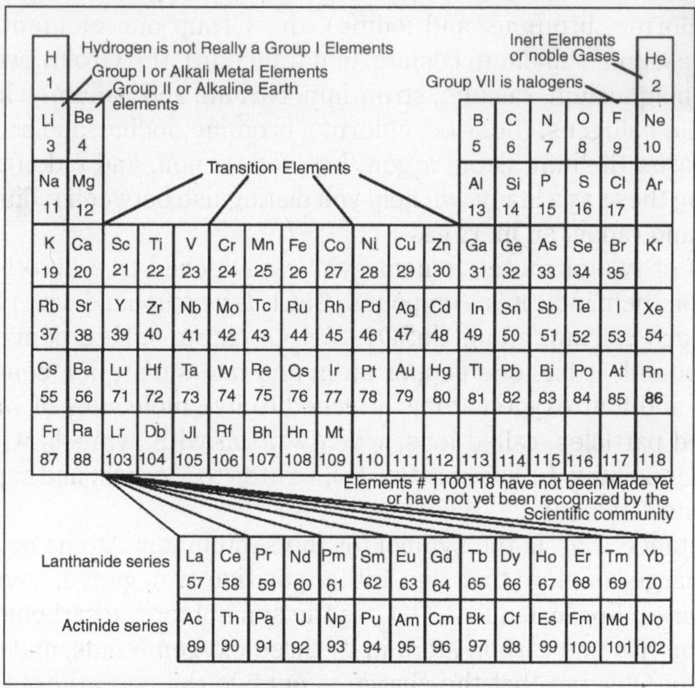

Fig. Periodic Chart of the Elements

None of the elements greater than #83 have any isotope that is completely stable. This means that all the elements larger than bismuth are naturally radioactive. The Lanthanide elements are so rare that you are not likely to run across them in most beginning chemistry classes.

Another oddity of the Periodic Chart is that hydrogen does not really belong to Group I — or any other group. Despite being over seventy per cent of the atoms in the known universe, hydrogen is a unique element.

ELEMENT, ION, AND COMPOUND SYMBOLS

For every element there is one and only one upper case letter. There may or may not be a lower case letter with it. When written in chemical equations, we represent the elements by the symbol alone with no charge attached.

The seven exceptions to that are the seven elements that are in gaseous form as a diatomic molecule, that is, two atoms of the same element attached to each other. The list of these elements is best memorized. They are: hydrogen, nitrogen, oxygen, fluorine, chlorine, bromine, and iodine. The chemical symbols for these diatomic gases are: H_2, N_2, O_2, F_2, Cl_2, Br_2, and I_2. Under some conditions oxygen makes a triatomic molecule, ozone, O_3. Ozone is not stable, so the oxygen atoms rearrange themselves into the more stable diatomic form.

Chemtutor highly recommends that a few short lists be well learned for immediate recognition. The diatomic gases (hydrogen, nitrogen, oxygen, fluorine, chlorine, bromine, and iodine), the Group one elements (lithium, sodium, potassium, rubidium, cesium, and francium), the Group two elements (beryllium, magnesium, calcium, strontium, barium, and radium), Group seven elements, the halogens, (fluorine, chlorine, bromine, iodine, and astatine), and the noble gases (helium, neon, argon, krypton, xenon, and radon). If nothing else, learning these as a litany will help you distinguish between radium, a Group 1 element, and radon, an inert gas.

Groups of two or more element symbols attached to each other without any charge on them indicate a compound. $CaCl_2$ is a compound with two chlorine atoms for each calcium atom. $CuSO_4 \cdot 5H_2O$, cupric sulfate pentahydrate, is also a compound. It has one copper atom and one sulfate ion consisting of a sulfur atom and four oxygen atoms attached to five molecules of water.

Charged particles, called ions, when written with symbols will have the charge, either positive (+) or negative (–), written to the right and superscripted to the chemical symbol.

For instance, Na^+ is the symbol for the sodium ion. Atoms or polyatomic ions with charges of more than one, either positive or negative, have a number with the charge. For instance $(CO_3)_2^-$ is the symbol for the carbonate ion. The carbonate ion has one carbon atom in it, three oxygen atoms, and a charge of negative two. Observe that the charge is outside the parentheses, indicating that the charge is from the polyatomic ion as a whole.

CATEGORIES OF ELEMENTS

What Chemtutor calls 'categories of elements' include; metals, non-metals, semi-metals, noble gases, and hydrogen.

H 1																	He 2
Li 3	Be 4											B 5	C 6	N 7	O 8	F 9	Ne 10
Na 11	Mg 12											Al 13	Si	P 15	S 16	Cl 17	Ar 18
K 19	Ca 20	Sc 21	Ti 22	V 23	Cr 24	Mn 25	Fe 26	Co 27	Ni 28	Cu 29	Zn 30	Ga 31	Ge 32	As 33	Se 34	Br 35	Kr 36
Rb 37	Sr 38	Y 39	Zr 40	Nb 41	Mo 42	Tc 43	Ru 44	Rh 45	Pd 46	Ag 47	Cd 48	In 49	Sn 50	Sb 51	Te 52	I 53	Xe 54
Cs 55	Ba 56	Lu 71	Hf 72	Ta 73	W 74	Re 75	Os 76	Ir 77	Pt 78	Au 79	Hg 80	Tl 81	Pb	Bi 83	Po 84	At 85	Rn 86
Fr 87	Ra 88	Lr 103	Db 104	Jl 105	Rf 106	Bh 107	Hn 108	Mt 109	110	111	112	113	114	115	116		

La 57	Ce 58	Pr 59	Nd 60	Pm 61	Sm 62	Eu 63	Gd 64	Tb 65	Dy 66	Ho 67	Er 68	Tm 69	Yb 70
Ac 89	Th 90	Pa 91	U 92	Np 93	Pu 94	Am 95	Cm 96	Bk 97	Cf 98	Es 99	Fm 100	Md 101	No 102

Fig. Periodic Chart of the Elements

Consider a staircase-shaped line on the Periodic Chart starting between boron and aluminum turns to be between aluminum and silicon then down between silicon and germanium, between germanium and arsenic, between arsenic and antimony, between antimony and tellurium, between tellurium and polonium, and between polonium and astatine. This is the line between metal and non-metal elements.

Metal elements are to the left and down from the line and non-metal elements are to the right and up from the line. Well, that's not exactly true. There is a line of non-metal elements, Group 8, or Group 18, or Group 0, whichever way you count them, the noble or inert gases that are really an entire Group and category to themselves. Hydrogen is a unique element, the only member of its own Group and category.

NOBLE GASES

The noble gases, or inert gases, have the following properties: For the most part, they do not make chemical combinations with any elements. There have been some compounds made with the noble gases, but only with difficulty.

There are certainly no natural compounds with this group. They are all gases at room temperature. They all have very low boiling and melting points. They all put out a colour in the visible wavelengths when a low pressure of the gas is put into a tube and a high voltage current is run through the tube. This type of tube is called a neon light whether the tube has neon in it or not. The inert gases are non-metals because they are not metals, but they are significantly different from the other non-metals. As closely akin as all the noble gases are to each other, they should surely be considered a separate group.

METALS

By far the largest category of elements on the Periodic Chart is the metal elements. Metals share a set of properties that are not as universal to them as the inert gases. Metal elements usually have the following properties: They have one, two, or three electrons on the outside electron shell. The outside electrons make it more likely that the metal will lose electrons, making positive ions. The ions of metals are usually plus one, plus two, or plus three in charge. Metals tend to lose electrons to become stable. They will attach to other elements with ionic bonds almost exclusively. When metal atoms are together in a group, there is a swarm of semi-loose electrons around the atoms. These electrons move about freely among the metal atoms making what is called an electron gas. The electron gas accounts for the shininess of metals. When there is a smooth surface on the metal it will reflect electromagnetic waves (to include visible light) in an organized manner. The shininess is also called metallic luster.

The same electron gas accounts for the cohesive tendencies of metals. Cohesive means the material clings to itself. This property can be easily seen with mercury. Mercury atoms cling to other mercury atoms or other metal atoms with an incredible tenacity. This same cohesion of metals occurs in the solid state. Silver is very malleable. That means that if you hit it, the material would more likely change shape than shatter.

At one time US half dollar coins were made of ninety per cent silver. It is illegal to deface money, but school children would take a spoon and beat the sides of the silver half dollars until the edges curled inward. When the center became the right size, it was taken out to make a silver ring beaten to fit your finger. Wire is made by pulling metals through a die. The metal coheres to itself so much that it will reshape itself to the shape of the die as it passes through the hole in the die. This property of being able to be pulled through a die to make wire is called ductility (from the Latin *ducere*, to pull). The presence of the electron gas makes metals good conductors of electricity. Again due to the cohesive property, metals have high melting and boiling points. Almost all metals are solids at room temperature. Metals are usually good conductors of heat. Active metals react with acids. Some very active metals will react with water. Metal elements tend to be denser than non-metals.

NON-METALS

The properties of non-metals are not as universal to them as the metals; there is a great deal of variation among this group. Non-metals have the following properties: Non-metals usually have four, five, six, or seven electrons in the outer shell. When they join with other elements non-metals can either share electrons in a covalent bond or gain electrons to become a negative ion and make an ionic bond. When non-metal elements join by covalent bonds, it is usually to other non-metals. Non-metals can attach together with covalent bonds to make a group of (usually non-metal) elements with a common charge called a radical or polyatomic ion. Elemental non-metals often have a dull appearance. They are more likely to be brittle, or shatter when struck. Although not a constant rule, non- metals tend to have lower melting and boiling points than metals and the solids tend to be less dense. Non-metals are not as cohesive as metals and certainly not ductile. Non-metals are not usually good conductors of heat or electricity. Many non-metals form diatomic or polyatomic molecules with other atoms of the same element. Many non-metals have more than one form of the free element, called allotropes, that appear in different conditions. (The word free here means that the element is unattached to other types of atom, not that it has a monetary value of zero.)

SEMI-METALS

We have pretended that there is a sharp dividing line between the metals and non-metals. This is not the case. The staircase-shaped line between metals and non-metals has several elements on or near it that have properties somewhere between the two categories.

By having three electrons in the outside shell, boron should be a metal element. It is not. Boron is more likely to form covalent bonds like a non-metal than donate electrons like aluminum, the next element down the chart in the same group. Aluminum is definitely a metal in most of its traits, but it has its own idiosyncrasy. Aluminum is amphoteric; it reacts with both acids and bases.

Silicon, germanium, arsenic, antimony, and tellurium are on the line between metals and non-metals and exhibit some of the qualities of both. These elements do not really comprise a clear-cut category, but, due to the mix of properties they show, they are often lumped into a classification called semi-metals. Many of the elements on the line are semiconductors of electricity, meaning that they have the ability to conduct electricity somewhere between almost none and full conduction. This property is useful in the electronics industry.

HYDROGEN

We have failed to include hydrogen in any of the categories, for good reasons. Hydrogen just does not match anything else. More than ninety-nine-point-nine per cent of hydrogen is just one proton and one electron. A very small proportion (one atom in several thousand) of hydrogen is deuterium, one proton, one neutron, and one electron. An even smaller portion (one hundred atoms per million billion) of hydrogen is tritium, one proton, two neutrons, and one electron. When a hydrogen atom gains an electron, it becomes a negative ion. The negative hydrogen ion, called hydride ion, can be attached to metals, but it is not seen in nature because it is not stable in water. The positive hydrogen ion is what is responsible for acids. There really is no such thing as a (positive) hydrogen ion. Having only a proton and an electron, hydrogen becomes only a proton if it loses its electron. Loose protons attach themselves to a water molecule to make H_3O^+ ion, a hydronium ion.

This hydronium is the real chemical that produces the properties of acids. Elemental hydrogen is a diatomic gas. Except for having a valence of +1, hydrogen has few other similarities with the Group 1 elements. Hydrogen makes covalent bonds between other hydrogen atoms or other non-metals.

GROUPS OR FAMILIES OF THE PERIODIC CHART

This section is not intended as an exhaustive study of the groups of the Periodic Chart, but a quick-and-dirty overview of the groups as a way to see the organization of the chart. Many texts and charts will label the groups with different names and numbers.

Chemtutor will attempt to give some standard numbers and identify the elements in those groups so there is no question about which ones we are describing. It is a good idea to have a copy of the Periodic Chart available as you go through this section.

Group I (1) elements, lithium, sodium, potassium, rubidium, cesium, and francium, are also called the alkali metal elements. They are all very soft metals that are not found free in nature because they react with water. In the element form they must be stored under kerosene to keep them from reacting with the humidity in the air. They all have a valence of plus one because they have one and only one electron in the outside shell.

All of the alkali metals show a distinctive colour when their compounds are put into a flame. Spectroscopy (dividing up the spectrum so you can see the individual frequencies) of the coloured light from the flame test shows strong emission lines from the elements. The lightest of them are the least reactive. Activity increases as the element is further down the Periodic Chart. Lithium reacts leisurely with water. Cesium reacts very violently. Very few of the salts of Group 1 elements are not soluble in water. The lightest of the alkali metals are very common in the earth's crust. Francium is both rare and radioactive.

Group II (2) elements, beryllium, magnesium, calcium, strontium, barium, and radium, all have two electrons in the outside ring, and so have a valence of two. Also called the alkaline earth metals, Group 2 elements in the free form are slightly soft metals. Magnesium and calcium are common in the crust of the earth. Group 3 elements, boron, aluminum, gallium, indium, and thallium, are a mixed group. Boron has mostly non- metal properties. Boron will bond covalently by preference. The rest of the group are metals. Aluminum is the only one common in the earth's crust. Group 3 elements have three electrons in the outer shell, but the larger three elements have valences of both one and three. Group 4 elements, carbon, silicon, germanium, tin, and lead, are not a coherent group either. Carbon and silicon bond almost exclusively with four covalent bonds. They both are common in the earth's crust.

Germanium is a rare semi-metal. Tin and lead are definitely metals, even though they have four electrons in the outside shell. Tin and lead have some differences in their properties from metal elements that suggest the short distance from the line between metals and non-metals (semi-metal weirdness). They both have more than one valence and are both somewhat common in the crust of the earth.

Group 5 is also split between metals and non-metals. Nitrogen and phosphorus are very definitely non-metals. The element nitrogen as a diatomic molecule forms about eighty per cent of the atmosphere. In the rare instances that nitrogen and phosphorus form ions, they form triple negative ions. Nitride (N^{-3}) and phosphide (P^{-3}) ions are unstable in water, and so are not found in nature. All of the Group 5 elements have five electrons in the outer shell. For the smaller elements it is easier to complete the shell to become stable, so they are non-metals and are more likely to form covalent bonds than ionic bonds.

The larger elements in the group, antimony and bismuth, tend to be metals because it is easier for them to donate the five electrons than to attract three more. Arsenic, antimony and bismuth have valences of +3 or +5. Arsenic is very much a semi-metal, but all three of them show some semi-metal weirdness, such as brittleness as a free element. Group VII (6 or 16) elements, oxygen, sulfur, selenium, and tellurium, have six electrons in the outside shell. We are not concerned with polonium as a Group 6 element. It is too rare, too radioactive, and too dangerous for us to even consider in a basic course. Tellurium is the only element in Group 6 that is a semi-metal. There are positive and negative ions of tellurium. Oxygen, sulfur, and selenium are true non-metals.

They have a valence of negative two as an ion, but they also bond covalently. Oxygen gas makes covalent double-bonded diatomic gas molecules that are about twenty per cent of the earth's atmosphere.. Oxygen and sulfur are common elements. Selenium has a property that may be from semi-metal weirdness; it conducts electricity much better when light is shining on it. Selenium is used in photocells for this property.

On some charts you will see hydrogen above fluorine in Group VII (7 or 17). Hydrogen does not belong there any more than it belongs above Group 1. Fluorine, chlorine, bromine, and iodine make up Group 7, the halogens. We can forget about astatine. It is too rare and radioactive to warrant any consideration here. Halogens have a valence of negative one when they make ions because they have seven electrons in the outer shell.

They are all diatomic gases as free elements near room temperature. They are choking poisonous gases. Fluorine and chlorine are yellow-green, bromine is reddish, and iodine is purple as a gas. All can be found attached to organic molecules. Chlorine is common in the earth's crust, much of it as the negative ion of salt, NaCl, in the oceans. Fluorine is the most active of them, and the activity decreases as the size of the halogen increases.

The inert gases or noble gases all have a complete outside shell of electrons. Helium is the only one that has only an "*s*" subshell filled, having only two electrons in the outer and only shell. All the others, neon, argon, krypton, xenon, and radon, have eight electrons in the outer shell. Since the electron configuration is most stable in this shape, the inert gases do not form natural compounds with other elements. The group is variously numbered as Group VIIIA, 8, 8A, 0, or 18. 'Group zero' seems to fit them nicely since it is easy to think of them as having a zero valence, that is no likely charge.

The Transition Elements make up a group between what Chemtutor has labeled Group 2 and Group 3. Transition elements are all metals. Very few of the transition elements have any non-metal properties. Within the transition elements many charts subdivide the elements into groups, but other than three horizontal groups, it is difficult to make meaningful distinctions among them. The horizontal groups are: iron, cobalt, and nickel; ruthenium, rhodium, and palladium; and osmium, iridium, and platinum. Iron is thought to be plentiful as a molten mass in the center of the earth.

Lanthanides and actinides are called the Inner Transition Elements. Lanthanides, elements 57 through 70, are also called the rare earth elements. They are all metal elements very similar to each other, but may be divided into a cerium and a yttrium group. They are often found in the same ores with other elements of the group. None are found in any great quantity in the earth's crust.

ENERGY SPECTRA AND NEUTRINO HYPOTHESIS

The hypothesis of tachyonic neutrinos deserves attention as being motivated by results of recent tritium decay experiments which are so far the most widely used kinematical method of direct determining the mass squared of the electron antineutrino, $m_{v_e}^2$. The energy spectral density describes how the energy (or variance) of a signal or a time series is distributed with frequency.

2

Theories of Relativity

INTRODUCTION

The theory of relativity, or simply relativity, generally encompasses two theories of Albert Einstein: special relativity and general relativity. (The word *relativity* can also be used in the context an older theory, that of Galilean invariance.)

Concepts introduced by the theories of relativity include:

- Measurements of various quantities are *relative* to the velocities of observers. In particular, space and time can dilate.
- Spacetime: space and time should be considered together and in relation to each other.
- The speed of light is nonetheless invariant, the same for all observers.

The term "theory of relativity" was based on the expression "relative theory" used by Max Planck in 1906, who emphasized how the theory uses the principle of relativity.

GENERAL RELATIVITY

EINSTEIN'S PARABLE

In Einstein's little book *Relativity: the Special and the General Theory*, he introduces general relativity with a parable.

He imagines going into deep space, far away from gravitational fields, where any body moving at steady speed in a straight line will continue in that state for a very long time. He imagines building a space station out there - in his words, "a spacious chest resembling a room with an observer inside who is equipped with apparatus."

Einstein points out that there will be no gravity, the observer will tend to float around inside the room.

But now a rope is attached to a hook in the middle of the lid of this "chest" and an unspecified "being" pulls on the rope with a constant force. The chest and its contents, including the observer, accelerate "upwards" at a constant rate. How does all this look to the man in the room? He finds himself moving towards what is now the "floor" and needs to use his leg muscles to stand.

If he releases anything, it accelerates towards the floor, and in fact all bodies accelerate at the same rate.

If he were a normal human being, he would assume the room to be in a gravitational field, and might wonder why the room itself didn't fall. Just then he would discover the hook and rope, and conclude that the room was suspended by the rope. Einstein asks: should we just smile at this misguided soul? His answer is no - the observer in the chest's point of view is just as valid as an outsider's.

In other words, being inside the (from an outside perspective) uniformly accelerating room is physically equivalent to being in a uniform gravitational field. This is the basic postulate of general relativity. Special relativity said that all inertial frames were equivalent. General relativity extends this to accelerating frames, and states their equivalence to frames in which there is a gravitational field. This is called the Equivalence Principle.

The acceleration could also be used to cancel an existing gravitational field—for example, inside a freely falling elevator passengers are weightless, conditions are equivalent to those in the unaccelerated space station in outer space.

It is important to realise that this equivalence between a gravitational field and acceleration is only possible because the gravitational mass is exactly equal to the inertial mass.

There is no way to cancel out electric fields, for example, by going to an accelerated frame, since many different charge to mass ratios are possible. As physics has developed, the concept of fields has been very valuable in understanding how bodies interact with each other. We visualize the electric field lines coming out from a charge, and know that something is there in the space around the charge which exerts a force on another charge coming into the neighbourhood.

We can even compute the energy density stored in the electric field, locally proportional to the square of the electric field intensity. It is tempting to think that the gravitational field is quite similar—after all, it's another inverse square field.

Evidently, though, this is not the case. If by going to an accelerated frame the gravitational field can be made to vanish, at least locally, it cannot be that it stores energy in a simply defined local way like the electric field. We should emphasize that going to an accelerating frame can only cancel a *constant* gravitational field, of course, so there is no accelerating frame in which the whole gravitational field of, say, a massive body is zero, since the field necessarily points in different directions in different regions of the space surrounding the body.

SOME CONSEQUENCES OF THE EQUIVALENCE PRINCIPLE

Consider a freely falling elevator near the surface of the earth, and suppose a laser fixed in one wall of the elevator sends a pulse of light horizontally across to the corresponding point on the opposite wall of the elevator. Inside the elevator, where there are no fields present, the environment is that of an inertial frame, and the light will certainly be observed to proceed directly across the elevator. Imagine now that the elevator has windows, and an outsider at rest relative to the earth observes the light.

As the light crosses the elevator, the elevator is of course accelerating downwards at g, so since the flash of light will hit the opposite elevator wall at precisely the height relative to the elevator at which it began, the outside observer will conclude that the flash of light also accelerates downwards at g. In fact, the light could have been emitted at the instant the elevator was released from rest, so we must conclude that light falls in an initially parabolic path in a constant gravitational field.

Of course, the light is traveling very fast, so the curvature of the path is small! Nevertheless, *the Equivalence Principle forces us to the conclusion that the path of a light beam is bent by a gravitational field*. The curvature of the path of light in a gravitational field was first detected in 1919, by observing stars very near to the sun during a solar eclipse.

The deflection for stars observed very close to the sun was 1.7 seconds of arc, which meant measuring image positions on a photograph to an accuracy of hundredths of a millimeter, quite an achievement at the time. One might conclude from the brief discussion above that a light beam in a gravitational field follows the same path a Newtonian particle would if it moved at the speed of light. This is true in the limit of small deviations from a straight line in a constant field, but is not true even for small deviations for a spatially varying field, such as the field near the sun the starlight travels through in the eclipse experiment mentioned above.

We could try to construct the path by having the light pass through a series of freely falling elevators, all falling towards the centre of the sun, but then the elevators are accelerating relative to each other (since they are all falling along *radii*), and matching up the path of the light beam through the series is tricky. If it is done correctly (as Einstein did) it turns out that the angle the light beam is bent through is twice that predicted by a naïve Newtonian theory.

What happens if we shine the pulse of light vertically *down* inside a freely falling elevator, from a laser in the centre of the ceiling to a point in the centre of the floor? Let us suppose the flash of light leaves the ceiling at the instant the elevator is released into free fall. If the elevator has height h, it takes time h/c to reach the floor. This means the floor is moving downwards at speed gh/c when the light hits.

Question: Will an observer on the floor of the elevator see the light as Doppler shifted? The answer has to be no, because inside the elevator, by the Equivalence Principle, conditions are identical to those in an inertial frame with no fields present. There is nothing to change the frequency of the light. This implies, however, that to an outside observer, stationary in the earth's gravitational field, the frequency of the light *will* change.

This is because he will agree with the elevator observer on what was the initial frequency f of the light as it left the laser in the ceiling (the elevator was at rest relative to the earth at that moment) so if the elevator operator maintains the light had the same frequency f as it hit the elevator floor, which is moving at gh/c relative to the earth at that instant, the earth observer will say the light has frequency $f(1 + v/c) = f(1 + gh/c^2)$, using the Doppler formula for very low speeds.

We conclude from this that light shining downwards in a gravitational field is shifted to a higher frequency. Putting the laser in the elevator floor, it is clear that light shining upwards in a gravitational field is red-shifted to lower frequency. Einstein suggested that this prediction could be checked by looking at characteristic spectral lines of atoms near the surfaces of very dense stars, which should be red-shifted compared with the same atoms observed on earth, and this was confirmed. This has since been observed much more accurately.

An amusing consequence, since the atomic oscillations which emit the radiation are after all just accurate clocks, is that *time passes at different rates at different altitudes*. The US atomic standard clock, kept at 5400 feet in Boulder, gains 5 microseconds per year over an identical clock almost at sea level in the Royal Observatory at Greenwich, England. Both clocks are accurate to one microsecond per year. This means you would age more slowly if you lived on the surface of a planet with a large gravitational field. Of course, it might not be very comfortable.

GENERAL RELATIVITY AND THE GLOBAL POSITIONING SYSTEM

Despite what you might suspect, the fact that time passes at different rates at different altitudes has significant practical consequences. An important *everyday* application of general relativity is the Global Positioning System. A GPS unit finds out where it is by detecting signals sent from orbiting satellites at precisely timed intervals.

If all the satellites emit signals simultaneously, and the GPS unit detects signals from four different satellites, there will be three relative time delays between the signals it detects. The signals themselves are encoded to give the GPS unit the precise position of the satellite they came from at the time of transmission. With this information, the GPS unit can use the speed of light to translate the detected time delays into distances, and therefore compute its own position on earth by triangulation.

But this has to be done very precisely! Bearing in mind that the speed of light is about one foot per nanosecond, an error of 100 nanoseconds or so could, for example, put an airplane off the runway in a blind landing. This means the clocks in the satellites timing when the signals are sent out must certainly be accurate to 100 nanoseconds a day.

That is one part in 10^{12}. It is easy to check that both the special relativistic time dilation correction from the speed of the satellite, and the general relativistic gravitational potential correction are much greater than that, so the clocks in the satellites must be corrected appropriately. (The satellites go around the earth once every twelve hours, which puts them at a distance of about four earth radii. The calculations of time dilation from the speed of the satellite, and the clock rate change from the gravitational potential, are left as exercises for the student In fact, Ashby reports that when the first Cesium clock was put in orbit in 1977, those involved were sufficiently skeptical of general relativity that the clock was not corrected for the gravitational redshift effect. But—just in case Einstein turned out to be right—the satellite was equipped with a synthesizer that could be switched on if necessary to add the appropriate relativistic corrections.

After letting the clock run for three weeks with the synthesizer turned off, it was found to differ from an identical clock at ground level by precisely the amount predicted by special plus general relativity, limited only by the accuracy of the clock. This simple experiment verified the predicted gravitational redshift to about one per cent accuracy! The synthesizer was turned on and left on..

SPECIAL RELATIVITY

Relativity is about the nature of space and time. It is not difficult mathematically; it makes us re-examine our ideas of space and time. Relativity is based upon two postulates. Accepting these postulates makes one examine their consequences and their agreement with experiments.

Einstein's special theory of relativity deals with how we observe events, particularly how we observe events from different frames of references.

POSTULATES OF RELATIVITY

The laws of physics are the same in all inertial frames of reference.

- This postulate does not say that the measured values of all physical quantities are the same for all reference frames. It is the laws of physics that relate these quantities to each other that are the same.

The speed of light in free space has the same value in all directions and in all inertial reference frames.

- Experimental proof (1964): Electrons were accelerated to various measured speeds. Their kinetic energies were determined by independent calorimetric methods. This experiment proved that an object with mass cannot travel at the speed of light. (Electrons have been accelerated to 0.999 999 999 95 c)

- If the speed of light is the same in all reference frames, then the speed of light emitted by a moving source should be the same as the speed of light emitted by a source that is at rest in a lab. This was proved in 1964 when the speed of light of gamma rays emitted during the decay of moving pions was determined to be the same as those emitted at rest.

Event A physical happening that occurs at a certain place and time. You can assign three space coordinants and one time coordinant to an event

SIMULTANEOUS PROCEEDINGS

One of the important consequences of the theory of relativity is that time can no longer be regarded as an absolute quantity. The time interval between two events depends upon the observer's reference frame. If two events occur at widely separated places, it is difficult to determine if they are simultaneous. One must take into account the time it takes for the light from the two events to reach the observer. Two events which are simultaneous to one observer are not necessarily simultaneous to another observer.

TIME DILATION AND EINSTEIN'S THEORY OF RELATIVITY PREDICTS

Einstein's theory of relativity predicts that time passes differently in one frame of reference than in another. If different observers measure the time interval between a pair of events, they will not agree about how long the time interval for the event was. The time dilation effect states that clocks moving relative to an observer are measured by that observer to run more slowly (as compared to clocks at rest). In other words, time is dilated (or stretched) for a moving object.

$$Dt = \gamma Dt_o$$

where Dt_o is the proper time interval, or that recorded at rest relative to an inertial reference frame

Dt is the dilated time interval

γ is the Lorentz factor

$$\gamma = [1 - (v/c)^2]^{-1/2}$$

An applet that will help you visualize space and time in special relativity.

RELATIVITY OF LENGTH MOVING

The length of a moving object, you must simultaneously measure both of its ends. The length of an object is measured to be shorter when it is moving relative to the observer than when it is at rest. In other words, the length of a moving object is always contracted, or shorter than its proper length (length contraction only occurs in the direction of motion).

$$L = L_o/\gamma$$

L_o is the proper length, or the length of the object at rest in an inertial reference frame

g is the Lorentz factor

MOMENTUM AND MASS ENERGY

We must redefine momentum as:

$$p = m_o v/\gamma$$

where m_o is the rest mass.

For very small values of v, this becomes simply $p=m_o v$

The mass of an object increases as its speed increases.

$$m = m_o/\gamma$$

A New Appear at Energy

While a steady force is applied to an object of rest mass, the object increases its speed. Work is done and its kinetic energy changes. As the speed of the object approaches c, the mass of the object increases. The work done on the object not only increases its speed but also its mass. Relativity thus predicts that mass is a form of energy. Einstein predicted that the kinetic energy of a particle is given by

$$KE = mc^2 - m_o c^2 \text{ or } mc^2 = m_o c^2 + KE$$

m is the mass of the particle traveling at speed v

mc^2 is the total energy E of the particle (assuming not potential energy)

We now have Einstein's famous formula

$$E = mc^2$$

Relativistic Calculation of Velocities

Einstein showed that since length and time are different in differnt reference frames, the old addition of velocities is no longer valid. For motion along a straight line,

$$v = (v' + u)/[1 + (uv'/c^2\}]$$

where v is the proper velocity (at rest)

v' is the velocity of the object at at velocity u

SCOPE OF RELATIVITY

The theory of relativity transformed theoretical physics and astronomy during the 20th century. When first published, relativity superseded a 200-year-old theory of mechanics stated by Isaac Newton. The theory of relativity overturned the concept of motion from Newton's day, by positing that all motion is relative. Time was no longer uniform and absolute. Physics could no longer be understood as space by itself, and time by itself. Instead, an added dimension had to be taken into account with curved spacetime. Time now depended on velocity, and contraction became a fundamental consequence at appropriate speeds.

In the field of physics, relativity catalyzed and added an essential depth of knowledge to the science of elementary particles and their fundamental interactions, along with ushering in the nuclear age. With relativity, cosmology and astrophysics predicted extraordinary astronomical phenomena such as neutron stars, black holes, and gravitational waves.

TWO-THEORY VIEW

The theory of relativity was representative of more than a single new physical theory. There are some explanations for this. First, special relativity was published in 1905, and the final form of general relativity was published in 1916. Second, special relativity fits with and solves for elementary particles and their interactions, whereas general relativity solves for the cosmological and astrophysical realm (including astronomy).

Third, special relativity was widely accepted in the physics community by 1920. This theory rapidly became a significant and necessary tool for theorists and experimentalists in the new fields of atomic physics, nuclear physics, and quantum mechanics.

Conversely, general relativity did not appear to be as useful. There appeared to be little applicability for experimentalists as most applications were for astronomical scales. It seemed limited to only making minor corrections to predictions of Newtonian gravitation theory. Its impact was not apparent until the 1930s. Finally, the mathematics of general relativity appeared to be incompre-hensibly dense.

Consequently, it was thought a small number of people in the world, at that time, could fully understand the theory in detail, but this has been discredited by Richard Feynman (video available on YouTube). Then, at around 1960 a critical resurgence in interest occurred which has resulted in making general relativity central to physics and astronomy. New mathematical techniques applicable to the study of general relativity substantially streamlined calculations.

From this, physically discernible concepts were isolated from the mathematical complexity. Also, the discovery of exotic astronomical phenomena in which general relativity was crucially relevant, helped to catalyze this resurgence.

The astronomical phenomena included quasars (1963), the 3-kelvin microwave background radiation (1965), pulsars (1967), and the discovery of the first black hole candidates (1971).

On the Theory of Relativity

Einstein stated that the theory of relativity belongs to a class of "principle-theories". As such it employs an analytic method. This means that the elements which comprise this theory are not based on hypothesis but on empirical discovery.

The empirical discovery leads to understanding the general characteristics of natural processes. Mathematical models are then developed which separate the natural processes into theoretical-mathematical descriptions. Therefore, by analytical means the necessary conditions that have to be satisfied are deduced. Separate events must satisfy these conditions. Experience should then match the conclusions.

The special theory of relativity and the general theory of relativity are connected. As stated below, special theory of relativity applies to all inertial physical phenomena except gravity. The general theory provides the law of gravitation, and its relation to other forces of nature.

Special Relativity

Special relativity is a theory of the structure of spacetime. It was introduced in Einstein's 1905 paper "On the Electrodynamics of Moving Bodies". Special relativity is based on two postulates which are contradictory in classical mechanics:

- The laws of physics are the same for all observers in uniform motion relative to one another (principle of relativity).
- The speed of light in a vacuum is the same for all observers, regardless of their relative motion or of the motion of the source of the light.

The resultant theory copes with experiment better than classical mechanics, *e.g.*, in the Michelson–Morley experiment that supports postulate 2, but also has many surprising consequences. Some of these are:

- Relativity of simultaneity: Two events, simultaneous for one observer, may not be simultaneous for another observer if the observers are in relative motion.
- Time dilation: Moving clocks are measured to tick more slowly than an observer's "stationary" clock.
- Length contraction: Objects are measured to be shortened in the direction that they are moving with respect to the observer.
- Mass–energy equivalence: $E = mc^2$, energy and mass are equivalent and transmutable.
- Maximum speed is finite: No physical object, message or field line can travel faster than the speed of light in a vacuum.

The defining feature of special relativity is the replacement of the Galilean transformations of classical mechanics by the Lorentz transformations.

General Relativity and Theory

General relativity is a theory of gravitation developed by Einstein in the years 1907–1915. The development of general relativity began with the equivalence principle, under which the states of accelerated motion and being at rest in a gravitational field (for example when standing on the surface of the Earth) are physically identical.

The upshot of this is that free fall is inertial motion; an object in free fall is falling because that is how objects move when there is no force being exerted on them, instead of this being due to the force of gravity as is the case in classical mechanics. This is incompatible with classical mechanics and special relativity because in those theories inertially moving objects cannot accelerate with respect to each other, but objects in free fall do so. To resolve this difficulty Einstein first proposed that spacetime is curved. In 1915, he devised the Einstein field equations which relate the curvature of spacetime with the mass, energy, and momentum within it.

Some of the consequences of general relativity are:

- Clocks run more slowly in deeper gravitational wells. This is called gravitational time dilation.
- Orbits process in a way unexpected in Newton's theory of gravity. (This has been observed in the orbit of Mercury and in binary pulsars).
- Rays of light bend in the presence of a gravitational field.
- Rotating masses "drag along" the spacetime around them; a phenomenon termed "frame-dragging".
- The Universe is expanding, and the far parts of it are moving away from us faster than the speed of light.

Technically, general relativity is a theory of gravitation whose defining feature is its use of the Einstein field equations. The solutions of the field equations are metric tensors which define the topology of the spacetime and how objects move inertially.

TESTS OF SPECIAL RELATIVITY

Like all falsifiable scientific theories, relativity makes predictions that can be tested by experiment. In the case of special relativity, these include the principle of relativity, the constancy of the speed of light, and time dilation. The predictions of special relativity have been confirmed in numerous tests since Einstein published his paper in 1905, but three experiments conducted between 1881 and 1938 were critical to its validation. These are the Michelson–Morley experiment, the Kennedy–Thorndike experiment, and the Ives–Stilwell experiment.

Einstein derived the Lorentz transformations from first principles in 1905, but these three experiments allow the transformations to be induced from experimental evidence.

Maxwell's equations – the foundation of classical electromagnetism – describe light as a wave which moves with a characteristic velocity. The modern view is that light needs no medium of transmission, but Maxwell and his contemporaries were convinced that light waves were propagated in a medium, analogous to sound propagating in air, and ripples propagating on the surface of a pond. This hypothetical medium was called the luminiferous aether, at rest relative to the "fixed stars" and through which the Earth moves. Fresnel's partial ether dragging hypothesis ruled out the measurement of first-order (v/c) effects, and although observations of second-order effects (v^2/c^2) were possible in principle, Maxwell thought they were too small to be detected with then-current technology.

The Michelson–Morley experiment was designed to detect second order effects of the "aether wind" – the motion of the aether relative to the earth. Michelson designed an instrument called the Michelson interferometer to accomplish this. The apparatus was more than accurate enough to detect the expected effects, but he obtained a null result when the first experiment was conducted in 1881, and again in 1887. Although the failure to detect an aether wind was a disappointment, the results were accepted by the scientific community.

In an attempt to salvage the aether paradigm, Fitzgerald and Lorentz independently created an *ad hoc* hypothesis in which motion the length of material bodies changes according to their motion through the aether. This was the origin of Fitzgerald-Lorentz contraction, and their hypothesis had no theoretical basis. The interpretation of the null result of the Michelson–Morley experiment is that the round-trip travel time for light is isotropic (independent of direction), but the result alone is not enough to discount the theory of the aether or validate the predictions of special relativity.

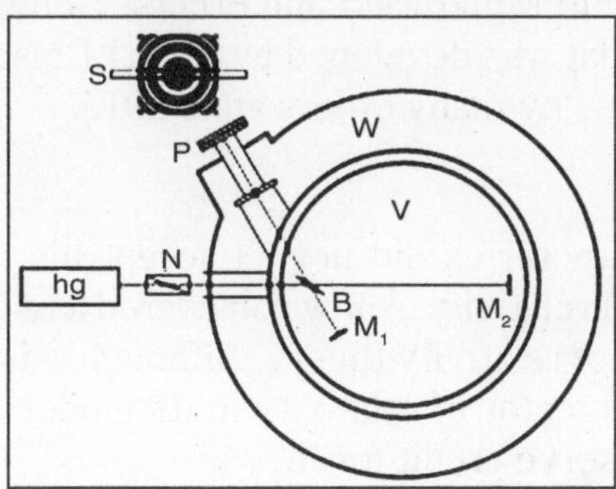

Fig. The Kennedy–Thorndike Experiment shown with Interference Fringes.

While the Michelson–Morley experiment showed that the velocity of light is isotropic, it said nothing about how the magnitude of the velocity changed (if at all) in different inertial frames. The Kennedy–Thorndike experiment was designed to do that, and was first performed in 1932 by Roy Kennedy and

Edward Thorndike. They obtained a null result, and concluded that "there is no effect ... unless the velocity of the solar system in space is no more than about half that of the earth in its orbit". That possibility was thought to be too coincidental to provide an acceptable explanation, so from the null result of their experiment it was concluded that the round-trip time for light is the same in all inertial reference frames.

The Ives–Stilwell experiment was carried out by Herbert Ives and G.R. Stilwell first in 1938 and with better accuracy in 1941. It was designed to test the transverse Doppler effect – the redshift of light from a moving source in a direction perpendicular to its velocity – which had been predicted by Einstein in 1905. The strategy was to compare observed Doppler shifts with what was predicted by classical theory, and look for a Lorentz factor correction. Such a correction was observed, from which was concluded that the frequency of a moving atomic clock is altered according to special relativity.

Those classic experiments have been repeated many times with increased precision. Other experiments include, for instance, relativistic energy and momentum increase at high velocities, time dilation of moving particles, and modern searches for Lorentz violations.

Tests of General Relativity

Also general relativity was confirmed many times, the classic experiments being the perihelion precession of Mercury's orbit, the deflection of light by the Sun, and the gravitational redshift of light. Other tests confirmed the equivalence principle and frame dragging. The history of special relativity consists of many theoretical results and empirical findings obtained by Albert Michelson, Hendrik Lorentz, Henri Poincaré and others. It culminated in the theory of special relativity proposed by Albert Einstein, and subsequent work of Max Planck, Hermann Minkowski and others. General relativity (GR) is a theory of gravitation that was developed by Albert Einstein between 1907 and 1915, with contributions by many others after 1915.

Minority Views

Einstein's contemporaries did not all accept his new theories at once. However, the theory of relativity is now considered as a cornerstone of modern physics, see Criticism of relativity theory. Although it is widely acknowledged that Einstein was the creator of relativity in its modern understanding, some believe that others deserve credit for it.

SPECIAL RELATIVITY POSTULATES A FOUR-DIMENSIONAL SPACE-TIME

Special Relativity postulates a four-dimensional space-time with a radically different spatio-temporal structure. Instead of having a pure temporal structure and a pure spatial structure, there is a single Relativistic "distance" between events, (The scare quotes around "distance" must be taken seriously, as the quantity is not at all like a spatial distance.) How can this spatio-temporal structure be specified?

The easiest method, albeit a bit roundabout, is by means of coordinates. Here we will take the analogy with Euclidean geometry quite seriously. As we saw, even though Euclidean geometry has no need of coordinate systems, still the spatial structure of a Euclidean space can be specified in this way: a Euclidean space is a space that admits of Cartesian coordinates. More specifically, a three-dimensional Euclidean space has a structure of distance relations among its points such that each point can be given coordinates (*x, y, z*) and the distance between any pair of points is

$$\sqrt{(x_p - x_q)^2 + (y_p - y_q)^2 + (z_p - z_q)^2}.$$

In exactly the same way, we can specify the spatio-temporal structure of Minkowski space-time, the space-time of Special Relativity. Minkowski space-time is a four-dimensional manifold that admits of Lorentz coordinates (or Lorentz frames). A Lorentz frame is a system of coordinates (*t, x, y, z*) such that the Relativistic spatio-temporal "distance" between any pair of events *p* and *q* is

$$\sqrt{\left(t_p - t_q\right)^2 - \left(x_p - x_q\right)^2 - \left(y_p - y_q\right)^2 - \left(z_p - z_q\right)^2}$$

Written this way, the similarity with the example of Cartesian coordinates on Euclidean space is manifest: the only difference is the minus signs in place of plus signs. The consequences of that small mathematical difference are profound.

Previous to investigating the nature of this spatio-temporal structure, we should renew some of our caveats. First, there is always the temptation to invest the *coordinates* with some basic physical significance. For example, it is very natural to regard the coordinate we are calling *t* as a *time* coordinate, and to suppose that it has something to do with what is measured by clocks. But as of yet, we have said nothing to justify that interpretation. The Lorentz coordinates are just some way or other of attaching numbers to points such that the quantity defined above is proportional to the spatio-temporal "distance" between events. Indeed, just as there are many ways to lay down Cartesian coordinates on a Euclidean plane, systems that differ with respect to the origin and orientation of the coordinate grid, so there are many ways to lay down Lorentz coordinates in Minkowski space-time. Different systems will assign different *t* values to the points, and will disagree about, for example, the difference in *t* value between two events. We do not invest these differences with any physical significance: since the various systems *agree* about the quantity defined above, they agree about all that is physically real. We have been speaking so far as if the spatio-temporal "distance" between events is itself a *number*, *viz.* the number that results when one plugs the coordinates of the events into the formula above. But it is easy to see that this is wrong even in the Euclidean case. Distances are only associated with numbers once one has chosen a *scale*, such as inches or meters. What exists as a purely geometrical, non-numerical structure is rather a system of *ratios of distances*.

Having chosen a particular geometrical magnitude as a unit, other magnitudes can be expressed as numbers, *viz.* the numbers that represent the ratio between the unit and the given magnitude. The Greeks had a deep insight when they divided mathematics into arithmetic (the theory of number) and geometry (the theory of magnitude). They recognized that the theory of ratios applied equally to each field, but kept the two subjects strictly separate. Our use of coordinates to associate curves in space with algebraic functions of numbers has blurred the distinction between magnitudes and numbers. To understand Relativity, it is important to recognize the conventions employed to associate geometrical structure with numerical structure.

Relativistic Spatio-temporal "Distance"

Holding these warnings in mind, let's turn to the Relativistic spatio-temporal "distance". What are the consequences of replacing the plus signs in the Euclidean distance function with minus signs? One understandable difference between the Euclidean structure and the Minkowski structure is this: in Euclidean space, the distance between any two distinct points is always positive, and only the only zero distance is between a point and itself. In mathematical terms, the Euclidean metrical structure is *positive definite*. But in the Minkowski structure, two distinct events can have zero "distance" between them. For example, the events with coordinates (0,0,0,0) and (1,1,0,0) have zero "distance" [where we list the coordinates in the order (t, x, y, z)]. Of course, this does not mean that these two events are the *same* event: assigning the numerical value zero to this sort of "distance" is just a product of the conventions we have used for assigning numbers to the "distances". But the fact that two events have a zero "distance" between then does show that they are related in a particular spatio-temporal way. In order to remind ourselves that these spatio-temporal "distances" do not behave like spatial distances, from now on we will call them spatio-temporal *intervals*.

If we choose a exacting event, the popping of a particular champagne bubble, and call the event p, then we can consider the entire locus of events that have zero interval from p. There will be infinitely many such events. If p happens to be at the origin of a Lorentz frame, assigned coordinates (0,0,0,0), then among the events at zero interval from it are (1,1,0,0), (1,0,1,0), (5,0,-3,4), and (-6,4,-4,2). To get a sense of how these events are distributed in space-time, we draw a *space-time diagram*, but again one must be very cautious when interpreting these diagrams. The diagrams must repress one or two dimensions of the space-time, since we can't draw four-dimensional pictures, but that is not the principle problem. The main problem is that the diagrams are drawn on a *Euclidean* sheet of paper, even though they represent events in *Minkowski* space-time. There is always the danger of investing some of the Euclidean structure of the representation with physical significance it does not have. Bearing that in mind, the natural thing to do is to suppress the z coordinate and draw the x, y, and t coordinates as the x, y and z coordinates of three-dimensional Euclidean space.

The points at zero interval from (0,0,0) will be points that solve the equation $t^2 - x^2 - y^2 = 0$, or $t^2 = x^2 + y^2$. The points that solve this equation form a double cone whose apex is at the origin. According to Relativity, the intrinsic spatio-temporal structure associates such a double cone with every event in the space-time. This locus of points is called the *light-cone* of the event p, and divides into two pieces, the two cones that meet at p. These cones are called the *future* light-cone and the *past* light cone of p.

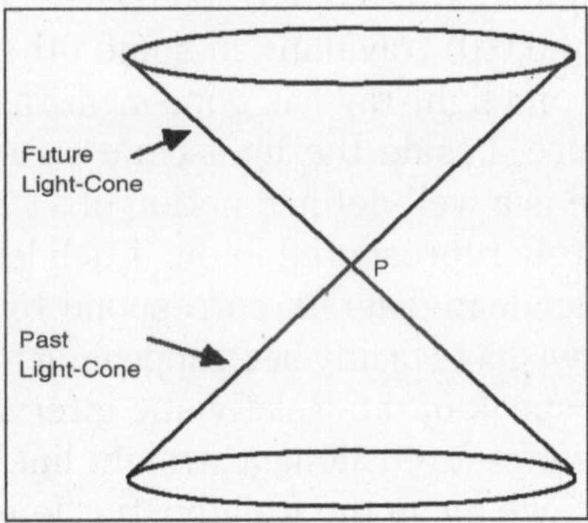

Spatio-temporal Structure

As the name "light-cone" suggests, we are now in a position to make contact between the spatio-temporal structure postulated by Relativity and the behaviour of physical entities. According to the laws of Relativistic physics, any light emitted at an event (in a vacuum) will propagate along the future light-cone of the event, and any light that arrives at an event (in a vacuum) arrives along the past light-cone. So the tiny flash of light emitted when our champagne bubble pops races away from the popping event along its future light-cone. One can think of the ever growing light-cone as representing the expanding circle (or, if we add back the z dimension, the expanding sphere) of light that originates at the bursting of the bubble.

Having associated the spatio-temporal structure with the behaviour of an observable phenomenon like light, we can now see how Relativistic physics gains empirical content. For example, it is an observable fact that any pair of light-rays travelling in parallel directions in a vacuum travel at the same speed: one light ray in a vacuum never overtakes another. This is not, of course, how material particles behave. One spaceship travelling in a vacuum can overtake another, or one electron in a vacuum can overtake another, because where a spaceship or an electron goes depends on more than the space-time location of the origin and direction of its journey. Two electrons can start out at the same place and time and set off in the same direction but end up in different locations because they were shot out at different speeds. Their trajectories depend on more than just the space-time structure. Light, in contrast, is intimately and directly tied to the Relativistic space-time structure. Space-time itself, as it were, tells light in a vacuum where to go. The assignment of zero Relativistic interval between the origin of a light-cone and any event on it has one other notable consequence. We have already said that when we assign numbers to magnitudes, we want the ratios between the numbers to be identical to the ratios between the magnitudes. Since 0:0 is not a proper ratio, the Relativistic interval does not license comparisons between the various intervals on a light-cone. If one light ray originates at (0,0,0,0) and travels to (1,1,0,0), and a second light ray originates at (0,0,0,0) travelling in some other direction, there is no fact about when the second light-ray has gone *as far* as the first.

What other structure, beside the light-cone structure, does Minkowski space-time have? There is a well-defined notion of a *straight line* in the space-time, and this is accurately represented in our Euclidean space-time diagram: straight lines in the Euclidean diagram correspond to straight trajectories in the space-time. Indeed, we have tacitly been appealing to the notion of a straight line all along: when we speak of the Relativistic interval between two events, we mean the interval as measured along a straight line connecting the events, or, even more precisely, we mean the Relativistic "length" of the straight line that connects the events. The straight-line structure (affine structure) of Minkowski space-time plays a central role in framing physical laws.

If a light ray is emitted from (0,0,0,0) into a vacuum, we already know that its trajectory through space-time will lie on the future light-cone of (0,0,0,0). But more than that, the trajectory will be a straight line on the light-cone. An analogous fact holds for material particles that travel below the speed of light. If a material particle is emitted from (0,0,0,0), its trajectory will lie entirely within the future light-cone of (0,0,0,0), which is to say that the particle can never travel at or above the speed of light. But more than that: if the particle is emitted into a vacuum, and is subject to no forces, then its trajectory will be a straight line in space-time.

ENORMOUSLY PREDATES THE THEORY OF RELATIVITY

This law, in abstract form, enormously predates the Theory of Relativity. For this is just the proper space-time formulation of Newton's First Law of

Motion: "Every body continues in its state of rest, or of uniform motion in a right line, unless compelled to change that state by forces impressed on it". The trajectory of a particle at rest or in uniform motion in Newtonian space-time is a straight line through the four-dimensional space-time.

Newton's first law, stated in terms of space-time trajectories, also retains the same form in Galilean space-time, and can be taken over without change into Minkowski space-time. As we will see, in this abstract space-time formulation, Newton's First Law also holds in the General Theory of Relativity. That is why we should try to formulate physical laws directly in terms of space-time structure.

Once we deal with material particles that travel below of the speed of light, the Relativistic interval takes on even greater significance. Consider a particle that travels from (0, 0, 0, 0) to (5, 4, 0, 0) along a straight trajectory, *i.e.,* a particle emitted from the origin of the coordinate system that arrives at the event (5, 4, 0, 0) without having any forces acting on it. The Relativistic interval along its space-time trajectory is

$$\sqrt{(5-0)^2-(4-0)^2-(0-0)^2-(0-0)^2}=3.$$

Clocks in the Theory of Relativity

The dimension of this interval has direct physical significant: *it is proportional to the amount of time that will elapse for a clock that travels along that trajectory*. Clocks in the Theory of Relativity are like odometers on cars: they measure the length of the path they take. But "length" here means the interval, and "path" the space-time trajectory of the clock. Events in space-time separated by positive intervals are called *time-like separated*.

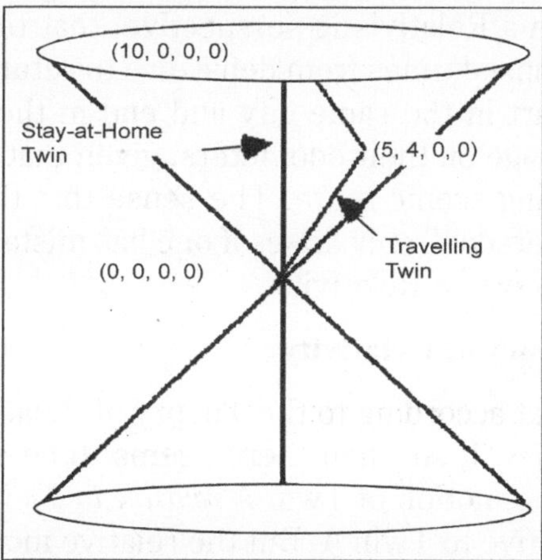

Postulate of Relativity that clocks measure the interval along their trajectory: clocks are physical mechanisms subject to physical analysis. But one can easily analyse how a simple clock will behave, such as a clock that counts the number of times a light ray gets reflected between two mirrors, and find that the reading on the clock will be proportional to the interval along the clock's trajectory.

With the clock postulate in hand, we can now analyse the notorious "twins paradox" of Relativity. One of a pair of twins takes a rocket from Earth and travels to a nearby star. Upon returning to Earth, the twin has aged less than the stay-at-home sister, and the clocks in the twin's space-ship show less elapsed time than those that remained on Earth. Why is that?

To be concrete, suppose the event of the rocket leaving Earth is at the point (0, 0, 0, 0) in our coordinate system, and the rocket travels inertially (without acceleration) to the point (5, 4, 0, 0). The rocket immediately turns around, and follows an inertial trajectory back Earth, arriving at the event (10, 0, 0, 0). The interval between (0, 0, 0, 0) and (5, 4, 0, 0) is as we have seen, 3. Suppose this corresponds to an elapse of three years according to the on-board clocks. The return trajectory from (5, 4, 0, 0) to (10, 0, 0, 0) also has interval length 3, corresponding to another 3 years elapsed. So the astronaut twin arrives back having aged 6 years, and having had all the experiences that correspond to 6 years of life.

The Relativistic Analysis of the Situation

The stay-at-home twin, however, always remained at the "spatial origin" of the coordinate system. Her trajectory through space-time is a straight line from (0, 0, 0, 0) to (10, 0, 0, 0). So the interval along her trajectory is 10, corresponding to an elapse of 10 years. She will have biologically aged 10 years at her sister's return, and had four more years of experience than her twin. The Relativistic analysis of the situation is quite straightforward. It is really no more surprising, from a Relativistic perspective, that the clocks of the twins will show different elapsed times from departure to return than it is surprising that two cars that start in the same city and end in the same city will show different elapsed mileage on their odometers, given that one took the freeway and the other a winding scenic route. The sense that there is a fundamental puzzle in the twins "paradox" only arises if one has mistaken views concerning the content of the Theory of Relativity.

According to the Theory of Relativity

It is often said that according to the Theory of Relativity, all motion is the *relative motion of bodies*. If so, then there seems to be a complete symmetry between the twins: the motion of Twin A *relative to Twin B* is identical to the motion of Twin B relative to Twin A. But the relative motion of the twins plays no role at all in the physical analysis of the situation. The amount of time that elapses for Twin B on her trip has nothing to do with what Twin A is doing, or even if there is a Twin A. The amount of time is just a function of the space-time interval along her trajectory.

It is also sometimes said that the Theory of Relativity gets rid of all Absolute spatio-temporal structure: all facts about space and time are ultimately understood in terms of relations between bodies, so in a world with only one body there could be no spatio-temporal facts. This is also incorrect. The Special Theory of Relativity postulates the existence of Minkowski space-time, whose intrinsic spatio-temporal structure is perfectly Absolute, in whatever sense one takes that term. It is not a *classical* space-time structure, but it is not just a system of relations between bodies.

One occasionally also hears that the resolution of the twins paradox rests on facts about *acceleration*: the situation of the two twins is not exactly symmetric since the astronaut twin must accelerate (when she turns around to come home) while the stay-at-home twin does not. That is true, but irrelevant: the difference in elapsed time is a function of the intervals along the trajectories, not a function of the accelerations that the twins experience. Indeed, in the General Theory of Relativity we will be able to construct a twins scenario in which neither twin accelerates at all, but still they suffer different elapsed times between parting and reunion.

It would be just as misleading to attribute the difference in elapsed time to the accelerations of the twins as it would the difference in odometer reading to the accelerations of the cars, even if the car that took the longer route did accelerate more.

The paradoxical or puzzling aspect of the twins paradox really arises from the difference between Euclidean geometry and Minkowski space-time geometry. If we draw the trajectories of the twins in space-time, we get a triangle whose corners lie a (0, 0, 0, 0), (5, 4, 0, 0) and (10, 0, 0, 0). The astronaut twin travels along two edges of this triangle, while the stay-at-home twin travels along the third. And in Euclidean geometry, the sum of the lengths of any two sides of a triangle are *greater* than the length of the remaining side. But in Minkowskian geometry, the opposite is true: the sum of the intervals of two sides is *less* that the interval along the remaining side. Indeed, for time-like separated events, a straight line is the *longest* path between the two points in space-time. This is one consequence of exchanging the plus signs in the Euclidean metric for minus signs in the Minkowski metric.

The Relativistic "clock postulate" has been most strikingly checked using natural clocks: unstable particles whose decay rate displays a known half-life in the laboratory. The muon, a sort of heavy electron, is unstable and will decay on average 10^{-6} seconds after having been created. Muons can be created in the upper atmosphere by collisions between molecules in the air and high-energy cosmic rays. According to clocks on Earth, it should take the muon about 10×10^{-6} second to reach the Earth, so very few should survive the trip without decaying. Nonetheless, many more muons than that calculation suggest do reach the Earth's surface. Calculation of the interval along the muon's trajectory predicts this, since that interval corresponds to less that 10^{-6} second.

Asymptotically Approaches the Light

If we romanticize muons a bit, and imagine that they all decay in exactly 10-6 seconds (according to their own "clocks"), then we can use them to map out the geometry of Minkowski space-time. Suppose we create a swarm of muons in space and send them out in all directions. Their decays will provide a map in space-time of events that are all the same interval from the point of creation. If we choose units so that the size of the interval corresponds to seconds, and we choose the creation of the muons as the origin of the coordinate system, then the coordinates of the decay events will satisfy:

$$\sqrt{t^2 - x^2 - y^2 - z^2} = 10^{-6}$$

This is the equation of a hyperboloid of revolution that asymptotically approaches the light-cone, as depicted below.

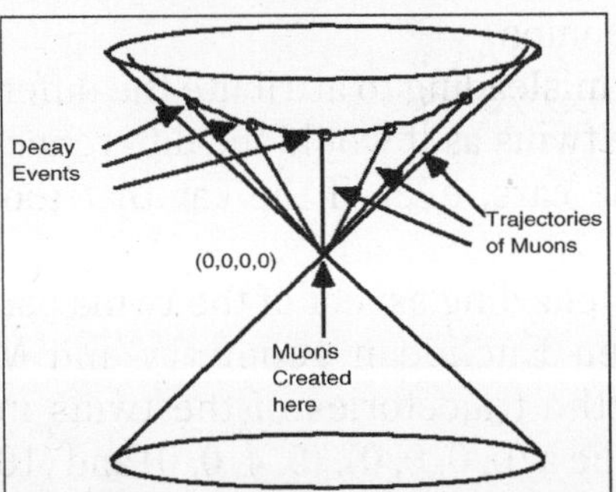

The hyperboloid represents events all at the same interval from (0, 0, 0, 0), and so corresponds to a *circle* or *sphere* of fixed radius in Euclidean geometry. There would be a corresponding hyperboloid in the past light-cone, representing places from which a muon could have been sent that would have decayed at (0, 0, 0, 0).

THE PRINCIPLE OF RELATIVITY (IN THE RESTRICTED SENSE)

IN order to attain the greatest possible clearness, let us return to our example of the railway carriage supposed to be travelling uniformly. We call its motion a uniform translation ("uniform" because it is of constant velocity and direction, "translation" because although the carriage changes its position relative to the embankment yet it does not rotate in so doing). Let us imagine a raven flying through the air in such a manner that its motion, as observed from the embankment, is uniform and in a straight line. If we were to observe the flying raven from the moving railway carriage, we should find that the motion of the raven would be one of different velocity and direction, but that it would still be uniform and in a straight line. Expressed in an abstract manner we may say: If a mass m is moving uniformly in a straight line with respect to a co-ordinate system K, then it will also be moving uniformly and in a straight line relative to a second co-ordinate system K', provided that the latter is executing a uniform translatory motion with respect to K. In accordance with the discussion contained in the preceding section, it follows that:

If K is a Galileian co-ordinate system, then every other co-ordinate system K' is a Galileian one, when, in relation to K, it is in a condition of uniform motion of translation. Relative to K' the mechanical laws of Galilei-Newton hold good exactly as they do with respect to K.

We advance a step farther in our generalisation when we express the tenet thus: If, relative to K, K' is a uniformly moving co-ordinate system devoid of rotation, then natural phenomena run their course with respect to K' according to exactly the same general laws as with respect to K. This statement is called the principle of relativity (in the restricted sense).

As long as one was convinced that all natural phenomena were capable of representation with the help of classical mechanics, there was no need to doubt the validity of this principle of relativity. But in view of the more recent development of electrodynamics and optics it became more and more evident that classical mechanics affords an insufficient foundation for the physical description of all natural phenomena. At this juncture the question of the validity of the principle of relativity became ripe for discussion, and it did not appear impossible that the answer to this question might be in the negative.

Nevertheless, there are two general facts which at the outset speak very much in favour of the validity of the principle of relativity. Even though classical mechanics does not supply us with a sufficiently broad basis for the theoretical presentation of all physical phenomena, still we must grant it a considerable measure of "truth," since it supplies us with the actual motions of the heavenly bodies with a delicacy of detail little short of wonderful. The principle of relativity must therefore apply with great accuracy in the domain of mechanics. But that a principle of such broad generality should hold with such exactness in one domain of phenomena, and yet should be invalid for another, is a priori not very probable.

We now proceed to the second argument, to which, moreover, we shall return later. If the principle of relativity (in the restricted sense) does not hold, then the Galileian co-ordinate systems K, K', K'', etc., which are moving uniformly relative to each other, will not be equivalent for the description of natural phenomena. In this case we should be constrained to believe that natural laws are capable of being formulated in a particularly simple manner, and of course only on condition that, from amongst all possible Galileian co-ordinate systems, we should have chosen one (K0) of a particular state of motion as our body of reference. We should then be justified (because of its merits for the description of natural phenomena) in calling this system "absolutely at rest," and all other Galileian systems K "in motion." If, for instance, our embankment were the system K0, then our railway carriage would be a system K, relative to which less simple laws would hold than with respect to K0. This diminished simplicity would be due to the fact that the carriage K would be in motion (i.e. "really") with respect to K0. In the general laws of natural which have been formulated with reference to K, the magnitude and direction of the velocity of the carriage would necessarily play a part. We should expect, for instance, that the note emitted by an organ-pipe placed with its axis parallel to the direction of travel would be different from that emitted if the axis of the pipe were placed perpendicular to this direction. Now in virtue of its motion in an orbit round the sun, our earth is comparable with a railway carriage travelling with a velocity of about 30 kilometres per second. If the principle of relativity were not valid we should therefore expect that the direction of motion of the earth at any moment would enter into the laws of nature, and also that physical systems in their behaviour would be dependent on the orientation in space with respect to the earth. For owing to the alteration in direction of the velocity of rotation of the earth in the course of a year, the earth cannot be at rest relative to the hypothetical system K0 throughout the whole year. However, the most careful observations have never revealed such anisotropic properties in terrestrial physical space, i.e. a physical non-equivalence of different directions. This is a very powerful argument in favour of the principle of relativity.

RELATIVITY OF THE CONCEPTION OF DISTANCE

LET us consider two particular points on the train travelling along the embankment with the velocity v, and enquire as to their distance apart. We already know that it is necessary to have a body of reference for the measurement of a distance, with respect to which body the distance can be measured up. It is the simplest plan to use the train itself as the reference-body (co-ordinate system). An observer in the train measures the interval by marking off his measuring-rod in a straight line (e.g. along the floor of the carriage) as many times as is necessary to take him from the one marked point to the other. Then the number which tells us how often the rod has to be laid down is the required distance.

It is a different matter when the distance has to be judged from the railway line. Here the following method suggests itself. If we call A' and B' the two points on the train whose distance apart is required, then both of these points are moving with the velocity v along the embankment. In the first place we require to determine the points A and B of the embankment which are just being passed by the two points A' and B' at a particular time t—judged from the embankment. The distance between these points A and B is then measured by repeated application of the measuring-rod along the embankment.

A priori it is by no means certain that this last measurement will supply us with the same result as the first. Thus the length of the train as measured from the embankment may be different from that obtained by measuring in the train itself. This circumstance leads us to a second objection which must be raised against the apparently obvious consideration. Namely, if the man in the carriage covers the distance w in a unit of time—measured from the train,—then this distance—as measured from the embankment—is not necessarily also equal to w.

EINSTEIN POSTULATES OF RELATIVITY

Einstein made two postulates on which he based his theory of special relativity:

1. *The principle of relativity*: The laws of physics are the same in all frames of reference moving at a constant velocity with respect to one another.
2. *The constancy of the speed of light*: The speed of light is the same to all observers.

The first postulate is basically the same as the Galilean principle of relativity. The second postulate is at odds with the Galilean transformation of velocities between two inertial frames. For example, in the case of a person shining a flashlight on a moving train, it would mean that if the train were to travel at 0.9c, then instead of the ground-based observer measuring the speed of the light as c + 0.9c = 1.9c, he would still measure the speed to be *c*.

Concept of Simultaneity

The concept of simultaneity – whether two events occur at the same time – is important when comparing measurements made in two different inertial systems. If the speed of light is the same in all moving coordinate systems, this means that events that occur simultaneously in one system may not be observed as being simultaneous in another coordinate system.

An example is illustrated in the figure below.

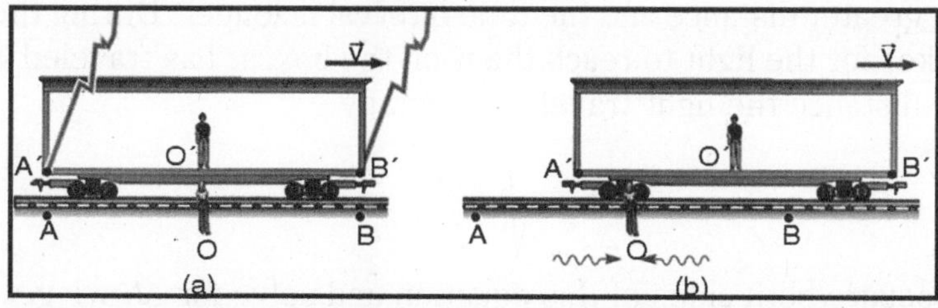

An observer *O'* stands in the middle of a moving boxcar and an observer *O* stands at rest beside the track. When the positions of the observers coincide, a lightning bolt strikes each end of the boxcar, leaving marks on the ground and marks on each end of the boxcar. The light from the lightning strikes at A and B reach observer O at the same time, so observer O concludes that the lightning strikes occurred simultaneously. But to observer O' in the moving boxcar, the lightning strikes do not appear to occur at the same time. The light travelling from A' to O' travels further than the light from B' to O'. Because of the motion, O' moves towards the incoming beam from B' and away from the incoming beam from A'. So to observer O' the strike at B' appeared to occur before the strike at A'.

TIME DILATION AND CONSTANCY OF THE SPEED OF LIGHT

One consequence of the constancy of the speed of light is that moving clocks run slower.

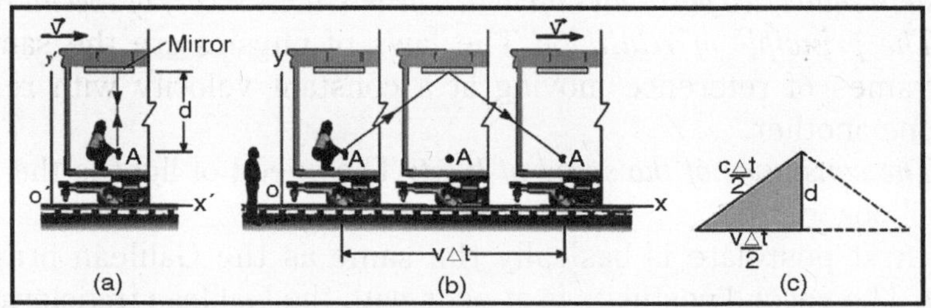

In the figure above a light clock consists of a laser beam being reflected from the roof of a boxcar of height d. The time interval of the clock is the time for the beam to go up and back down. To an observer at rest in the boxcar, this time interval is

$$\Delta t_p = \frac{2d}{c}.$$

This time is referred to as the 'proper' time since the clock is at rest with respect to the observer and the time interval is measured between two events, sending and receiving the laser pulse, that occur at the same position to the observer. To an observer at rest beside the track, the light must travel at an angle to reflect from the moving roof and back to the moving observer. Thus, it travels a greater distance and the time interval is longer. During the time $\Delta t/2$ that it takes for the light to reach the roof, the boxcar has traveled a distance $v\Delta t/2$. The distance the light travels is given by

$$\frac{c\Delta t}{2} = \sqrt{\left(\frac{v\Delta t}{2}\right)^2 + d^2}.$$

If we square both sides of this equation and solve for Δt, we get

$$\Delta t = \frac{2d / c}{\sqrt{1 - v^2 / c^2}}.$$

Comparing this expression with that for Δt_p, we see that

$$\Delta t = \frac{\Delta t_p}{\sqrt{1 - v^2 / c^2}} = \gamma \Delta t_p,$$

where

$$\gamma = \frac{1}{\sqrt{1 - v^2 / c^2}}.$$

Length and Space Contraction

Time dilation leads to the concept of space contraction. Consider an observer in a rocket ship travelling with speed v from earth to a distant planet. To an observer at rest on earth, this distance is $L_p = v\Delta t$. The distance the observer measures is considered the *proper* length, L_p, since it is the length of something that is at rest with respect to the observer. (The planet is at rest with respect to the earth.) The time that the earth observer measures for the trip, Δt, is not the proper time since the two events that are timed, leaving earth and arriving at the planet, do not occur at the same position. Now the guy in the rocket ship measures the time for the trip as Δt_p. This would be considered the proper time since two events occur at the same position with respect to him. He sees the planet as moving towards him with speed v, and he is measuring a moving length, L. So,

$$L_p = v\Delta t \text{ and } L = v\Delta t_p.$$

Using the time dilation formula relating Δt and Δt_p, we obtain

$$L = L_p\sqrt{1 - v^2 / c^2} = \frac{L_p}{\gamma}.$$

Thus, an object moving with respect to an observer appears have its dimension in the direction of the motion shortened.

MUON DECAY IN THE EARTH'S ATMOSPHERE

Evidence of time dilation and space contraction has been obtained by measuring the number of muons in the earth's atmosphere as a function of elevation. Muons are unstable particles that are produced in the earth's atmosphere by the collision of high-energy cosmic rays with atmospheric molecules. They are created in large numbers at an altitude of about 4,800 m with enough energy to travel close to the speed of light. When at rest they have a half-life of 2.2 ms and decay into an electron (or positron) and two neutrinos. A particle with a speed c would travel in 2.2 ms a distance

$$d = (3 \times 10^8 \text{ m/s})(2.2 \times 10^{-6} \text{ s}) = 660 \text{ m}.$$

This would mean that the muons should never reach the surface of the earth. However, a considerable number of muons do indeed reach earth. So, how is this possible?

The result can be explained by time dilation. To an observer at rest on earth, the half-life of the moving muons is larger by an amount that depends on their speed. If their speed was v = 0.99c, then the half-life observed by someone on earth would be

$$\Delta t = \frac{\Delta t_p}{\sqrt{1 - v^2 / c^2}} = \frac{2.2\mu s}{\sqrt{1-(0.99)^2}} = 16\,\mu s\,.$$

So, the person on earth would calculate the distance traveled by the dying muon to be

$$d' » c\Delta t = (3 \times 10^8 \text{ m/s})(16 \times 10^{-6} \text{ s}) = 4{,}800 \text{ m},$$

so they would reach the surface of earth before decaying.

An alternative view is that the muon sees the distance to earth contracted because of the motion of the earth respect to the muon. The contracted distance would be

$$d = d'\sqrt{1 - v^2 / c^2} = (4800\,\text{m})\sqrt{1-(0.99)^2} = 660\,\text{m}..$$

The Concept of Motion and Twin Paradox

The concept of motion is relative to the observer. So when thinking about time dilation, a person at rest on earth would say that the clocks in a moving spacecraft run more slowly. But on the other hand a person in the spacecraft could say that he was at rest and the person on earth was moving and say that the earth clocks were running more slowly. Actually, both observers are correct. A conflict would appear to arise if the space traveller were to go on a long trip and return back to earth so they could compare their clocks. This situation occurs in the so-called *twin paradox*.

In the twin paradox, at age 20 twin Speedo leaves earth in a spacecraft for a round trip to Planet X while twin Goslo stays behind on earth. The planet is 20 light-years from earth and the spacecraft travels with a speed of 0.95c. What are the ages of the twins when Speedo returns?

To Goslo the time of travel is

$$\Delta t = 2d / v = 2(20c \cdot yr) / (0.95c) = 42\,yr$$

(Note: A light year is the distance that light travels in one year. So 1 light-yr = speed × time = 1 c × yr.) Goslo perceives Speedo's time of travel to be shorter, so he determines that Goslo has aged by

$$\Delta t_p = \Delta t / \gamma = \Delta t\sqrt{1 - v^2 / c^2}$$

$$= (42yr)\sqrt{1 - (0.95)^2} = (42yr)(0.31) = 13\,yr$$

However, one might think that Speedo could consider that he was as rest and Goslo was moving, so that Goslo aged by 13 yrs and Speedo aged by 44 yrs. However, when applying the concepts of relativity, we must compare inertial frames of reference. Goslo is in a fixed inertial frame of reference, whereas Speedo is in one inertial frame going and a different inertial frame returning. Thus, Goslo's calculations are correct. Upon Speedo's return, Speedo is 33 yrs old and Goslo is 62 yrs old. Another way to envision this is that from Speedo's point of view the distance that he has to travel to Planet X is less than 20 light-years because of space contraction. The distance he travels is

$$d' = d\sqrt{1 - v^2 / c^2} = (20\,c \cdot yr)(0.31) = 6.2\,c \cdot yr$$

So the time for him to travel to Planet X and back is

$$\Delta t_p = 2d'/ v = 2(6.2) / 0.95 = 13\,yr\,.$$

Velocity Addition and Galilean Relativity

In Galilean relativity it is possible for velocities to add such that objects can exceed the speed of light. In special relativity, the addition rule for velocities makes this impossible. Consider a spaceship (system 2) moving at a speed v_{21} with respect to an observer at rest on earth (system 1). If an observer at rest in the spaceship fires a projectile (object 3) with speed v_{32} relative to the spaceship, then the speed of the projectile as measured by the earth observer is

$$v_{31} = \frac{v_{32} + v_{21}}{1 + v_{32}v_{21} / c^2}\,.$$

Special Theory of Relativistic Momentum

In the special theory of relativity, linear momentum is given by

$$p = \frac{mv}{\sqrt{1 - v^2 / c^2}} = \gamma\, mv\,.$$

If we use the classical definition of momentum, which is $p = mv$, then momentum is not conserved in high speed collisions. The relativistic momentum approaches infinity as the speed of a particle approaches the speed of light.

Classical Relativistic Energy

Classically, kinetic energy is given by $KE = ½mv^2$. The relativistic formula for kinetic energy is

$$KE = \frac{mc^2}{\sqrt{1-v^2/c^2}} - mc^2 = \gamma mc^2 - mc^2 .$$

The quantity mc^2 is the rest mass energy,

$$E_0 = mc^2 .$$

So we can think of $g\ mc^2$ as the total energy,

$$E = \gamma mc^2 = \frac{mc^2}{\sqrt{1-v^2/c^2}} .$$

The kinetic energy and the total energy of a particle approach infinity as its speed approaches the speed of light. This means that it is impossible to accelerate a particle with finite mass to the speed of light.

Expressions for Relativistic Energy and Momentum Relationship

From the expressions for the relativistic energy and momentum, an expression relating the two can be obtained. The result is

$$E^2 = p^2c^2 + (mc^2)^2 .$$

For a photon, $m = 0$, so

$$E = pc.$$

Since for a photon $E = hf$, we have

$$p = \frac{E}{c} = \frac{hf}{c} = \frac{h}{\lambda} .$$

RELATIVITY OF LIGHT

Light is arguably the phenomenon of nature with which we have the most conscious experience, by means of our sense of vision, and yet throughout most of human history very little seems to have been known about how vision works. Interestingly, from the very beginning there were at least two distinct concepts of light, existing side by side, as can be seen in some of the earliest known writings.

For example, the description of creation in the biblical book of Genesis says light was created on the first day, and yet the sun, moon, and stars were not created until the fourth day "to give light upon the earth". Evidently the word "light" is being used to signify two different things on the first and fourth days. For another example, Plato argued in Timaeus that there are two kinds of "fire" involved in our sense of vision, one coming from inside ourselves, emanating as visual rays from our eyes to make contact with distant objects, and another, which he called "daylight", that (when present) surrounds the visual rays from our eyes and facilitates the conveyance of the visual images.

These two kinds of "fire" correspond roughly with the later scholastic concepts of *lux* and *lumen*. The word lux was used to signify our visual sensations, whereas the word lumen referred to an external agent (such as light from the sun) that somehow participates in our sense of vision.

There was also, in ancient times, a competing theory of vision, according to which all objects naturally emit whole "images" (eidola) of themselves in small packets, and these enter our souls by way of our eyes. To account for our inability to see at night, it was thought that light from the sun or moon struck the objects and caused them to emit their images. This model of vision still entailed two distinct kinds of light: the facilitating illumination from the sun or moon, and the eidola emitted by ordinary objects. This somewhat awkward conception of vision was improved by Ibn al-Haitham and later by Kepler, who argued that it is not necessary to assume whole objects emit multiple copies of themselves; we can simply consider each tiny part of an object as the source of rays emanating in all directions, and a sub-set of these rays intersecting in the eye can be re-assembled into an image of the object.

Until the end of the 17th century there was no evidence to indicate that rays of light propagated at a finite speed, and they were often assumed to be instantaneous. Only in 1689 with Roemer's observations of the moons of Jupiter, and even more convincingly in 1728 with Bradley's discovery of stellar aberration, did it become clear that the rays of lumen propagate through space with a characteristic finite speed. This suggested that light, and the energy it conveys, must have some mode of existence during the interval of time between its emission and its absorption.

Hence light became an entity or process in itself, rather than just a relation between entities, but again there were two competing notions as to the mode of existence. Two different analogies were conceived, based on the behaviour of ordinary material substances. Some thought light could be regarded as a stream of material corpuscles moving through empty space, whereas other believed light consists of undulations or waves in a pervasive material medium.

Each of these analogies was consistent with some of the attributes of light, but neither could be reconciled fully with all the attributes. For example, if light consists of material corpuscles, then according to Galilean relativity there should be an inertial reference frame with respect to which light is at rest in a vacuum, whereas in fact we never observe light in a vacuum to be at rest, nor even noticeably slow, with respect to any inertial reference frame. On the other hand, if light is a wave propagating through a material medium, then the constituent parts of that medium should, according to Galilean relativity, behave inertially, and in particular should have a definite rest frame, whereas we find that light propagates best through regions (vacuum) in which there is no detectable material with a definite rest frame, and again we cannot conceive of light at rest in any inertial frame. Thus the behaviour of light defies realistic representation in terms of the behaviour of material substances within the framework of Galilean space and time, even if we consider just the classical attributes, let alone quantum phenomena.

By the end of the 19th century the inadequacy of both of the materialistic analogies for explaining the behaviour of light had become acute, because there was strong evidence that light exhibits two seemingly mutually exclusive properties. First, Maxwell showed how light can be regarded as a propagating electromagnetic wave, and as such the speed of propagation is obviously independent of the speed of the source.

Second, numerous experiments showed that light propagates at the same speed in all directions relative to the source, just as we would expect for streams of inertial corpuscles. Hence some of the attributes of light seemed to unequivocally support an emission theory, while others seemed just as unequivocally to support a wave theory. In retrospect it's clear that there was an underlying confusion regarding the terms of description, *i.e.*, the systems of inertial coordinates, but this was far from clear at the time.

One of the first clues to unraveling the mystery was found in 1887, when Woldemar Voigt made a remarkable discovery concerning the ordinary wave equation. Recall that the wave equation for a time-dependent scalar field $\phi(x, t)$ in one dimension is

$$\frac{\partial^2\phi}{\partial x^2} = \frac{1}{u^2}\frac{\partial^2\phi}{\partial t^2}$$

where u is the propagation speed of the wave. This equation was first studied by Jean d'Alembert in the 18th century, and it applies to a wide range of physical phenomena.

In fact it seems to represent a fundamental aspect of the relationship between space, time, and motion, transcending any particular application. Traditionally it was considered to be valid only for a coordinate system x, t with respect to which the wave medium (presumed to be an inertial substance) is at rest and has isotropic properties, because if we apply a Galilean transformation to these coordinates, the wave equation is not satisfied with respect to the transformed coordinates. However, Galilean transformations are not the most general possible linear transformations. Voigt considered the question of whether there is *any* linear transformation that leaves the wave equation unchanged.

The general linear transformation between (X, T) and (x, t) is of the form

$$x = AX + BT \qquad t = CX + DT$$

for constants A, B, C, D. If we choose units of space and time so that the acoustic speed u equals 1, the wave equation in terms of (X, T) is simply

$$\partial^2\phi/\partial X^2 = \partial^2\phi/\partial T^2.$$

To express this equation in terms of the transformed (x, t) coordinates, recall that the total differential of f can be written in the form

$$d\phi = \frac{\partial\phi}{\partial x}dx + \frac{\partial\phi}{\partial t}dt$$

Also, at any constant T, the value of ϕ is purely a function of X, so we can divide through the above equation by dX to give

$$\frac{\partial\phi}{\partial X} = \left(\frac{\partial\phi}{\partial X}\right)_T = \frac{\partial\phi}{\partial x}\left(\frac{dx}{dX}\right)_T + \frac{\partial\phi}{\partial t}\left(\frac{dt}{dX}\right)_T$$

$$= A\frac{\partial\phi}{\partial x} + C\frac{\partial\phi}{\partial t}$$

Taking the partial derivative of this with respect to X then gives

$$\frac{\partial^2\phi}{\partial X^2} = A\frac{\partial^2\phi}{\partial X\partial x} + C\frac{\partial^2\phi}{\partial X\partial t}$$

Since partial differentiation is commutative, this can be written as

$$\frac{\partial^2\phi}{\partial X^2} = A\frac{\partial}{\partial x}\left(\frac{\partial\phi}{\partial X}\right) + C\frac{\partial}{\partial t}\left(\frac{\partial\phi}{\partial X}\right)$$

Substituting the prior expression for $\partial\phi/dX$ and carrying out the partial differentiations gives an expression for $\partial^2\phi/\partial X^2$ in terms of partials of ϕ with respect to x and t. Likewise we can derive an expression for $\partial^2\phi/\partial T^2$. Substituting into the wave equation gives

$$A^2\frac{\partial^2\phi}{\partial x^2} + 2AC\frac{\partial^2\phi}{\partial x\partial t} + C^2\frac{\partial^2\phi}{\partial t^2} = B^2\frac{\partial^2\phi}{\partial x^2} + 2BD\frac{\partial^2\phi}{\partial x\partial t} + D^2\frac{\partial^2\phi}{\partial t^2}$$

This is equivalent to the condition that $\phi(X, T)$ is a solution of the wave equation with respect to the X, T coordinates. Since the mixed partial generally varies along a path of constant second partial with respect to x or t, it follows that a necessary and sufficient condition for $\phi(x, t)$ to also be a solution of the wave equation in terms of the x, t coordinates is that the constants A, B, C, D of our linear transformation satisfy the relations

$$A^2 + C^2 = B^2 + D^2 \qquad AC = BD$$

Furthermore, the differential of the space transformation is $dx = AdX + BdT$, so an increment with $dx = 0$ satisfies $dX/dT = -B/A$. This represents the velocity at which the spatial origin of the *x*, *t* coordinates is moving relative to the *X*, *T* coordinates. We will refer to this velocity as v. We also have the inverse transformation from (X, T) to (x, t):

$$X = \frac{D}{AD - BC^x} + \frac{-B}{AD - BC}t$$

$$T = \frac{-C}{AD - BC^x} + \frac{A}{AD - BC}t$$

Proceeding as before, the differential of this space transformation gives $dx/dt = B/D$ for the velocity of the spatial origin of the *X*, *T* coordinates with respect to the *x*, *t* coordinates, and this must equal $-v$. Therefore we have $B = -Av = -Dv$, and so $A = D$. It follows from the condition imposed by the wave equation that $B = C$, so both of these equal $-Av$. Our transformation can then be written in the form

$$x = A(X - vT) \quad x = A(T - vX)$$

The same analysis shows that the perpendicular coordinates *y* and *z* of the transformed system must be given by

$$y = A\sqrt{1 - v^2}\,Y \quad z = A\sqrt{1 - v^2}\,Z$$

In order to make the transformation formula for x agree with the Galilean transformation, Voigt chose $A = 1$, so he did not actually arrive at the Lorentz transformation, but nevertheless he had shown roughly how the wave equation could actually be relativistic – just like the dynamic behaviour of inertial particles– provided we are willing to consider a transformation of the space and time coordinates that differs from the Galilean transformation.

Had he considered the inverse transformation

$$X = \frac{1}{A(1 - v^2)}(x + vt) \quad T = \frac{1}{A(1 - v^2)}(t + vx)$$

he might have noticed that the determinant is $A^2(1 - v^2)$, so to make this equal to 1 we must have $A = 1/(1 - v^2)^{1/2}$, which not only implies $y = Y$ and $z = Z$, but also makes the transformation formally identical to its inverse. In other words, he would have arrived at a completely relativistic framework for the wave equation. However, this was not Voigt's objective, and he evidently regarded the transformed coordinates *x*, *y*, *z* and *t* as merely a convenient parameterization for purposes of calculation, without attaching any greater significance to them.

Voigt's transformation was the first hint of how a wavelike phenomenon could be compatible with the principle of relativity, which is that there exist inertial coordinate systems in terms of which free motions are linear, inertia is isotropic, and every material object is instantaneously at rest with respect to one of these systems.

None of this conflicts with the observed behaviour of light, because the motion of light is observed to be both linear and isotropic with respect to inertial coordinate systems.

The fact that light is not at rest with respect to any system of inertial coordinates does not conflict with the principle of relativity if we agree that light is not a material object.

The incompatibility of light with the Galilean framework arises not from any conflict with the principle of relativity, but from the tacitly adopted empirical conclusion that two relatively moving systems of inertial coordinates are related to each other by Galilean transformations, so that the composition of co-linear speeds is simply additive. We aren't free to impose this assumption on the class of inertial coordinate systems, because they are fully determined by the requirement for inertia to be homogeneous and isotropic. There are no more adjustable parameters (aside from insignificant scale factors), so the composition of velocities with respect to relatively moving inertial coordinate systems is a matter to be determined empirically.

The basis of slowly moving reference frames, Galileo and Newton had inferred that the composition of speeds was simply additive. In other words, if a material object B is moving at the speed v in terms of inertial rest frame coordinates of a material object A, and if an object C is moving in the same direction at the speed u in terms of inertial rest frame coordinates of B, then Newton found that object C has the speed $v + u$ in terms of the inertial rest frame coordinates of A. Towards the end of the nineteenth century, more precise observations revealed that is not quite correct. It was found that the speed of object C in terms of inertial rest frame coordinates of A is not $v + u$, but rather $(v + u)/(1 + uv/c^2)$, where c is the speed of light in a vacuum.

Obviously these conclusions would be identical if the speed of light was infinitely great, which was still considered a real possibility in Galileo's day. Many people, including Descartes, regarded rays of light as instantaneous. Even Newton's Opticks, published in 1704, made allowances for the possibility that "light be propagated in an instant" (although Newton himself was persuaded by Roemer's observations that light has a finite speed).

Hence it can be argued that the principles of Galileo and Einstein are essentially identical in both form and content. The only difference is that Galileo assessed the propagation of light to be "if not instantaneous then extraordinarily fast", and thus could neglect the term uv/c^2, especially since he restricted his considerations to the movements of material objects, whereas subsequently it became clear that the speed of light has a finite value, and it was necessary to take account of the uv/c^2 term when attempting to incorporating the motions of light and high-speed particles into the framework of mechanics. The empirical correspondence between inertial isotropy and lightspeed isotropy can be illustrated by a simple experiment.

Three objects, *A*, *B*, and *C*, at rest with respect to each other can be arranged so that one of them is at the midpoint between the other two (the midpoint having been determined using standard measuring rods at rest with respect to those objects). The two outer objects, *A* and *C*, are equipped with identical clocks, and the central object, *B*, is equipped with two identical cannons. Let the two cannons in the centre be fired simultaneously in opposite directions towards the two outer objects, and then at a subsequent time let object B emit a flash of light.

If the arrivals of the cannonball and light coincide at *A*, then they also coincide at *C*, signifying that the propagation of light is isotropic with respect to the same system of coordinates in terms of which mechanical inertia is isotropic, as illustrated in the figure below.

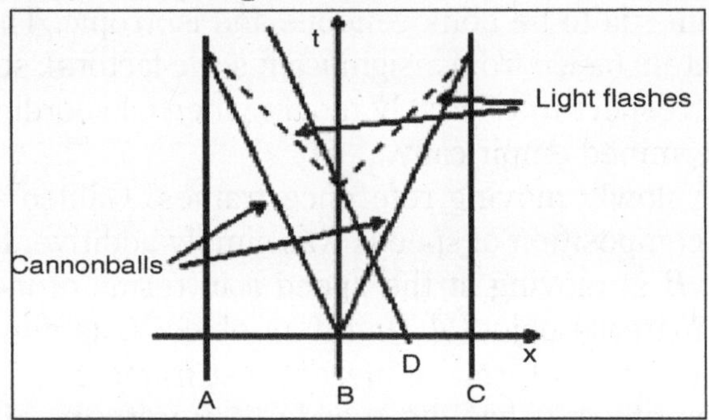

The fact that light emitted from object B propagates isotropically with respect to B's inertial rest frame might seem to suggest that light can be treated as an inertial object within the Galilean framework, just like cannon-balls. However, we also find that if the light is emitted at the same time and place from an object *D* that is moving with respect to *B*, the light's speed is still isotropic with respect to *B*'s inertial rest frame.

Now, this might seem to suggest that light is a disturbance in a material medium in which the objects *A, B, C* just happen to be at rest, but this is ruled out by the fact that it applies regardless of the state of (uniform) motion of those objects.

Naturally this implies that the flash of light propagates isotropically with respect to the inertial rest coordinates of object *D* as well.

To demonstrate this, we could arrange for two other bodies, denoted by *E* and *F*, to be moving at the same speed as *D*, and located an equal distance from *D* in opposite directions. Then we could fire two identically constructed cannons (at rest with respect to *D*) in opposite directions, towards *E* and *F*. The results are illustrated below.

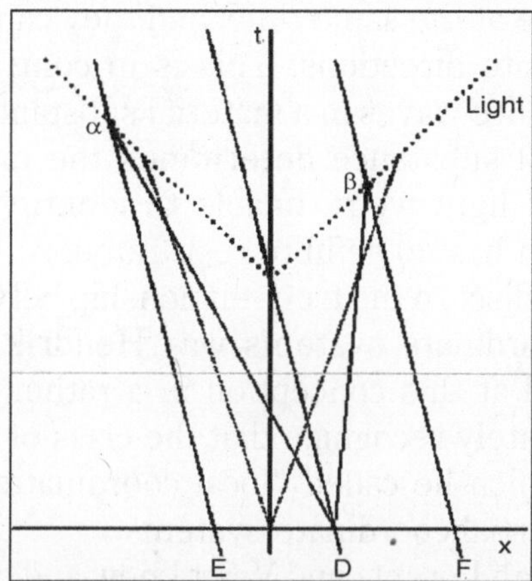

The cannons are fired from *D* when it crosses the *x* axis, and the cannon-balls strike *E* and *F* at the events marked α and β, coincident with the arrival of the light pulse from *D*. Obviously the time axis for the inertial rest frame coordinates of object *D* is the worldline of *D* itself (rather than the original "*t*" axis shown on the figure).

In addition, since inertial coordinates are defined such that mechanical inertia is isotropic, it follows that the cannon-balls fired from identical cannons at rest with *D* are moving with equal and opposite speeds with respect to *D*'s inertial rest coordinates, and since E and F are at equal distances from *D*, it also follows that the events a and b are simultaneous with respect to the inertial rest coordinates of *D*. Hence, not only is the time axis of *D*'s rest frame slanted with respect to *B*'s time axis, the spatial axis of *D*'s rest frame is equally slanted with respect to *B*'s spatial axis.

Several other important conclusions can be deduced from this figure. For example, with respect to the original *x, t* coordinate system, the speeds of the cannon-balls from *D* are not given by simply adding (or subtracting) the speed of the cannon-balls with respect to *D*'s rest frame to (or from) the speed of *D* with respect to the *x, t* coordinates. Since momentum is explicitly conserved, this implies that the inertia of a body increases with it's velocity (*i.e.*, kinetic energy).

We should also note that although the *speed* of light is isotropic with respect to any inertial spacetime coordinates, independent of the motion of the source, it is not correct to say that the light itself is isotropic. The relationship between the frequency (and energy) of the light with respect to the rest frame of the emitting body and the frequency (and energy) of the light with respect to the rest frame of the receiving body *does* depend on the relative velocity between those two massive bodies.

Incidentally, notice that we can rule out the possibility of object *B* and *D* dragging the light medium along with them, because they are moving through the *same* region of space at the same time, and they can't *both* be dragging the same medium in opposite directions. This is in contrast to the case of (for example) acoustic pressure waves in a material substance, because in that case a recognizable material substance determines the unique isotropic frame, whereas in the case of light we're unable to identify any definite material medium, so the medium has no definite rest frame.

The first person to discern the true relationship between relatively moving systems of inertial coordinate systems was Hendrik Antoon Lorentz. Not surprisingly, he arrived at this conception in a rather indirect and laborious way, and didn't immediately recognize that the class of coordinate systems he had discovered (and which he called "local coordinate" systems) were none other than Galileo's inertial coordinate systems.

Incidentally, although Lorentz and Voigt knew and corresponded with each other, Lorentz apparently was not aware of Voigt's earlier work on coordinate transformations that leave the wave equation invariant, and so that work had no influence on Lorentz's search for coordinate systems in terms of which Maxwell's equations are invariant. Unlike Voigt, Lorentz derived the transformation in two separate stages. He first developed the "local time" coordinate, and only years later came to the conclusion (after, but independently of, Fitzgerald) that a "contraction" of spatial length was also necessary in order to account for the absence of second-order effects in Michelson's experiment.

Lorentz began with the absolute ether frame coordinates t and x, in terms of which every event can be assigned a unique space-time position (t, x), and then he considered a system moving with the velocity v in the positive x direction. He applied the traditional Galilean transformation to assign a new set of coordinates to every event.

Thus an event with ether-frame coordinates t, x is assigned the new coordinates $x'' = x - vt$ and $t'' = t$. Then he tentatively proposed an additional transformation that must be applied to x'', t'' in order to give coordinates in terms of which Maxwell's equations apply in their standard form. Lorentz was not entirely clear about the physical significance of these "local" coordinates, but it turns out that *all* physical phenomena conform to the same isotropic laws of physics when described in terms of these coordinates. (Lorentz's notation made use of the parameter $\beta = 1/\gamma = 1/(1 - v^2)^{1/2}$ and another constant which he later determines to be 1). Taking units such that $c = 1$, his equations for the local coordinates x' and t' in terms of the Galilean coordinates which we are calling x' and t'' are

$$x' = \frac{x''}{\gamma} \qquad t' = \gamma t'' - \frac{vx''}{\gamma}$$

Recall that the traditional Galilean transformation is $x'' = x - vt$ and $t'' = t$, so we can make these substitutions to give the complete transformation from the original ether rest frame coordinates x, t to the local coordinates moving with speed v

$$x' = \frac{x - vt}{\gamma} \qquad t' = \frac{t - vx}{\gamma}$$

These effective coordinates enabled Lorentz to explain how two relatively moving observers, each using his own local system of coordinates, both *seem* to remain at the centre of expanding spherical light waves originating at their point of intersection.

The x and x'' axes represent the respective spatial coordinates (say, in the east/west direction), and the t and t' axes represent the respective time coordinates. One observer is moving through time along the t axis, and the other has some relative westward velocity as he moves through time along the t' axis. The two observers intersect at the event labeled O, where they each emit a pulse of light.

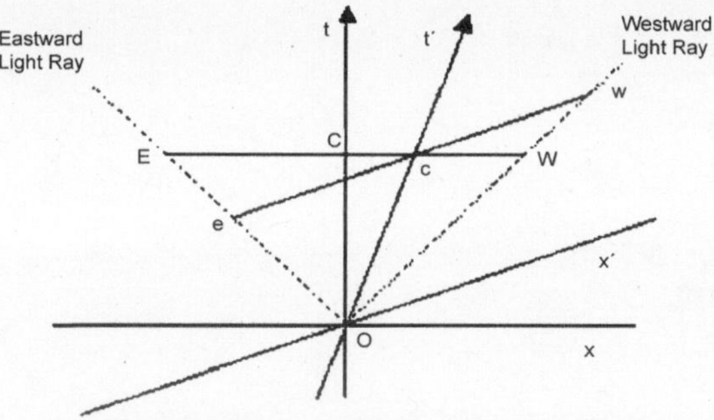

Those light pulses emanate away from O along the dotted lines. Subsequently the observer moving along the t axis finds himself at C, and according to his measures of space and time the outward going light waves are at E and W at that same instant, which places him at the midpoint between them. On the other hand, the observer moving along t' axis finds himself at point c, and according to his measures of space and time the outward going light waves are at e and w at this instant, which implies that *he* is at the midpoint between them.

Thus Lorentz discovered that by means of the "fictitious" coordinates x', t' it was possible to conceive of a class of relatively moving coordinate systems with respect to which the speed of light is invariant. He went beyond Voigt in the realization that the existence of this class of coordinate systems ensures the appearance of relativity, at least for optical phenomena, and yet, like Voigt, he still tended to regard the "local coordinates" as artificial.

Having been derived specifically for electromagnetism, it was not clear that the same transformations should apply to all physical phenomena, including inertia, gravity, and whatever forces are responsible for the stability of matter – at least not without simply hypothesizing this to be the case.

However, Lorentz was dissatisfied with the proliferation of hypotheses that he had made in order to arrive at this theory. The same criticism was made in a contemporary review of Lorentz's work by Poincare, who chided him with the remark "hypotheses are what we lack least". The most glaring of these was the hypothesis of contraction, which seemed distinctly "ad hoc" to most people, including Lorentz himself originally, but gradually he came to realise that the contraction hypothesis was not as unnatural as it might seem. Surprising as this hypothesis may appear at first sight, yet we shall have to admit that it is by no means far-fetched, as soon as we assume that molecular forces are also transmitted through the ether, like the electric and magnetic forces…

He set about trying to show (admittedly after the fact) that the Fitzgerald contraction was to be expected based on what he called the Molecular Force Hypothesis and his theorem of Corresponding States.

3

Electromagnetic Theory of Radiation

ELECTROMAGNETIC ULTRAVIOLET RADIATION

Ultraviolet Radiation, electromagnetic radiation that has wavelengths in the range between 4000 angstrom units (Å), the wavelength of violet light, and 150 Å, the length of X-rays. Natural ultraviolet radiation is produced principally by the sun. Ultraviolet radiation is produced artificially by electric-arc lamps. Ultraviolet radiation is often divided into three categories based on wavelength, *UV-A, UV-B*, and *UV-C.* In general shorter wavelengths of ultraviolet radiation are more dangerous to living organisms. *UV-A* has a wavelength from 4000 Å to about 3150 Å. *UV-B* occurs at wavelengths from about 3150 Å to about 2800 Å and causes sunburn; prolonged exposure to *UV-B* over many years can cause skin cancer. *UV-C* has wavelengths of about 2800 Å to 150 Å and is used to sterilize surfaces because it kills bacteria and viruses.

The earth's atmosphere protects living organisms from the sun's ultraviolet radiation. If all the ultraviolet radiation produced by the sun were allowed to reach the surface of the earth, most life on earth would probably be destroyed. Fortunately, the ozone layer of the atmosphere absorbs almost all of the short-wavelength ultraviolet radiation, and much of the long-wavelength ultraviolet radiation. However, ultraviolet radiation is not entirely harmful; a large portion of the vitamin *D* that humans and animals need for good health is produced when the human's or animal's skin is irradiated by ultraviolet rays.

When exposed to ultraviolet light, many substances behave differently than when exposed to visible light. For example, when exposed to ultraviolet radiation, certain minerals, dyes, vitamins, natural oils, and other products become *fluorescent*—that is, they appear to glow. Molecules in the substances absorb the invisible ultraviolet light, become energetic, then shed their excess energy by emitting visible light.

As another example, ordinary window glass, transparent to visible light, is opaque to a large portion of ultraviolet rays, particularly ultraviolet rays with short wavelengths. Special-formula glass is transparent to the longer ultraviolet wavelengths, and quartz is transparent to the entire naturally occurring range.

SUN

Sun, closest star to Earth. The Sun is a huge mass of hot, glowing gas. The strong gravitational pull of the Sun holds Earth and the other planets in the solar system in orbit. The Sun's light and heat influence all of the objects in the solar system and allow life to exist on Earth. The Sun is an average star its size, age, and temperature fall in about the middle of the ranges of these properties for all stars. Astronomers believe that the Sun is about 4.6 billion years old and will keep shining for about another 7 billion years.

For humans, the Sun is beautiful and useful, but also powerful and dangerous. As Earth turns, the Sun rises over the eastern horizon in the morning, passes across the sky during the day, and sets in the west in the evening. This movement of the Sun across the sky marks the passage of time during the day. The Sun's movement can produce spectacular sunrises and sunsets under the right atmospheric conditions. At night, reflected sunlight makes the Moon and planets bright in the night sky.

The Sun provides Earth with vast amounts of energy every day. The oceans and seas store this energy and help keep the temperature of Earth at a level that allows a wide variety of life to exist. Plants use the Sun's energy to make food, and plants provide food for other organisms. The Sun's energy also creates wind in Earth's atmosphere. This wind can be harnessed and used to produce power. While it lights our day and provides energy for life, sunlight can also be harmful to people. Human skin is sensitive to ultraviolet light emitted from the Sun. Earth's atmosphere blocks much of the harmful light, but sunlight is still strong enough to burn skin under some conditions. Sunburn is one of the most important risk factors in the development of skin cancers, which can be fatal. Sunlight is also very harmful to human eyes.

A person should never look directly at the Sun, even with sunglasses or during an eclipse. The Sun influences Earth with more than just light. Particles flowing from the Sun can disrupt Earth's magnetic field, and these disruptions can interfere with electronic communications.

ELECTROMAGNETIC SPECTRUM

VISIBLE SPECTRUM

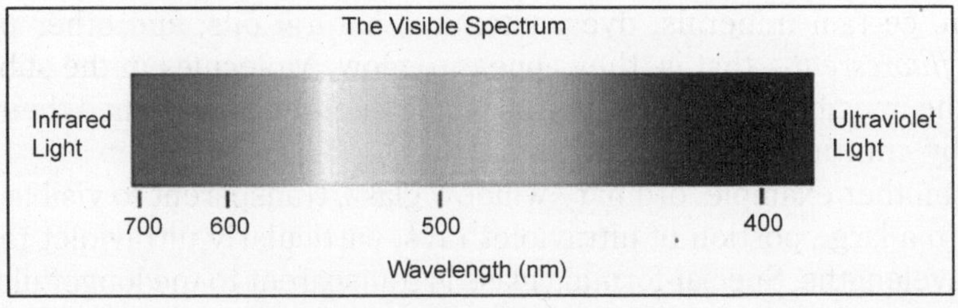

Fig. Electromagnetic Spectrum

Type of Radiation	*Frequency* Range (Hz)	*Wavelength* Range	*Type of Transition*
gamma-rays	$10^{20} - 10^{24}$	$<10^{-12}$ m	Nuclear
x-rays	$10^{17} - 10^{20}$	1 nm – 1 pm	inner electron
ultraviolet	$10^{15} - 10^{17}$	400 nm – 1 nm	outer electron
visible	$4 - 7.5 \times 10^{14}$	750 nm – 400 nm	outer electron
near-infrared	$1 \times 10^{14} - 4 \times 10^{14}$	2.5 um – 750 nm	outer electron molecular vibrations
infrared	$10^{13} - 10^{14}$	25 um – 2.5 um	molecular vibrations
microwaves	$3 \times 10^{11} - 10^{13}$	1 mm – 25 um	molecular rotations, electron spin flips*
radio waves	$<3 \times 10^{11}$	>1 mm	nuclear spin flips*

*energy levels split by a magnetic field.

X-RAY PHOTOELECTRON SPECTROSCOPY (XPS, ESCA)

X-ray photoelectron spectroscopy (XPS, also called electron spectroscopy for chemical analysis, ESCA) is a electron spectroscopic method that uses x-rays to eject electrons from inner-shell orbitals. The kinetic energy, E_k, of these photoelectrons is determined by the energy of the x-ray radiation, *h*, ν and the electron binding energy, E_b, as given by:

$$E_k = h\nu - E_b$$

The experimentally measured energies of the photoelectrons are given by:

$$E_k = h\nu - E_b - E_w$$

where E_w is the work function of the spectrometer.

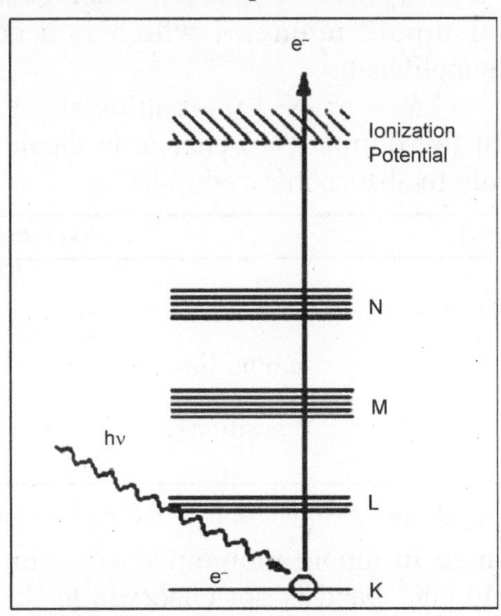

Fig. Energy-level Diagram for XPS

The electron binding energies are dependent on the chemical environment of the atom, making XPS useful to identify the oxidation state and ligands of an atom.

XPS instruments consist of an x-ray source, an energy analyser for the photoelectrons, and an electron detector. The analysis and detection of photoelectrons requires that the sample be placed in a high-vacuum chamber. Since the photoelectron energy depends on x-ray energy, the excitation source must be monochromatic. The energy of the photoelectrons is analysed by an electrostatic analyser, and the photoelectrons are detected by an electron multiplier tube or a multichannel detector such as a microchannel plate.

INFRARED ABSORPTION SPECTROSCOPY (IR)

IR spectroscopy is the measurement of the wavelength and intensity of the absorption of mid-infrared light by a sample. Mid-infrared light (2.5 – 50 μm, 4000 – 200 cm^{-1}) is energetic enough to excite molecular vibrations to higher energy levels. The wavelength of IR absorption bands are characteristic of specific types of chemical bonds, and IR spectroscopy finds its greatest utility for identification of organic and organometallic molecules.

MECHANISM OF IR ABSORPTION

The transition moment for infrared absorption is:

$$R = \langle X_i \mid u \mid X_j dt \rangle$$

where X_i and X_j are the initial and final states, respectively, and u is the electric dipole moment operator:

$$u = u_o + (r - r_e)(du/dr) + \text{... higher terms.}$$

u_o is the permanent dipole moment, which is a constant, and since $\langle X_i \mid X_j \rangle = 0$, R simplifies to:

$$R = \langle X_i \mid (r - r_e)(du/dr) \mid X_j \rangle$$

The result is that there must be a change in dipole moment during the vibration for a molecule to absorb infrared radiation.

Symmetric stretch		Asymmetric stretch
1340 cm^{-1}		2350 cm^{-1}
O=C=O		O = C=O
O = C = O	equilibrium ← → position	O = C = O
O = C = O		O=C = O

Figur Examples of Infrared Active and Inactive Absorption Bands in CO_2.

There is no change in dipole moment during the symmetric stretch vibration and the 1340 cm^{-1} band is not observed in the infrared absorption spectrum (the symmetric stretch is called infrared inactive).

There is a change in dipole moment during the asymmetric stretch and the 2350 cm^{-1} band does absorb infrared radiation (the asymmetric stretch in infrared active). A related vibrational spectroscopic method is Raman spectroscopy, which has a different mechanism and therefore provides complementary information to infrared absorption.

INFRARED ABSORPTION BANDS

IR absorption spectroscopy uses mid-infrared light (2.5 – 50 μm, 4000 – 200 cm^{-1}) to detect specific types of chemical bonds in a sample for identification of organic and organometallic molecules.

Table of Characteristic IR Bands

Group	*Bond*	*Appox. Energy (cm^{-1})*
Hydroxyl	O–H	3610 – 3640
Amines	N–H	3300 – 3500
Aromatic rings	C–H	3000 – 3100
Alkenes	C–H	3020 – 3080
Alkanes	C–H	2850 – 2960
Nitriles	C=–N	2210 – 2260
Carbonyl	C=O	1650 – 1750
Amines	C–N	1180 – 1360

INFRARED ABSORPTION SPECTROMETERS

This document describes dispersive and Fourier-transform spectrometers that are used in infrared absorption spectroscopy.

DISPERSIVE INFRARED SPECTROMETERS

Common light sources are tungsten lamps, Nernst glowers, or glowbars. Dispersive IR spectrometers use a grating monochromator to select wavelengths and are commonly used when a single wavelength is desired to monitor the kinetics of a reaction or as a GC or LC detector.

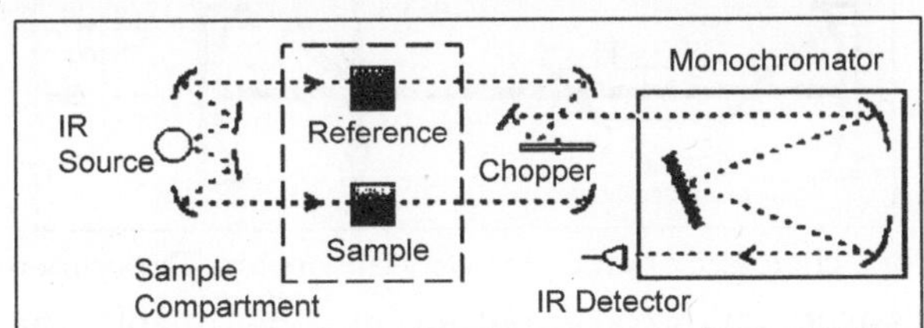

Fig. Figure of a Dispersive IR Absorption Spectrometer

CAVITY-RINGDOWN LASER ABSORPTION SPECTROSCOPY (CRLAS)

Cavity-ringdown laser absorption spectroscopy (CRLAS) is an ultrasensitive method to make quantitative absorption measurements of very low concentrations of analytes. The technique uses a laser pulse that is reflected back and forth between two highly reflecting mirrors. This procedure results in a very long path length, to which the measured absorbance is directly proportional as described by the Beer-Lambert law. An optical detector is placed behind one of the mirrors to detect the small amount of the light that passes through the mirror.

With no absorbing analyte present, the laser pulse will decrease in intensity after each round trip due to the loss of light through the monitoring mirror and other losses. When an absorbing species is present between the mirrors, the intensity of the laser pulse decreases more rapidly. The analyte concentration is determined by calibrating this decay time with known concentration of analyte. CRLAS can be used from the near-UV to the mid-IR. The wavelength range is only limited by instrumental constraints, such as the availability of high reflectivity mirrors and suitable pulsed laser sources. Accessing the near and mid-IR is accomplished with frequency-conversion techniques such as Raman shifting or optical parametric oscillators. In principle, the analyte can be in a solid, liquid, or gas. In practice, CRALS is primarily used to measure gas-phase species due to scattering losses in solids or liquids.

INTRACAVITY-ABSORPTION SPECTROSCOPY

Intracavity-absorption spectroscopy is an ultrasensitive method for measuring very small absorptions. Some examples are very low concentrations of analytes, and very weak absorption such as infrared overtone bands. The technique takes advantage of the very high light intensities within a laser resonance cavity, and the sensitivity of the lasing output to losses within the cavity. Small losses within the laser cavity due to absorption results in large changes in the intensity of the laser output.

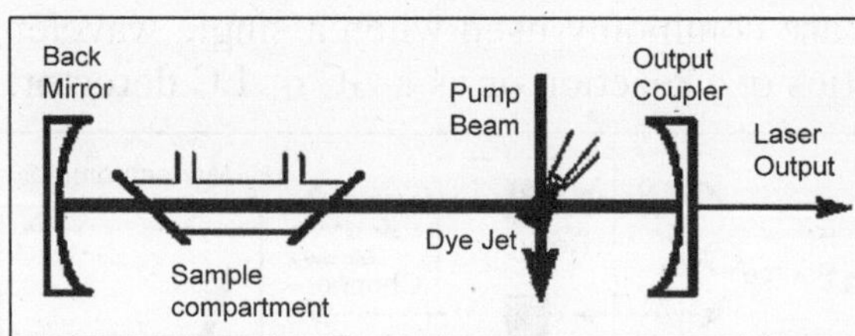

Fig. Schematic of an Intracavity-absorption Experiment

To record a spectrum the laser output must be tunable, therefore the most common lasers used in intracavity-absorption measurements are dye lasers. Samples can be solids, liquids, or gases. Liquids and gases are contained in a sample holder that has windows at Brewster's angle to minimize losses due to reflection. Small absorptions are detected by monitoring the output of the laser beam with a detector. Since the laser light is relatively intense, the main requirement for the detector is that it not be destroyed by the laser beam.

LASER-INDUCED FLUORESCENCE (LIF)

Laser-induced fluorescence (LIF) is the optical emission from molecules that have been excited to higher energy levels by absorption of electromagnetic radiation. The main advantage of fluorescence detection compared to absorption measurements is the greater sensitivity achievable because the fluorescence signal has a very low background. For molecules that can be resonant excitated, LIF provides selective excitation of the analyte to avoid interferences. LIF is useful to study the electronic structure of molecules and to make quantitative measurements of analyte concentrations. Analytical applications include monitoring gas-phase concentrations in the atmosphere, flames, and plasmas; and remote sensing using light detection and ranging (LIDAR). Because of the differences in the nature of the energy-level structure between atoms and molecules, the discussion on atomic fluorescence spectroscopy is in a separate document.

The excitation source for molecular LIF is typically a tunable dye laser in the visible spectral region. Studies in the near-ultraviolet and near-infrared are becoming more common as near-infrared lasers and frequency-doubling methods improve. High-resolution studies require cooling of the molecules to remove spectral congestion and to reduce the Doppler width of the transitions. A separate document on high-resolution spectroscopy describes cooling methods such as molecular beams, free-jet expansions, and cryogenic glass or crystalline matrices.

EMISSION OF ENERGY OF INFRARED RADIATION

Infrared Radiation, emission of energy as electromagnetic waves in the portion of the spectrum just beyond the limit of the red portion of visible radiation. The wavelengths of infrared radiation are shorter than those of radio waves and longer than those of light waves. They range between approximately 10^{-6} and 10^{-3} (about 0.0004 and 0.04 in). Infrared radiation may be detected as heat, and instruments such as bolometers are used to detect it.

Infrared radiation is used to obtain pictures of distant objects obscured by atmospheric haze, because visible light is scattered by haze but infrared radiation is not. The detection of infrared radiation is used by astronomers to observe stars and nebulas that are invisible in ordinary light or that emit radiation in the infrared portion of the spectrum.

An opaque filter that admits only infrared radiation is used for very precise infrared photographs, but an ordinary orange or light-red filter, which will absorb blue and violet light, is usually sufficient for most infrared pictures.

Developed about 1880, infrared photography has today become an important diagnostic tool in medical science as well as in agriculture and industry. Use of infrared techniques reveals pathogenic conditions that are not visible to the eye or recorded on X-ray plates.

Remote sensing by means of aerial and orbital infrared photography has been used to monitor crop conditions and insect and disease damage to large agricultural areas, and to locate mineral deposits. In industry, infrared spectroscopy forms an increasingly important part of metal and alloy research, and infrared photography is used to monitor the quality of products.

Infrared devices such as those used during World War II enable sharpshooters to see their targets in total visual darkness. These instruments consist essentially of an infrared lamp that sends out a beam of infrared radiation, often referred to as black light, and a telescope receiver that picks up returned radiation from the object and converts it to a visible image.

SPECTRUM

Spectrum, rainbowlike series of colours, in the order violet, blue, green, yellow, orange, and red, produced by splitting a composite light, such as white light, into its component colours. Indigo was formerly recognized as a distinct spectral colour. The rainbow is a natural spectrum, produced by meteorological phenomena. A similar effect can be produced by passing sunlight through a glass prism. The first correct explanation of the phenomenon was advanced in 1666 by the English mathematician and physicist Sir Isaac Newton.

When a ray of light passes from one transparent medium, such as air, into another, such as glass or water, it is bent; upon re-emerging into the air, it is bent again. This bending is called refraction; the amount of refraction depends on the wavelength of the light. Violet light, for example, is bent more than red light in passing from air to glass or from glass to air. A mixture of red and violet light is thus dispersed into the two colours when it passes through a wedge-shaped glass prism.

A device for producing and observing a spectrum visually is called a spectroscope; a device for observing and recording a spectrum photographically is called a spectrograph; a device for measuring the brightness of the various portions of spectra is called a spectrophotometer; and the science of using spectroscopes, spectrographs, and spectrophotometers to study spectra is called spectroscopy. For extremely accurate spectroscopic measurements, an interferometer is used. During the 19th century, scientists discovered that beyond the violet end of the spectrum, radiations could be detected that were invisible to the human eye but that had marked photochemical action; these radiations were termed ultraviolet.

Similarly, beyond the red end of the spectrum, infrared radiations were detected that, although invisible, transmitted energy, as shown by their ability to raise the temperature of a thermometer. The definition of spectrum was then revised to include these invisible radiations, and has since been extended to include radio waves beyond the infrared, and X rays and gamma rays beyond the ultraviolet. The term *spectrum* is often loosely applied today to any orderly array produced by analysis of a complex phenomenon. A complex sound such as noise, for example, may be analysed into an audio spectrum of pure tones of various pitches. Similarly, a complex mixture of elements or isotopes of different atomic weights can be separated into an orderly sequence called a mass spectrum in order of their atomic weights.

Spectroscopy has not only provided an important and sensitive method of chemical analysis, but has also been the chief tool for discoveries in the apparently unrelated fields of astrophysics and atomic theory.

In general, changes in motions of the outer electrons of atoms produce spectra in the visible, infrared, and ultraviolet regions. Changes in motions of the inner electrons of heavy atoms produce X-ray spectra. Changes in the configurations of the nucleus of an atom produce gamma-ray spectra. Changes in the configurations of molecules produce visible and infrared spectra.

Different colours of light are similar in consisting of electromagnetic radiations that travel at a speed of approximately 300,000 km per sec (about 186,000 mi per sec). They differ in having varying frequencies and wavelengths, the frequency being equal to the speed of light divided by wavelength. Two rays of light having the same wavelength also have the same frequency and the same colour. The wavelength of light is so small that it is conveniently expressed in nanometers (nm), which are equal to one-billionth of a meter. The wavelength of violet light varies from about 400 to 450 nm, and of red light from about 620 to 760 nm, or from about 0.000016 to 0.000018 in for violet, and from 0.000025 to 0.000030 in for red.

QUANTA OF RADIATION

At the beginning of the 20th century, however, physicists found that the wave theory did not account for all the properties of radiation. In 1900 the German physicist Max Planck demonstrated that the emission and absorption of radiation occur in finite units of energy, known as *quanta.* In 1904, German-born American physicist Albert Einstein was able to explain some puzzling experimental results on the external photoelectric effect by postulating that electro-magnetic radiation can behave like a particle. Other phenomena, which occur in the interaction between radiation and matter, can also be explained only by the quantum theory. Thus, modern physicists were forced to recognize that electromagnetic radiation can sometimes behave like a particle, and sometimes behave like a wave. The parallel concept—that matter also exhibits the same duality of having particlelike and wavelike characteristics—was developed in 1923 by the French physicist Louis Victor, Prince de Broglie.

X-RAY

X-Ray, penetrating electromagnetic radiation, having a shorter wavelength than light, and produced by bombarding a target, usually made of tungsten, with high-speed electrons. X-rays were discovered accidentally in 1895 by the German physicist Wilhelm Conrad Roentgen while he was studying cathode rays in a high-voltage, gaseous-discharge tube.

Despite the fact that the tube was encased in a black cardboard box, Roentgen noticed that a barium-platinocyanide screen, inadvertently lying nearby, emitted fluorescent light whenever the tube was in operation.

After conducting further experiments, he determined that the fluorescence was caused by invisible radiation of a more penetrating nature than ultraviolet rays. The invisible radiation "X-ray" because of its unknown nature. Subsequently, X-rays were known also as Roentgen rays in his honour.

Nature of X-Rays

X-rays are electromagnetic radiation ranging in wavelength from about 100 A to 0.01 A. The shorter the wavelength of the X-ray, the greater is its energy and its penetrating power. Longer wavelengths, near the ultraviolet-ray band of the electromagnetic spectrum, are known as soft X-rays.

The shorter wavelengths, closer to and overlapping the gamma-ray range, are called hard X-rays. A mixture of many different wavelengths is known as "white" X-rays, as opposed to "monochromatic" X-rays, which represent only a single wavelength. Both light and X-rays are produced by transitions of electrons that orbit atoms, light by the transitions of outer electrons and X-rays by the transitions of inner electrons. In the case of bremsstrahlung radiation X-rays are produced by the retardation or deflection of free electrons passing through a strong electrical field. Gamma rays, which are identical to X-rays in their effect, are produced by energy transitions within excited nuclei.

X-rays are produced whenever high-velocity electrons strike a material object. Much of the energy of the electrons is lost in heat; the remainder produces X-rays by causing changes in the target's atoms as a result of the impact. The X-rays emitted can have no more energy than the kinetic energy of the electrons that produce them. Moreover, the emitted radiation is not monochromatic but is composed of a wide range of wavelengths with a sharp, lower wavelength limit corresponding to the maximum energy of the bombarding electrons.

This continuous spectrum is referred to by the German name *bremsstrahlung,* which means "braking," or slowing down, radiation, and is independent of the nature of the target. If the emitted X-rays are passed through an X-ray spectrometer, certain distinct lines are found superimposed on the continuous spectrum; these lines, known as the characteristic X-rays, represent wavelengths that depend only on the structure of the target atoms. In other words, a fast-moving electron striking the target can do two things: It can excite X-rays of any energy up to its own energy; or it can excite X-rays of particular energies, dependent on the nature of the target atom.

X-Ray Production

The first X-ray tube was the Crookes tube, a partially evacuated glass bulb containing two electrodes, named after its designer, the British chemist and physicist Sir William Crookes. When an electric current passes through such a tube, the residual gas is ionized and positive ions, striking the cathode, eject electrons from it.

These electrons, in the form of a beam of cathode rays, bombard the glass walls of the tube and produce X rays. Such tubes produce only soft X rays of low energy. An early improvement in the X-ray tube was the introduction of a curved cathode to focus the beam of electrons on a heavy-metal target, called the anticathode, or anode.

This type generates harder rays of shorter wavelengths and of greater energy than those produced by the original Crookes tube, but the operation of such tubes is erratic because the X-ray production depends on the gas pressure within the tube. The next great improvement was made in 1913 by the American physicist William David Coolidge.

The Coolidge tube is highly evacuated and contains a heated filament and a target. It is essentially a thermionic vacuum tube in which the cathode emits electrons because the cathode is heated by an auxiliary current and not because it is struck by ions as in the earlier types of tubes. The electrons emitted from the heated cathode are accelerated by the application of a high voltage across the tube. As the voltage is increased, the minimum wavelength of the radiation decreases.

Most of the X-ray tubes in present-day use are modified Coolidge tubes. The larger and more powerful tubes have water-cooled anticathodes to prevent melting under the impact of the electron bombardment. The widely used shockproof tube is a modification of the Coolidge tube with improved insulation of the envelope (by oil) and grounded power cables. Such devices as the betatron are used to produce extremely hard X-rays, of shorter wavelength than the gamma rays emitted by naturally radioactive elements.

Properties of X-Rays

X-rays affect a photographic emulsion in the same way light does. Absorption of X-radiation by any substance depends upon its density and atomic weight. The lower the atomic weight of the material, the more transparent it is to X-rays of given wavelengths. When the human body is X-rayed, the bones, which are composed of elements of higher atomic weight than the surrounding flesh, absorb the radiation more effectively and therefore cast darker shadows on a photographic plate. Another type of radiation, which is known as neutron radiation and is now used in some types of radiography, produces almost opposite results. Objects that cast dark shadows in an X-ray picture are almost always light in a neutron radiograph.

Fluorescence

X-rays also cause fluorescence in certain materials, such as barium platinocyanide and zinc sulfide. If a screen coated with such fluorescent material is substituted for the photographic films, the structure of opaque objects may be observed directly. This technique is known as fluoroscopy.

Ionization

Another important characteristic of X-rays is their ionizing power, which depends upon their wavelength. The capacity of monochromatic X-rays to ionize is directly proportional to their energy. This property provides a method for measuring the energy of X-rays. When X-rays are passed through an ionization chamber an electric current is produced that is proportional to the energy of the incident beam. In addition to ionization chambers, more sensitive devices, such as the Geiger-Müller counter and the scintillation counter, can measure the energy of X-rays on the basis of ionization. In addition, the path of X-rays, by virtue of their capacity to ionize, can be made visible in a cloud chamber.

X-Ray Diffraction

X-rays may be diffracted by passage through a crystal or by reflection (scattering) from a crystal, which consists of regular lattices of atoms that serve as fine diffraction gratings. The resulting interference patterns may be photographed and analysed to determine the wavelength of the incident X-rays or the spacings between the crystal atoms, whichever is the unknown factor. X-rays may also be diffracted by ruled gratings if the spacings are approximately equal to the wavelengths of the incident X-rays.

Interaction with Matter

In the interaction between matter and X-rays, three mechanisms exist by which X-rays are absorbed; all three mechanisms demonstrate the quantum nature of X-radiation.

Photoelectric Effect

When a quantum of radiation, or a photon, in the X-ray portion of the electromagnetic spectrum strikes an atom, it may impinge on an electron within an inner shell and eject it from the atom. If the photon carries more energy than is necessary to eject the electron, it will transfer its residual energy to the ejected electron in the form of kinetic energy. This phenomenon, called the photoelectric effect, occurs primarily in the absorption of low-energy X-rays.

Compton Effect

The Compton effect, discovered in 1923 by the American physicist and educator Arthur Holly Compton, is an important manifestation of the absorption of X-rays of shorter wavelengths. When a high-energy photon collides with a stationary electron, both particles may be deflected at an angle to the direction of the path of the incident X-ray. The incident photon, having delivered some of its energy to the electron, emerges from the impact with a lower frequency and a correspondingly longer wavelength. These deflections, accompanied by a change of wavelength, are known as Compton scattering.

Pair Production

In the third type of absorption, especially evident when elements of high atomic weight are irradiated with extremely high-energy X-rays, the phenomenon of pair production occurs. When a high-energy photon penetrates the electron shell close to the nucleus, it may create a pair of electrons, one of negative charge and the other positive; a positively charged electron is also known as a positron.

This pair production is an example of the conversion of energy into mass. The photon requires at least 1.2 MeV of energy to yield the mass of the pair. If the incident photon possesses more energy than is required for pair production, the excess energy is imparted to the electron pair as kinetic energy. The paths of the two particles are divergent.

APPLICATIONS OF X-RAYS

The principal uses of X-radiation are in the field of scientific research, industry, and medicine. The study of X-rays played a vital role in theoretical physics, especially in the development of quantum mechanics. As a research tool, X-rays enabled physicists to confirm experimentally the theories of crystallography. By using X-ray diffraction methods, crystalline substances may be identified and their structure determined. Virtually all present-day knowledge in this field was either discovered or verified by X-ray analysis. X-ray diffraction methods can also be applied to powdered substances that are not crystalline but that display some regularity of molecular structure. By means of such methods, chemical compounds can be identified and the size of ultramicroscopic particles can be established. Chemical elements and their isotopes may be identified by X-ray spectroscopy, which determines the wavelengths of their characteristic line spectra.

Several elements were discovered by analysis of X-ray spectra.A number of recent applications of X-rays in research are assuming increasing importance. Microradiography, for instance, produces fine-grain images that can be enlarged considerably. Two radiographs can be combined in a projector to produce a three-dimensional image called a stereoradiogram.

Colour-radiography is also used to enhance the detail of X-ray photographs; in this process, differences in the absorption of X-rays by a specimen are shown as different colours. Extremely detailed and analytical information is provided by the electron microprobe, which uses a sharply defined beam of electrons to generate X-rays in an area of specimen as small as 1 micrometer (about 1/ 25,000 in) square.

Industry

In addition to the research applications of X-rays in physics, chemistry, mineralogy, metallurgy, and biology, X-rays are used in industry as a research tool and for many testing processes. They are valuable in industry as a means of testing objects such as metallic castings without destroying them. X-ray images on photographic plates reveal the presence of flaws, but a disadvantage of such inspection is that the necessary high-powered X-ray equipment is bulky and expensive. In some instances, therefore, radioisotopes, which emit highly penetrating gamma rays, are used instead of X-ray equipment. These isotope sources can be housed in relatively light, compact, and shielded containers. Cobalt-60 and cesium -137 have been used widely for industrial radiography. Thulium-70 has been used in small, convenient, isotope projectors for some medical and industrial applications.

Many industrial products are inspected routinely by means of X-rays so that defective products may be eliminated at the point of production. Other applications include the detection of fake gems and the detection of smuggled goods in customs examinations.

Ultrasoft X-rays are used to determine the authenticity of works of art and for art restoration.

Medicine

X-ray photographs, called radiographs, and fluoroscopy are used extensively in medicine as diagnostic tools. In radiotherapy, X-rays are used to treat certain diseases, notably cancer, by exposing tumors to X-radiation. The use of radiographs for diagnostic purposes was inherent in the penetrating properties of X-rays. Within a few years of their discovery, X-rays were being used to locate foreign bodies, such as bullets, within the human body.

With the development of improved X-ray techniques, minute differences in tissues were revealed by radiographs, and many pathological conditions could be diagnosed by means of X-rays. X-rays provided the most important single method of diagnosing tuberculosis when that disease was prevalent. Pictures of the lungs were easy to interpret because the air spaces are more transparent to X-rays than the lung tissues.

Various other cavities in the body can be filled artificially with contrasting media, either more transparent or more opaque to X-rays than the surrounding tissue, so that a particular organ is brought more sharply into view. Barium sulfate, which is highly opaque to X-rays, is used for the X-ray examination of the gastrointestinal tract.

Certain opaque compounds are administered either by mouth or by injection into the bloodstream in order to examine the kidneys or the gallbladder. Such dyes can have serious side effects, however, and should be used only after careful consultation. The routine use of X-ray diagnosis has in fact been discouraged—by the American College of Radiology in 1982, for example—as of questionable usefulness.

A recent X-ray device, used without dyes, offers clear views of any part of the anatomy, including soft organ tissues. Called the body scanner, or computerized axial tomography (CAT or CT) scanner, it rotates 180° around a patient's body, sending out a pencil-thin X-ray beam at 160 different points. Crystals positioned at the opposite points of the beam pick up and record the absorption rates of the varying thicknesses of tissue and bone.

These data are then relayed to a computer that turns the information into a picture on a screen. Using the same dosage of radiation as that of the conventional X-ray machine, an entire "slice" of the body is made visible with about 100 times more clarity. The scanner was invented in 1972 by the British electronics engineer Godfrey N. Hounsfield, and was in general use by 1979.

4

Quantum Mechanics

INTRODUCTION

Quantum mechanics (QM) is a relatively new ground of physics that was developed at the beginning of the last century as the common effort of different scientists. Up to that time, it was possible to describe all known physical phenomena with the laws of classical physics as either *particles* (classical mechanics, kinematics) or *waves* (optics, classical electro-magnetism). The detection of new phenomena such as *e.g.* the *photoelectric effect* has lead to the development of a new physics that has revolutionized the way we look at the world around us. Many insights from quantum mechanics are at primary sight counterintuitive (*i.e.* they differ substantially from what we would expect in a classical picture!) which makes this field one of the most fascinating areas of modern physics. Quantum mechanics is a branch of physics dealing with physical phenomena where the action is on the arrange of the Planck constant. Quantum mechanics departs from classical mechanics primarily at the *quantum realm* of atomic and subatomic length scales. QM provides a mathematical description of much of the dual *particle-like* and *wave-like* behaviour and interactions of energy and matter.

In advanced topics of quantum mechanics, some of these behaviours are macroscopic and only emerge at extreme (*i.e.*, very low or very high) energies or temperatures. The name *quantum mechanics* derives from the observation that some physical quantities can change only in *discrete* amounts (Latin *quanta*), and not in a continuous (*cf.* analog) way. For example, the angular momentum of an electron bound to an atom or molecule is quantized. In the context of quantum mechanics, the wave-particle duality of energy and matter and the uncertainty principle provide a unified view of the behaviour of photons, electrons, and other atomic-scale objects.

The mathematical formulations of quantum mechanics are abstract. A mathematical purpose called the wavefunction provides information about the probability amplitude of position, momentum, and other physical properties of a particle. Mathematical manipulations of the wavefunction usually involve the braket notation, which requires an understanding of complex numbers and linear functionals. The wavefunction treats the object as a quantum harmonic oscillator, and the mathematics is akin to that describing acoustic resonance. Many of the results of quantum mechanics are not easily visualized in terms of classical mechanics - for instance, the ground state in a quantum mechanical model is a non-zero energy state that is the lowest permitted energy state of a system, as opposed a more "traditional" system that is thought of as simply being at rest, with zero kinetic energy. Instead of a traditional static, unchanging zero state, quantum mechanics allows for far more dynamic, chaotic possibilities, according to John Wheeler.

The earliest versions of quantum mechanics were formulated in the first decade of the 20th century. At around the same time, the atomic theory and the corpuscular theory of light (as updated by Einstein) first came to be widely accepted as scientific fact; these latter theories can be viewed as quantum theories of matter and electromagnetic radiation, respectively. Early quantum theory was significantly reformulated in the mid-1920s by Werner Heisenberg, Max Born, Wolfgang Pauli and their collaborators, and the Copenhagen interpretation of Niels Bohr became widely accepted. By 1930, quantum mechanics had been further unified and formalized by the work of Paul Dirac and John von Neumann, with a greater emphasis placed on measurement in quantum mechanics, the statistical nature of our knowledge of reality, and philosophical speculation about the role of the observer.

Quantum mechanics has since branched out into almost every aspect of 20th century physics and other disciplines, such as quantum chemistry, quantum electronics, quantum optics, and quantum information science. Much 19th century physics has been re-evaluated as the "classical limit" of quantum mechanics, and its more advanced developments in terms of quantum field theory, string theory, and speculative quantum gravity theories.

Particles and Waves

All phenomena in classical physics can be described either as a *particle* or a *wave*.

Wave-functions and Operators

The theory of quantum mechanics is built upon the fundamental concepts of *wave-functions* and *operators*. The wave-function is a single-valued square-integrable function of the system parameters and time which provides a complete description of the system. Linear Hermitian operators act on the wave-function and correspond to the physical *observables*, those dynamical variables which can be measured, *e.g.* position, momentum and energy.

For systems of atomic nuclei (Eq. 5.1) and electrons, which are the subject of this dissertation, the system parameters might be taken to be a set of position variables of the constituent particles *i.e.*, $\{(r_i),(r_\alpha)\}$ their momenta $\{(p_i),(p_\alpha)\}$ or even a mixture of the two *e.g.* $\{(r_i),(p_\alpha)\}$. In contrast to a Newtonian system which is completely described by the positions *and* momenta of its constituents, the quantum-mechanical wave-function is a function of only one of these parameters per particle (Eq. 5.2). The wave-function for the system is thus typically denoted by $\Psi\{(r_i),(r_\alpha),t\}$.

A notation due to Dirac is often employed, which reflects the fact that this wave-function is simply one of many representations of a single *state-vector* in a Hilbert space, which is written as $|\Psi\rangle$, known as a *ket*. There also exists a *dual space* containing a set of *bra* vectors, denoted $|\Psi\rangle$, defined by their scalar products and in one-to-one correspondence with the kets.

The scalar product is written as a *braket* and is anti-linear in the first argument and linear in the second: thus $|\Psi\,|\,\Phi\rangle = (\langle\Psi\,|\,\Phi\rangle)^*$.

It is worth noting here that state-vectors which differ only by a multiplicative non-zero complex constant describe the same state: we can thus restrict our interest to the set of *normalised* vectors defined such that the scalar product of the vector with its own conjugate equals unity:

$$\langle\Psi\,|\,\Psi\rangle = \int\prod_j dr_\beta\,\Psi^*(\{r_i\},\{r_a\},t)\,\Psi(\{r_i\},\{r_\alpha\},t) = 1. \quad (3.1)$$

The operator corresponding to some observable O is often written $\hat{O}$, and in general when this operator acts on some state-vector $|\Psi\rangle$, a different (not necessarily normalised) state-vector $|\Phi\rangle$ results:

$$\hat{O}\,|\,\Psi\rangle = |\,\Phi\rangle. \quad (3.2)$$

However, for each operator there exists a set of normalised *eigenstates*, say $\{|\,\chi_n\rangle\}$, which remain unchanged by the action of the operator *i.e.*

$$= 6.58217\times 10^{-16}\ \mathrm{e\,V\,s}. \quad (3.3)$$

in which the constant (always real for Hermitian operators) is the *eigenvalue*.

The postulates of quantum mechanics state that for a system in state $|\,\Psi\rangle$:

- The outcome of a measurement of a dynamical variable is always one of the eigenvalues l_n of the corresponding operator,
- Immediately following a measurement, the state-vector collapses to the eigenstate $|\,\chi_n\rangle$ corresponding to the measured eigenvalue,
- The probability of such a measurement is

$$P(\lambda_n) = |\langle \chi_n | \Psi \rangle|^2. \quad (3.4)$$

Wave Phenomena

Not all phenomena of classical physics are describable in a particle picture. A second large class of physical events can be described as *waves*. A *wave* is a disturbance that propagates through space or space and time, often transferring energy.

Surface Waves in Water with Highs and Lows.
Planck's Quantum Theory.

From every day live, you are already familiar with many wave phenomena, such as for instance *water waves* or *sound waves* (periodical compressions of air that travel through space and can be detected by a mechanosensitive device in your inner ear).

Water waves and sound waves are examples for *mechanical waves* that exist in a medium (which on deformation is capable of producing elastic restoring forces).

Wave Phenomena: Interference and Diffraction

Waves can exhibit phenomena that cannot be explain in a classical particle pictures. Typical such wave phenomena are *interference* and *diffraction*. Two or more waves can superimpose and form a new wave, this effect is called *interference*. When two sinusoidal waves place over, the resulting waveform depends on the frequency, the amplitude and in particular on the relative difference in phase (the *phase shift*) of the two waves.

If the two waves have the same amplitude A and wavelength the resultant waveform will have an amplitude between 0 and $2A$ depending on whether the two waves are *in phase* (*i.e.* both waves have their minima and maxima at exactly the same time) or *out of phase* (*i.e.* one wave has a maximum when the other one has a minimum). Consider two waves that are in phase, with amplitudes A_1 and A_2. Their troughs and peaks line up and the resultant wave will have an amplitude $A = A_1 + A_2$. This is known as *constructive interference*.

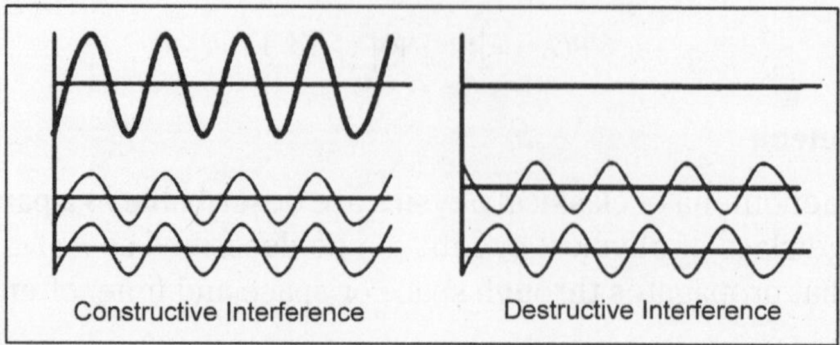

Fig. Constructive and Destructive Interference of Two Waves.

If the two waves are π radians, or 180°, out of phase, then one wave's crests will coincide with another wave's troughs and so will tend to cancel out. The resultant amplitude is $A = |A1 - A2|$. If $A1 = A2$, the resultant amplitude will be zero. This is known as *destructive interference*.

PRINCIPLES OF QUANTUM MECHANICS

One of the most important concepts of quantum mechanics is the uncertainty principle, which states that there is always some uncertainty when trying to measure the posistion and momentum of a particle. More specifically, the product of the uncertainty in position, D *x*, and the uncertainty in momentum, D*m*u, is always greater than or equal to *h* over 2π, where *h* is planck's constant.

When trying to determine the position of an electron, for example, we cannot know *exactly* where it is and where it's going at the same time. The more accurately one is measured, the less accurately the other is known. To we measure the position of the electron, for example, we can only obtain the information through the collision of the electron with another subatomic particle, such as a photon. After the collision, the momentum of the photon changes—it "bounces" away, and from this change in momentum the position of the electron is calculated. However, the photon also gives the electron a change in momentum, so that now the position and velocity of the electron has changed from what it was. Thus, we can't know precisely the position and velocity simultaneously.

ELECTRON IN THE BOHR ATOM

In the Bohr model, electrons revolved around the nucleus in orbits. According to classical physics, they should radiate energy as they move and ultimately fall to the centre of the atom. Yet this is not so. The Quantum explanation is that electrons can only exist in predefined, discrete energy levels, and nowhere in between. The lowest energy level is not at the centre of the atom, and electrons may go no closer than the lowest level. This explains why they don't collide into the nucleus. To move to another energy level, they must gain or lose *exactly* the energy difference between the levels. This explains spectral lines. Another explanation is that, if electrons crashed into the nucleus, their velovity would be zero and the uncertainty would also be zero. This would be a direct violation of the uncertainty principle, a fundamental property of nature.

SPIN AND QUANTUM NUMBERS

It was discovered experimentally that subatomic particles have intrinsic angular momentum. That is, that angular momentum is not given by some interaction, but a property of the particles. This is a quantum property that classical physics has no explanation for. The closest thing on the macroscopic level is rotation (since angular momentum leads to rotating), so this property is called "spin". Subatomic particles, however, don't really spin; they just have the angular momenta. This property is quantatized since the particles may only have certain discrete values of spin and not any value in between. Electrons have spins of either +1/2 or –1/2. In the quantum model, electrons in the atom are given four quantum numbers to distinguish them. The first three distinguish the energy level, the shape of the orbital, and in which orbital the electron is located (has 90 per cent chance of being found), and the last one gives the spin. The Pauli Exclusion Principle states that no 2 electrons in the atom can have the same 4 quantum numbers. If they occupy the same energy level and orbital, then they must have opposite spin.

NUCLEUS

The nucleus is a compact and dense region in the centre of the atom that makes up for most of its mass. It is made up of a group of subatomic particles known collectively as nucleons. The two states of nucleons are protons and neutrons. Although the electron, and not the nucleus, is responsable for chemical reactions, the nucleus still holds much information about the atom. Since nucleons are about 2000 times heavier than electrons, the mass of the nucleus is approximately equal to the mass of the atom. The number of protons in the nucleus determines the type of atom, and nuclear charge is the property by which the elements are arranged in the periodic table.

NUCLEAR ENERGY

A tremendous amount of energy is used to hold the nucleus together, since the repulsion between the positively charged protons is great. The bonding forces that hold the nucleus together are collectively known as the nucleur forces. When heavy element nuclei are split into smaller nuclei, or light element nuclei combined to make one nucleus, tremendous energies are released. The former is known fission, and the latter, fusion, and they are the driving forces behind stars and nuclear weopons.

HISTORY

From Dalton's theory, it was thought that atoms were indivisible particles that made up everything in the world. When Rutherford made his gold foil experiment, however, he found that when high-velocity positive alpha particles bombarded a thing metal foil, while most passed straight through, some were reflected back or deflected. This led him to postulate that most of the atom was empty space, with most of the mass concentrated in one dense region in the centre of the arom. Thus, most of the positive alpha particles passed through the empty space, but some hit the dense core and were reflected back by electric repulsion. He called this the nucleus of the atom. Subsequently, improvements were made on the Rutherford model. The nucleu was theorized to contain both positive and neutral particles.

APPLICATION

The interactions in the nucleus, the strong and the weak forces, are the strongest forces in nature. Processes such as fission and fusion release much of this energy, and make possible new energy sources. In fact, fusion is the process by which stars generate light and heat, and thus the source from which all life on earth is derived. Other processes, such as radiation, have medical applications.

DE BROGLIE HYPOTHESIS

After Albert Einstein's photon theory became accepted, the query became whether this was true only for light or whether material objects also exhibited wave-like behaviour. In quantum mechanics (QM) particles and waves have identity crises because one can act as a particle or a wave; and when not a particle, its "matter-wave" has a frequency proportional to its kinetic and relativistic energy. Motivated by this, and a number of similarities between QM and time-frequency analysis (TFA), comes the idea that signals can be represented and synthesized by an dynamic system of particles construed as waves. Vice versa, a sound might be "materialized" into its corresponding system of particles, and modifications made in that domain to synthesize variants.

Essentially what is developed here is a technique of sound work of art using classical many-body mechanics with the most natural mapping of parameters. Such a direct mapping of both multi-dimensional fields not only allows an interchange of concepts to enrich both, for instance sonifying particle collisions for musical purposes, but also enables an additional level of comprehension of the underlying physical concepts—an important goal of sonification. Consequently, when using this technique for composition and synthesis, the composer must also possess skill in physics to do an effective job. Interestingly, it has been observed that, when composing with these techniques, the compositional and scientific concerns merge into the "composerscientist"—a state of thought where physics and music become identical. It has also been found that the audience need not be versed in physics to appreciate or enjoy what they hear; however, more compositions need to be developed to explore this experience. Separately from these concepts, some problematic issues currently exist in the system and its execution.

The computational complexity required for physical subdivision simulations can be enormous, and when compounded with requirements for audio, there is a large latency period between execution and resultant signal. If precaution is not taken, one is at risk of spending several hours for a sound that could have more easily been produced, albeit without a "physical" correspondence. It can be said that even if a simple sound were created, one would hear it differently because of what it represents. Thus there exists a programmatic issue, whereby a composition developed through this technique might only be interesting within this context. In addition to these, psychoacoustic principles have yet to be integrated to facilitate a more perceivable and precise audification of the system. These issues, and more, will obviously be addressed in future work.

De Broglie's Thesis

In his 1923 (or 1924, depending on the source) doctoral dissertation, the French physicist Louis de Broglie made a bold assertion. Considering Einstein's relationship of wavelength *lambda* (λ) to momentum p, de Broglie proposed that this relationship would determine the wavelength of any matter, in the relationship:

$$\lambda = h/p$$

recall that h is Planck's constant.

This wavelength is called the *de Broglie wavelength*. The reason he chose the momentum equation over the energy equation is that it was unclear, with matter, whether E should be total energy, kinetic energy, or total relativistic energy. For photons they are all the same, but not so for matter.

Assuming the momentum relationship, however, allowed the derivation of a similar de Broglie relationship for frequency f using the kinetic energy E_k:

$$f = E_k/h$$

Alternate Formulations

De Broglie's relationships are sometimes expressed in terms of Dirac's constant, $\hbar = h/(2\pi)$, and the angular frequency w and wavenumber k:

$$p = \hbar k$$

$$E_k = \hbar w.$$

Experimental Confirmation

In 1927, physicists Clinton Davisson and Lester Germer, of Bell Labs, performed an experiment where they passionate electrons at a crystalline nickel target. The resulting diffraction pattern matched the predictions of the de Broglie wavelength. De Broglie received the 1929 Nobel Prize for his theory (the first time it was ever awarded for a Ph.D. thesis) and Davisson/Germer jointly won it in 1937 for the experimental discovery of electron diffraction (and thus the proving of de Broglie's hypothesis).

Further experiments have held de Broglie's hypothesis to be true, including the quantum variants of the double slit experiment.

Diffraction experiments in 1999 confirmed the de Broglie wavelength for the behaviour of molecules as large as buckeyballs (complex molecules made up of 60 or more carbon atoms).

Significance of the de Broglie Hypothesis

The de Broglie hypothesis showed that wave particle duality was not merely an aberrant behaviour of light, but rather was a fundamental principle exhibited by both radiation and matter. As such, it becomes possible to use wave equations to describe material behaviour, so long as one properly applies the de Broglie wavelength. This would prove crucial to the development of quantum mechanics.

Macroscopic Objects and Wavelength

Though de Broglie's hypothesis predicts wavelengths for matter of any size, there are realistic limits on when it's helpful. A baseball thrown at a pitcher has a de Broglie wavelength that is smaller than than the diameter of a proton... by about 20 orders of magnitude. The wave aspects of a macroscopic object are so tiny as to be unobservable in any useful sense.

Photoelectric Effect

Though at first observed in 1839, the photoelectric effect was documented by Heinrich Hertz in 1887 in a paper to the *Annalen der Physik*. It was originally called the Hertz effect, in fact, though this name fell out of use. When a light source (or, more generally, electromagnetic radiation) is incident upon a metallic surface, the surface can emit electrons. Electrons emitted in this fashion are called *photoelectrons* (although they are still just electrons).

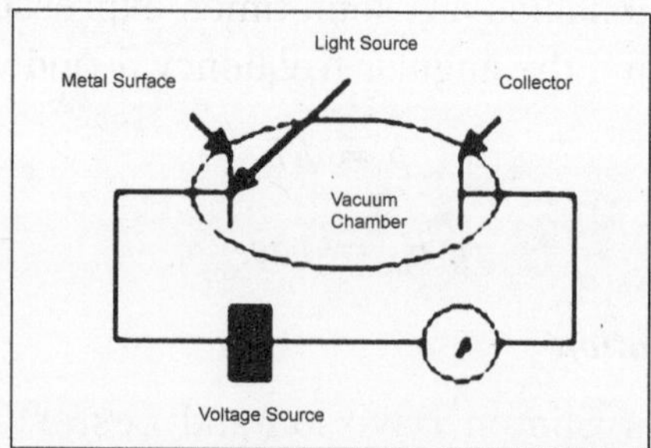

Fig. A Basic Set of Photoelectric Effect.

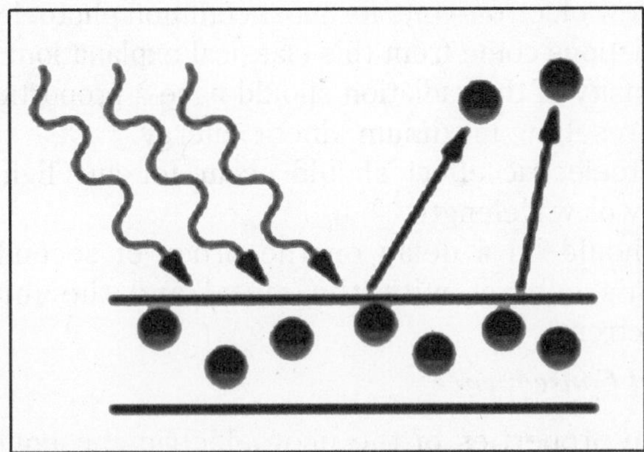

Fig. Light Collides with a Metal Surface, Releasing Electrons.

Setting Up the Photoelectric Effect

To observe the photoelectric result, you create a vacuum chamber with the photoconductive metal at one end and a collector at the other. When a light shines on the metal, the electrons are released and move through the vacuum towards the collector.

This creates a current in the wires connecting the two ends, which can be measured with an ammeter. A basic example of the experiment in shown in Figure .

By administering a negative voltage potential (the black box in the picture) to the collector, it takes more energy for the electrons to complete the journey and initiate the current. The point at which no electrons make it to the collector is called the *stopping potential* V_s, and can be used to determine the maximum kinetic energy K_{max} of the electrons (which have electronic charge e) by using the following equation:

$$K_{max} = eV_s$$

It is significant to note that not all of the electrons will have this energy, but will be emitted with a range of energies based upon the properties of the metal being used. The above equation allows us to calculate the maximum kinetic energy or, in other words, the energy of the particles knocked free of the metal surface with the greatest speed, which will be the trait that is most useful in the rest of this analysis.

THE CLASSICAL WAVE EXPLANATION

In classical wave theory, the energy of electromagnetic radiation is carried within the wave itself. As the electromagnetic wave (of intensity I) collides with the surface, the electron absorbs the energy from the wave until it exceeds the binding energy, releasing the electron from the metal. The minimum energy needed to remove the electron is the *work function phi* (ϕ) of the material. (ϕ is in the range of a few electron-volts for most common photoelectric materials.) Three main predictions come from this classical explanation:

- The intensity of the radiation should have a proportional relationship with the resulting maximum kinetic energy.
- The photoelectric effect should occur for any light, regardless of frequency or wavelength.
- There should be a delay on the order of seconds between the radiation's contact with the metal and the initial release of photoelectrons.

The Experimental Consequence

By 1902, the properties of the photoelectric consequence were well documented. Experiment showed that:

- The intensity of the light source had no effect on the maximum kinetic energy of the photoelectrons.
- Below a certain frequency, the photoelectric effect does not occur at all.
- There is no significant delay (less than 10^{-9} s) between the light source activation and the emission of the first photoelectrons.

As you can tell, these three results are the exact opposite of the wave theory predictions. Not only that, but they are all three completely counter-intuitive. Why would low-frequency light not trigger the photoelectric effect, since it still carries energy? How do the photoelectrons release so quickly? And, perhaps most curiously, why does adding more intensity not result in more energetic electron releases? Why does the wave theory fail so utterly in this case, when it works so well in so many other situation

Einstein's Wonderful Time

In 1905, Albert Einstein published four papers in the *Annalen der Physik* journal, each of which was significant enough to warrant a Nobel Prize in its own right. The first paper (and the only one to actually be recognized with a Nobel) was his explanation of the photoelectric effect.

Building on Max Planck's blackbody radiation theory, Einstein proposed that radiation energy is not continuously distributed over the wavefront, but is instead localized in small bundles (later called photons). The photon's energy would be associated with its frequency (ν), through a proportionality constant known as *Planck's constant* (h), or alternately, using the wavelength (*lambda*) and the speed of light (c):

$$E = h\nu = hc/\lambda$$

or the momentum equation: $p = h/\lambda$

In Einstein's theory, a photoelectron releases as a effect of an interaction with a single photon, rather than an interaction with the wave as a whole. The energy from that photon transfers instantaneously to a single electron, knocking it free from the metal if the energy (which is, recall, proportional to the frequency ν) is high enough to overcome the work function (ϕ) of the metal. If the energy (or frequency) is too low, no electrons are knocked free.

If, however, there is excess energy, beyond ϕ, in the photon, the excess energy is converted into the kinetic energy of the electron:

$$K_{max} = h\nu - \phi$$

Therefore, Einstein's theory predicts that the maximum kinetic energy is completely independent of the intensity of the light (because it doesn't show up in the equation anywhere). Shining twice as much light results in twice as many photons, and more electrons releasing, but the maximum kinetic energy of those individual electrons won't change unless the energy, not the intensity, of the light changes.

The maximum kinetic energy results when the least-tightly-bound electrons break free, but what about the most-tightly-bound ones; The ones in which there is *just* enough energy in the photon to knock it loose, but the kinetic energy that results in zero? Setting K_{max} equal to zero for this *cutoff frequency* (ν_c), we get:

$$\nu_c = \phi/h$$

or the cutoff wavelength: $\lambda_c = hc/\phi$

These equations indicate why a low-frequency light source would be unable to free electrons from the metal, and thus would produce no photoelectrons.

After Einstein

Experimentation in the photoelectric result was carried out extensively by Robert Millikan in 1915, and his work confirmed Einstein's theory. Einstein won a Nobel Prize for his photon theory (as applied to the photoelectric effect) in 1921, and Millikan won a Nobel in 1923 (in part due to his photoelectric experiments). Most significantly, the photoelectric effect, and the photon theory it inspired, crushed the classical wave theory of light. Though no one could deny that light behaved as a wave, after Einstein's first paper, it was undeniable that it was also a particle.

Matter Waves

We have seen that in certain experiments electrons produce patterns attributed to waves. From this observation, we concluded that electrons have wave-like properties. We also concluded that as the energy of electrons increases their wavelength decreases.

Electrons are a form of matter, so these waves are called matter waves. In this tutorial we will relate quantitative features of matter waves with familiar, measurable physical quantities such as energy, mass, and momentum. We will apply these relationships to forms of matter other than electrons, and see how these results can be applied to the electron microscope.

Historical context

At the end of the 19th century, light was thought to consist of waves of electromagnetic fields which propagated according to Maxwell's equations, while matter was thought to consist of localized particles. This division was challenged when, in his 1905 paper on the photoelectric effect, Albert Einstein postulated that light was emitted and absorbed as localized packets, or "quanta" (now called photons). These quanta would have an energy

$$E = h\nu$$

where ν is the incidence of the light and h is Planck's constant. .

Einstein's postulate was confirmed experimentally by Robert Millikan and Arthur Compton over the next two decades. Thus it became apparent that light has both wave-like and particle-like properties. De Broglie, in his 1924 PhD thesis, sought to expand this wave-particle duality to all particles:

"When I conceived the first basic ideas of wave mechanics in 1923-24, I was guided by the aim to perform a real physical synthesis, valid for all particles, of the coexistence of the wave and of the corpuscular aspects that Einstein had introduced for photons in his theory of light quanta in 1905." —De Broglie

In 1926, Erwin Schrödinger published an equation describing how this matter wave should evolve-the matter wave equivalent of Maxwell's equations-and used it to derive the energy spectrum of hydrogen. That same year Max Born published his now-standard interpretation that the square of the amplitude of the matter wave gives the probability to find the particle at a given place. This interpretation was in contrast to De Broglie's own interpretation, in which the wave corresponds to the physical motion of a localized particle.

Experimental Confirmation

Elementary Particles

In 1927 at Bell Labs, Clinton Davisson and Lester Germer fired slow-moving electrons at a crystalline nickel target. The angular dependence of the reflected electron intensity was measured, and was determined to have the same diffraction pattern as those predicted by Bragg for x-rays.

Before the acceptance of the de Broglie hypothesis, diffraction was a property that was thought to be only exhibited by waves. Therefore, the presence of any diffraction effects by matter demonstrated the wave-like nature of matter. When the de Broglie wavelength was inserted into the Bragg condition, the observed diffraction pattern was predicted, thereby experimentally confirming the de Broglie hypothesis for electrons.

This was a pivotal result in the development of quantum mechanics. Just as the photoelectric effect demonstrated the particle nature of light, the Davisson-Germer experiment showed the wave-nature of matter, and completed the theory of wave-particle duality. For physicists this idea was important because it means that not only can any particle exhibit wave characteristics, but that one can use wave equations to describe phenomena in matter if one uses the de Broglie wavelength.

Since the original Davisson-Germer experiment for electrons, the de Broglie hypothesis has been confirmed for other elementary particles. The wavelength of a thermalized electron in a non-metal at room temperature is about 8 nm.

Neutral Atoms

Experiments with Fresnel diffraction and specular reflection of neutral atoms confirm the application of the de Broglie hypothesis to atoms, *i.e.* the existence of atomic waves which undergo diffraction, interference and allow quantum reflection by the tails of the attractive potential. Advances in laser cooling have allowed cooling of neutral atoms down to nanokelvin temperatures. At these temperatures, the thermal de Broglie wavelengths come into the micrometre range. Using Bragg diffraction of atoms and a Ramsey interferometry technique, the de Broglie wavelength of cold sodium atoms was explicitly measured and found to be consistent with the temperature measured by a different method.

This effect has been used to demonstrate atomic holography, and it may allow the construction of an atom probe imaging system with nanometer resolution. The description of these phenomena is based on the wave properties of neutral atoms, confirming the de Broglie hypothesis.

Waves of Molecules

Recent experiments even corroborate the relations for molecules and even macromolecules, which are normally considered too large to undergo quantum mechanical effects. In 1999, a research team in Vienna demonstrated diffraction for molecules as large as füllerenes. The researchers calculated a De Broglie wavelength of the most probable C_{60} velocity as 2.5 pm. More recent experiments prove the quantum nature of molecules with a mass up to 6910 amu. In general, the De Broglie hypothesis is expected to apply to any well isolated object.

Davisson and Germer Experiment

The Davisson-Germer experiment demonstrated the wave nature of the electron, confirming the earlier hypothesis of deBroglie. Putting wave-particle duality on a firm experimental footing, it represented a major step forward in the development of quantum mechanics. The Bragg law for diffraction had been applied to x-ray diffraction, but this was the first application to particle waves.

Davisson and Germer intended and built a vacuum apparatus for the purpose of measuring the energies of electrons scattered from a metal surface. Electrons from a heated filament were accelerated by a voltage and allowed to strike the surface of nickel metal. The electron beam was directed at the nickel target, which could be rotated to observe angular dependence of the scattered electrons. Their electron detector (called a Faraday box) was mounted on an arc so that it could be rotated to observe electrons at different angles. It was a great surprise to them to find that at certain angles there was a peak in the intensity of the scattered electron beam. This peak indicated wave behaviour for the electrons, and could be interpreted by the Bragg law to give values for the lattice spacing in the nickel crystal.

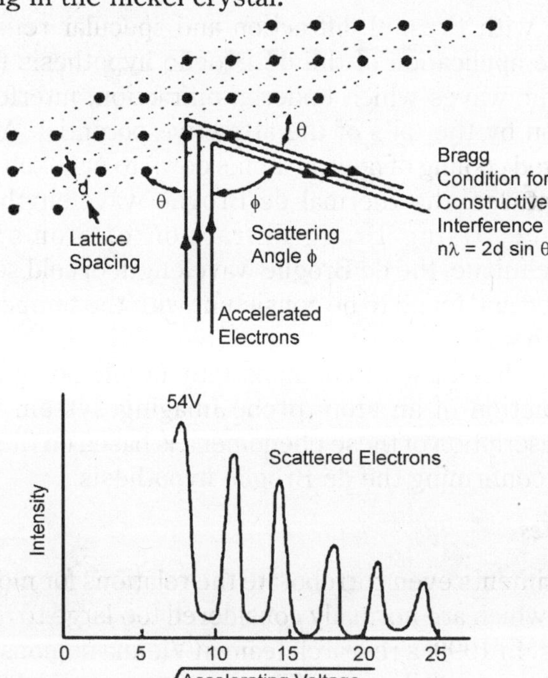

The experimental data above, reproduced above Davisson's article, shows repeated peaks of scattered electron intensity with increasing accelerating voltage. This data was collected at a fixed scattering angle. Using the Bragg law, the de Broglie wavelength expression, and the kinetic energy of the accelerated electrons gives the relationship.

$$\underbrace{\frac{1}{\lambda}}_{\text{Electron wavelength}} = \underbrace{\frac{n}{2d\sin\theta}}_{\text{Bragg law}} = \underbrace{\frac{p}{h}}_{\text{deBroglie relationship}} = \frac{\sqrt{2mE}}{h} = \underbrace{\frac{\sqrt{2meV}}{h}}_{\text{Acceleration through voltage } V}$$

In the historical data, an accelerating voltage of 54 volts gave a definite peak at a scattering angle of 50°. The angle theta in the Bragg law corresponding to that scattering angle is 65°, and for that angle the calculated lattice spacing is 0.092 nm. For that lattice spacing and scattering angle, the relationship for wavelength as a function of voltage is empirically.

$$\frac{1}{\lambda(nm)} = \frac{n}{2d\sin\theta} = 0.815\sqrt{V}$$

Trying this relationship for n = 1,2,3 gives values for the square root of voltage 7.36, 14.7 and 22, which appear to agree with the first, third and fifth peaks above. Then what gives the second, fourth and sixth peaks? Perhaps they originate from a different set of planes in the crystal. Those peaks satisfy a sequence 2,3,4, suggesting that the first peak of that series would have been at 5.85. That corresponds to an electron wavelength of 0.21 nm and a lattice spacing of 0.116 nm ??

History and Overview

According to Maxwell's equations in the late 19th century, light was thought to consist of waves of electromagnetic fields and matter consist of localized particles. However this was challenged in Albert Einstein's 1905 paper on the photoelectric effect, which described light as discrete and localized quanta of energy (now called photons), and won him the Nobel Prize in Physics in 1921. In 1927 Louis de Broglie presented his thesis concerning the wave-particle duality theory, which proposed the idea that all matter displays the wave-particle duality of photons. According to de Broglie for all matter and for radiation alike, the energy E of the particle was related to the frequency of its associated wave v by the Planck relation:

$$E = h\nu$$

And that the momentum of the particle p was related to its wavelength by what is now known as the de Broglie relation:

$$p = \frac{h}{\lambda},$$

where h is Planck's constant.

An important contribution to the Davisson–Germer experiment was made by Walter M. Elsasser in Göttingen in the 1920s, who remarked that the wave-like nature of matter might be investigated by electron scattering experiments on crystalline solids, just as the wave-like nature of X-rays had been confirmed through X-ray scattering experiments on crystalline solids.

This suggestion of Elsasser was then communicated by his senior colleague (and later Nobel Prize recipient) Max Born to physicists in England. When the

Davisson and Germer experiment was performed, the results of the experiment were explained by Elsasser's proposition. However the initial intention of the Davisson and Germer experiment was not to confirm the de Broglie hypothesis, but rather to study the surface of nickel.

In 1927 at Bell Labs, Clinton Davisson and Lester Germer fired slow moving electrons at a crystalline nickel target. The angular dependence of the reflected electron intensity was measured and was determined to have the same diffraction pattern as those predicted by Bragg for X-rays. This experiment was independently replicated by George Paget Thomson, and Davisson and Thomson shared the Nobel Prize in Physics in 1937. The Davisson -Germer experiment confirmed the de Broglie hypothesis that matter has wave-like behaviour. This, in combination with the Compton effect discovered by Arthur Compton (who won the Nobel Prize for Physics in 1927), established the wave–particle duality hypothesis which was a fundamental step in quantum theory.

Experiment

Davisson and Germer's actual objective was to study the surface of a piece of nickel by directing a beam of electrons at the surface and observing how many electrons bounced off at various angles. They expected that for electrons even the smoothest crystal surface would be too rough and so the electron beam would experience diffuse reflection.

The experiment consisted of firing an electron beam from an electron gun directed to a piece of nickel crystal at normal incidence (*i.e.* perpendicular to the surface of the crystal). The experiment included an electron gun consisting of a heated filament that released thermally excited electrons, which were then accelerated through a potential difference giving them a certain amount of kinetic energy towards the nickel crystal. To avoid collisions of the electrons with other molecules on their way towards the surface, the experiment was conducted in a vacuum chamber. To measure the number of electrons that were scattered at different angles, an electron detector that could be moved on an arc path about the crystal was used. The detector was designed to accept only elastically scattered electrons.

During the experiment an accident occurred and air entered the chamber, producing an oxide film on the nickel surface. To remove the oxide, Davisson and Germer heated the specimen in a high temperature oven, not knowing that this affected the formerly polycrystalline structure of the nickel to form large single crystal areas with crystal planes continuous over the width of the electron beam.

When they started the experiment again and the electrons hit the surface, they were scattered by atoms which originated from crystal planes inside the nickel crystal. As Max von Laue proved in 1912 the crystal structure serves as a type of three dimensional diffraction grating. The angles of maximum reflection are given by Bragg's condition for constructive interference from an array, Bragg's law

$$n\lambda = 2d\sin\left(90^\circ - \frac{\theta}{2}\right),$$

for $n = 1$, $\theta = 50°$, and for the spacing of the crystalline planes of nickel (d = 0.091 nm) obtained from previous X-ray scattering experiments on crystalline nickel. By varying the applied voltage to the electron gun, the maximum intensity of electrons diffracted by the atomic surface was found at different angles. The highest intensity was observed at an angle $\theta = 50°$ with a voltage of 54 V, giving the electrons a kinetic energy of 54 eV.

According to the de Broglie relation and Bragg's law, a beam of 54 eV had a wavelength of 0.165 nm. The experimental outcome was 0.167 nm, which closely matched the predictions.

Davisson and Germer's accidental discovery of the diffraction of electrons was the first direct evidence confirming de Broglie's hypothesis that particles can have wave properties as well.

THE UNCERTAINTY PRINCIPLE

The position and momentum of a particle cannot be simultaneously measured with arbitrarily high precision. There is a minimum for the product of the uncertainties of these two measurements. There is likewise a minimum for the product of the uncertainties of the energy and time.

$$\Delta \times \Delta p > \frac{\hbar}{2}$$

$$\Delta E \Delta t > \frac{\hbar}{2}$$

This is not a declaration about the inaccuracy of measurement instruments, nor a reflection on the quality of experimental methods; it arises from the wave properties inherent in the quantum mechanical description of nature. Even with perfect instruments and technique, the uncertainty is inherent in the nature of things.

Matter Wave Interpretation

According to the de Broglie hypothesis, every object in our Universe is a wave, a situation which gives rise to this phenomenon. Consider the measurement of the position of a particle. The particle's wave packet has non-zero amplitude, meaning that the position is uncertain – it could be almost anywhere along the wave packet. To obtain an accurate reading of position, this wave packet must be 'compressed' as much as possible, meaning it must be made up of increasing numbers of sine waves added together. The momentum of the particle is proportional to the wavenumber of one of these waves, but it *could be* any of them. So a more precise position measurement-by adding together more waves–means that the momentum measurement becomes less precise (and *vice versa*). The only kind of wave with a definite position is concentrated at one point, and such a wave has an indefinite wavelength (and therefore an indefinite momentum). Conversely, the only kind of wave with a definite wavelength is an infinite regular periodic oscillation over all space, which has no definite position. So in quantum mechanics, there can be no states that describe a particle with both a definite position and a definite momentum. The more precise the position, the less precise the momentum.

A mathematical statement of the principle is that every quantum state has the property that the root mean square (RMS) deviation of the position from its mean (the standard deviation of the *x*-distribution):

$$\sigma_x = \sqrt{\langle (x - \langle x \rangle)^2 \rangle}$$

times the RMS deviation of the momentum from its mean (the standard deviation of *p*):

$$\sigma_p = \sqrt{\langle (p - \langle p \rangle)^2 \rangle}$$

can never be smaller than a fixed fraction of Planck's constant:

$$\sigma_x \sigma_p \geq \hbar / 2.$$

The uncertainty principle can be restated in terms of other measurement processes, which involves collapse of the wavefunction. When the position is initially localized by preparation, the wavefunction collapses to a narrow bump in an interval $\Delta x > 0$, and the momentum wavefunction becomes spread out (cf. wave packet). The particle's momentum is left uncertain by an amount inversely proportional to the accuracy of the position measurement:

If the initial preparation in Δx is understood as an observation or disturbance of the particles then this means that the uncertainty principle is related to the observer effect. However, this is not true in the case of the measurement process corresponding to the former inequality but only for the latter inequality.

Robertson-Schrödinger Uncertainty Relations

The most common general form of the doubt principle is the *Robertson uncertainty relation*. For an arbitrary Hermitian operator $\hat{O}$, we can associate a standard deviation

$$\sigma_O = \sqrt{\langle \hat{O}^2 \rangle - \langle \hat{O} \rangle^2},$$

where the brackets $\langle O \rangle$ indicate an expectation value. For a pair of operators $\hat{A}$ and $\hat{B}$, we may define their *commutator* as

$$[\hat{A}, \hat{B}] = \hat{A}\hat{B} - \hat{B}\hat{A},$$

In this notation, the Robertson uncertainty relation is given by

$$\sigma_A \sigma_B \geq \left| \frac{1}{2i} \langle [\hat{A}, \hat{B}] \rangle \right|.$$

The Robertson uncertainty relation immediately follows from a slightly stronger inequality, the *Schrödinger uncertainty relation*,

$$\sigma_A^2 \sigma_B^2 \geq \left(\frac{1}{2} \langle \{\hat{A}, \hat{B}\} \rangle - \langle \hat{A} \rangle \langle \hat{B} \rangle \right)^2 + \left(\frac{1}{2i} \langle [\hat{A}, \hat{B}] \rangle \right)^2,$$

where we have introduced the *anticommutator*,

$$\{\hat{A}, \hat{B}\} = \hat{A}\hat{B} + \hat{B}\hat{A}.$$

Proof of the Schrödinger Uncertainty Relation

Since the Robertson and Schrödinger relations are for general operators, the relations can be applied to any two observables to obtain specific uncertainty relations. A few of the most common relations found in the literature are given below.

- For position and linear momentum, the canonical commutation relation $[\hat{x}, \hat{p}] = i\hbar$ implies the Kennard inequality from above:

$$\sigma_x \sigma_p \geq \frac{\hbar}{2}$$

- For two orthogonal components of the total angular momentum operator of an object:

$$\sigma_{J_i} \sigma_{J_j} \geq \frac{\hbar}{2} \left| \langle J_k \rangle \right|,$$

where i, j, k are distinct and J_i denotes angular momentum along the x_i axis.

This relation implies that only a single component of a system's angular momentum can be defined with arbitrary precision, normally the component parallel to an external (magnetic or electric) field. Moreover, for $[J_x, J_y] = i\hbar\epsilon_{xyz} J_z$, a choice $\hat{A} = J_x$, $\hat{B} = J_y$, in angular momen-tum multiplets, $\psi = |j, m\rangle$, bounds the Casimir invariant (angular momentum squared, $\langle J_x^2 + J_y^2 + J_z^2 \rangle$) from below and thus yields useful constraints such as j $(j + 1) \geq m\ (m + 1)$, and hence $j \geq m$, among others.

- For the number of electrons in a superconductor and the phase of its Ginzburg–Landau order parameter

$$\Delta N \Delta \phi \geq 1 .$$

History

Werner Heisenberg formulated the Uncertainty Principle at Niels Bohr's institute in Copenhagen, while working on the mathematical foundations of quantum mechanics. In 1925, following pioneering work with Hendrik Kramers, Heisenberg developed matrix mechanics, which replaced the ad-hoc old quantum theory with modern quantum mechanics. The central assumption was that the classical concept of motion does not fit at the quantum level, and that electrons in an atom do not travel on sharply defined orbits. Rather, the motion is smeared out in a strange way: the Fourier transform of time only involve those frequencies that could be seen in quantum jumps.

Heisenberg's paper did not admit any unobservable quantities like the exact position of the electron in an orbit at any time; he only allowed the theorist to talk about the Fourier components of the motion. Since the Fourier components were not defined at the classical frequencies, they could not be used to construct an exact trajectory, so that the formalism could not answer certain overly precise questions about where the electron was or how fast it was going.

The most striking property of Heisenberg's infinite matrices for position and momentum is that they do not commute. Heisenberg's canonical commutation relation indicates by how much:

and this result did not have a clear physical interpretation in the beginning.

In March 1926, working in Bohr's institute, Heisenberg realised that the non-commutativity implies the uncertainty principle. This implication provided a clear physical interpretation for the non-commutativity, and it laid the foundation for what became known as the Copenhagen interpretation of quantum mechanics. Heisenberg showed that the commutation relation implies an uncertainty, or in Bohr's language a complementarity. Any two variables that do not commute cannot be measured simultaneously — the more precisely one is known, the less precisely the other can be known. Heisenberg wrote: It can be expressed in its simplest form as follows: One can never know with perfect accuracy both of those two important factors which determine the movement of one of the smallest particles—its position and its velocity. It is impossible to determine accurately *both* the position and the direction and speed of a particle *at the same instant*.

One way to understand the complementarity between position and momentum is by wave-particle duality. If a particle described by a plane wave passes through a narrow slit in a wall like a water-wave passing through a narrow channel, the particle diffracts and its wave comes out in a range of angles.

The narrower the slit, the wider the diffracted wave and the greater the uncertainty in momentum afterwards. The laws of diffraction require that the spread in angle $\Delta\theta$ is about λ / d, where d is the slit width and λ is the wavelength. Reasoning based on the de Broglie relation shows that the size of the slit and the range in momentum of the diffracted wave are related by Heisenberg's rule:

$$\Delta x \Delta p \approx h.$$

In his celebrated 1927 paper, "Über den anschaulichen Inhalt der quantentheoretischen Kinematik und Mechanik" ("On the Perceptual Content of Quantum Theoretical Kinematics and Mechanics"), Heisenberg established this expression as the minimum amount of unavoidable momentum disturbance caused by any position measurement, but he did not give a precise definition for the uncertainties Δx and Δp.

Instead, he gave some plausible estimates in each case separately. In his Chicago lecture he refined his principle:

$$\Delta x \Delta p \geq h$$

Kennard in 1927 first proved the modern inequality:

$$\sigma_x \sigma_p \geq \frac{\hbar}{2}$$

where $\hbar = h/2\pi$, and σ_x, σ_p are the standard deviations of position and momentum. Heisenberg himself only proved relation for the special case of Gaussian states.

Terminology and Conversion

Throughout the main body of his original 1927 paper, written in German, Heisenberg used the word "Unbestimmtheit" ("indeterminacy") to describe the basic theoretical principle. Only in the endnote did he switch to the word "Unsicherheit" ("uncertainty").

However, when the English-language version of Heisenberg's textbook, *The Physical Principles of the Quantum Theory*, was published in 1930, the translation "uncertainty" was used, and it became the more commonly used term in the English language thereafter.

Heisenberg's Microscope

The principle is quite counter-intuitive, so the early students of quantum theory had to be reassured that naive measurements to violate it were bound to be always unworkable. One way in which Heisenberg originally illustra-ted the intrinsic impossibility of violating the uncertainty principle is by using an imaginary microscope as a measuring device. He imagines an experimenter trying to calculate the position and momentum of an electron by shooting a photon at it.

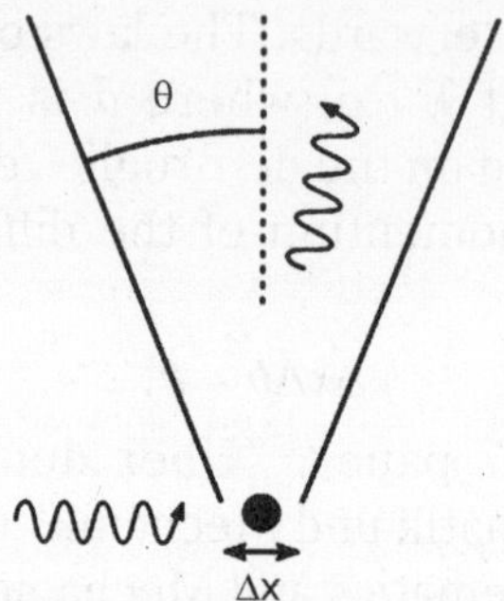

Figure 3.6: Heisenberg's Gamma-ray Microscope for Locating an Electron. The Incoming Gamma Ray is Scattered by the Electron up into the Microscope's Aperture Angle q. The Scattered Gamma-ray. Classical Optics Shows that the Electron Position can be Resolved only up to an Uncertainty Dx that Depends on q and the Wavelength l of the Incoming Light.

Problem 1 : If the photon has a short wavelength, and therefore a large momentum, the position can be measured accurately. But the photon scatters in a random direction, transferring a large and uncertain amount of momentum to the electron. If the photon has a long wavelength and low momentum, the collision does not disturb the electron's momentum very much, but the scattering will reveal its position only vaguely.

Problem 2 : If a large aperture is used for the microscope, the electron's location can be well resolved ; but by the principle of conservation of momentum, the transverse momentum of the incoming photon and hence the new momentum of the electron resolves poorly. If a small aperture is used, the accuracy of both resolutions is the other way around.

The combination of these trade-offs imply that no matter what photon wavelength and aperture size are used, the product of the uncertainty in measured position and measured momentum is greater than or equal to a lower limit, which is (up to a small numerical factor) equal to Planck's constant. Heisenberg did not care to formulate the uncertainty principle as an exact limit, and preferred to use it instead as a heuristic quantitative statement, correct up to small numerical factors, which makes the radically new noncommutativity of quantum mechanics inevitable.

Critical Reactions

The Copenhagen interpretation of quantum mechanics and Heisenberg's Uncertainty Principle were in fact seen as twin targets by detractors who believed in an underlying determinism and realism. According to the Copenhagen interpretation of quantum mechanics, there is no fundamental reality that the quantum state describes, just a prescription for calculating experimental results. There is no way to say what the state of a system fundamentally is, only what the result of observations might be.

Albert Einstein believed that randomness is a reflection of our ignorance of some fundamental property of reality, while Niels Bohr believed that the probability distributions are fundamental and irreducible, and depend on which measurements we choose to perform. Einstein and Bohr debated the uncertainty principle for many years.

Einstein's Slit

The first of Einstein's thought experiments challenging the uncertainty principle went as follows:

Consider a particle passing through a slit of width d. The slit introduces an uncertainty in momentum of approximately h/d because the particle passes through the wall. But let us determine the momentum of the particle by measuring the recoil of the wall. In doing so, we find the momentum of the particle to arbitrary accuracy by conservation of momentum.

Bohr's response was that the wall is quantum mechanical as well, and that to measure the recoil to accuracy Δp the momentum of the wall must be known to this accuracy before the particle passes through. This introduces an uncertainty in the position of the wall and therefore the position of the slit equal to $h / \Delta p$, and if the wall's momentum is known precisely enough to measure the recoil, the slit's position is uncertain enough to disallow a position measurement.

A similar analysis with particles diffracting through multiple slits is given by Richard Feynman.

Einstein's Box

Bohr was present when Einstein future the thought experiment which has become known as Einstein's box. Einstein argued that "Heisenberg's uncertainty equation implied that the uncertainty in time was related to the uncertainty in energy, the product of the two being related to Planck's constant." Consider, he said, an ideal box, lined with mirrors so that it can contain light indefinitely. The box could be weighed before a clockwork mechanism opened an ideal shutter at a chosen instant to allow one single photon to escape. "We now know, explained Einstein, precisely the time at which the photon left the box." "Now, weigh the box again. The change of mass tells the energy of the emitted light. In this manner, said Einstein, one could measure the energy emitted and the time it was released with any desired precision, in contradiction to the uncertainty principle."

Bohr spent a sleepless night considering this argument, and eventually realised that it was flawed. He pointed out that if the box were to be weighed, say by a spring and a pointer on a scale, "since the box must move vertically with a change in its weight, there will be uncertainty in its vertical velocity and therefore an uncertainty in its height above the table.... Furthermore, the uncertainty about the elevation above the earth's surface will result in an uncertainty in the rate of the clock," because of Einstein's own theory of gravity's effect on time. "Through this chain of uncertainties, Bohr showed that Einstein's light box experiment could not simultaneously measure exactly both the energy of the photon and the time of its escape."

EPR Paradox for Tangled Particles

Bohr was compelled to modify his understanding of the uncertainty principle after another thought experiment by Einstein. In 1935, Einstein, Podolsky and Rosen published an analysis of widely separated entangled particles. Measuring one particle, Einstein realised, would alter the probability distribution of the other, yet here the other particle could not possibly be disturbed. This example led Bohr to revise his understanding of the principle, concluding that the uncertainty was not caused by a direct interaction.

But Einstein came to much more far-reaching conclusions from the same thought experiment. He believed the "natural basic assumption" that a complete description of reality would have to predict the results of experiments from "locally changing deterministic quantities", and therefore would have to include more information than the maximum possible allowed by the uncertainty principle. In 1964 John Bell showed that this assumption can be falsified, since it would imply a certain inequality between the probabilities of different experiments. Experimental results confirm the predictions of quantum mechanics, ruling out Einstein's basic assumption that led him to the suggestion of his *hidden variables*.

While it is possible to assume that quantum mechanical predictions are due to *nonlocal* hidden variables, and in fact David Bohm invented such a formulation, this resolution is not satisfactory to the vast majority of physicists. The question of whether a random outcome is predetermined by a nonlocal theory can be philosophical, and it can be potentially intractable. If the hidden variables are not constrained, they could just be a list of random digits that are used to produce the measurement outcomes. To make it sensible, the assumption of nonlocal hidden variables is sometimes augmented by a second assumption — that the size of the observable universe puts a limit on the computations that these variables can do. A nonlocal theory of this sort predicts that a quantum computer encounters fundamental obstacles when it tries to factor numbers of approximately 10,000 digits or more; an achievable task in quantum mechanics.

Popper's Criticism

Karl Popper approached the difficulty of indeterminacy as a logician and metaphysical realist. He disagreed with the application of the uncertainty relations to individual particles rather than to ensembles of identically prepared particles, referring to them as "statistical scatter relations". In this statistical interpretation, a *particular* measurement may be made to arbitrary precision without invalidating the quantum theory. This directly contrasts the Copenhagen interpretation of quantum mechanics, which is non-deterministic but lacks local hidden variables.

In 1934 Popper published *Zur Kritik der Ungenauigkeitsrelationen* (*Critique of the Uncertainty Relations*) in *Naturwissenschaften*, and in the same year *Logik der Forschung* (translated and updated by the author as *The Logic of Scientific Discovery* in 1959), outlining his arguments for the statistical interpretation. In 1982, he further developed his theory in *Quantum theory and the schism in Physics*, writing:

[Heisenberg's] formulae are, beyond all doubt, derivable *statistical formulae* of the quantum theory. But they have been *habitually misinterpreted* by those quantum theorists who said that these formulae can be interpreted as determining some upper limit to the *precision of our measurements*. Popper proposed an experiment to falsify the uncertainty relations, though he later withdrew his initial version after discussions with Weizsäcker, Heisenberg, and Einstein; this experiment may have influenced the formulation of the EPR experiment.

Many-worlds Uncertainty

The many-worlds interpretation originally outlined by Hugh Everett III in 1957 is partly meant to reconcile the differences between the Einstein and Bohr's views by replacing Bohr's wave function collapse with an ensemble of deterministic and independent universes whose *distribution* is governed by wave functions and the Schrödinger equation. Thus, uncertainty in the many-worlds interpretation follows from each observer within any universe having no knowledge of what goes on in the other universes.

ALGEBRA FOR QUANTUM MECHANICS

We've seen that in quantum mechanics, the state of an electron in some potential is given by a wave function $\psi(\vec{x},t)$, and physical variables are represented by operators on this wave function, such as the momentum in the x-direction $P_x = -i\hbar\partial/\partial x$. The Schrodinger wave equation is a *linear* equation, which means that if ψ_1 and ψ_2 are solutions, then so is $c_1\psi_2 + c_2\psi_2$, where c_1, c_2 are arbitrary complex numbers. This linearity of the sets of possible solutions is true generally in quantum mechanics, as is the representation of physical variables by operators on the wave functions. The mathematical structure this describes, the linear set of possible states and sets of operators on those states, is in fact a *linear algebra* of operators acting on a *vector space*. From now on, this is the language we'll be using most of the time. To clarify, we'll give some definitions.

Vector Space

The prototypical vector space is of course the set of real vectors in ordinary three-dimensional space, these vectors can be represented by trios of real numbers (v_1,v_2,v_3) measuring the components in the x, y and z directions respectively.

The basic properties of these vectors are:

Any vector multiplied by a number is another vector in the space, $a(v_1,v_2,v_3)=(av_1,av_2,av_3)$;
the sum of two vectors is another vector in the space, that given by just adding the corresponding components together: $(v_1,w_1,v_2+w_2,v_3+w_3)$.

These two properties together are referred to as "*closure*": adding vectors and multiplying them by numbers cannot get you out of the space.

A further property is that there is a unique null vector (0,0,0) and each vector has an additive inverse $(-v_1,-v_2,-v_3)$ which added to the original vector gives the null vector.

Mathematicians have generalized the definition of a vector space: a general vector space has the properties we've listed above for three-dimensional real vectors, but the operations of addition and multiplication by a number are generalized to more abstract operations between more general entities. The operators are, however, restricted to being commutative and associative.

Notice that the list of necessary properties for a general vector space does not include that the vectors have a magnitude—that would be an additional requirement, giving what is called a *normed* vector space.

To go from the familiar three-dimensional vector space to the vector spaces relevant to quantum mechanics, first the real numbers (components of the vector and possible multiplying factors) are to be generalized to complex numbers, and second the three-component vector goes an n component vector.

The consequent n-dimensional complex space is sufficient to describe the quantum mechanics of angular momentum, an important subject. But to describe the wave function of a particle in a box requires an *infinite* dimensional space, one dimension for each Fourier component, and to describe the wave function for a particle on an infinite line requires the set of all normalizable continuous differentiable functions on that line.

Fortunately, all these generalizations are to finite or infinite sets of complex numbers, so the mathematicians' vector space requirements of commutativity and associativity are always trivially satisfied.

We use Dirac's notation for vectors, $|1\rangle, |2\rangle$ and call them "kets", so, in his language, if $|1\rangle, |2\rangle$ belong to the space, so does $c_1|1\rangle + c_2,|2\rangle$ for arbitrary complex constants c_1, c_2. Since our vectors are made up of complex numbers, multiplying any vector by zero gives the null vector, and the additive inverse is given by reversing the signs of all the numbers in the vector. Clearly, the set of solutions of Schrodinger's equation for an electron in a potential satisfies the requirements for a vector space: $\psi(\vec{x},t)$ is just a complex number at each point in space, so only complex numbers are involved in forming $c_1\psi_1 + c_2\psi_2$, and commutativity, associativity, etc., follow at once.

Vector Space Dimensionality

The vectors $|1\rangle, |2\rangle, |3\rangle$ are *linearly independent* if

$$c_1|1\rangle + c_2|2\rangle + c_3|3\rangle = 0 \text{ implies } c_1 = c_2 = c_3 = 0.$$

A vector space is n-dimensional if the maximum number of linearly independent vectors in the space is n.

Such a space is often called $V^n(C)$, or $V^n(R)$ if only real numbers are used.

Now, vector spaces with finite dimension n are clearly insufficient for describing functions of a continuous variable x. But they are well worth reviewing here: they are fine for describing quantized angular momentum, and they serve as a natural introduction to the infinite-dimensional spaces needed to describe spatial wavefunctions.

A set of n linearly independent vectors in n-dimensional space is a *basis*—any vector can be written *in a unique way* as a sum over a basis:

$$|V\rangle = \sum v_i |i\rangle$$

You can check the uniqueness by taking the difference between two supposedly distinct sums: it will be a linear relation between independent vectors, a contradiction.

Since all vectors in the space can be written as linear sums over the elements of the basis, the sum of multiples of any two vectors has the form:

$$a|V\rangle + b\sum(av_i + bv_i)|i\rangle$$

Inner Product Spaces

The vector spaces of relevance in quantum mechanics also have an operation associating a number with a pair of vectors, a generalization of the dot product of two ordinary three-dimensional vectors,

$$\vec{a}.\vec{b}\sum a_i b_i$$

Following Dirac, we write the inner product of two ket vectors $|V\rangle, |W\rangle$ as $|W\rangle|V\rangle$. Dirac refers to this $\langle | \rangle$ form as a "bracket" made up of a "bra" and a "ket".

An *important difference* from the simple three-dimensional real vector inner product is that we're usually dealing with *complex* vector spaces, and *we require*

$$\langle W|V\rangle = \langle V|W\rangle^*$$

where * denotes complex conjugate. This means that $\langle V|V\rangle$ is real. We require $\langle V|V\rangle$ to be *positive*, except for the null vector, for which it is zero. (These requirements do restrict the classes of vector spaces we are considering, but they are all satisfied by the spaces relevant to quantum mechanics.)

The *norm* of $|V\rangle$ is defined by

$$|V| = \sqrt{\langle V|V\rangle}$$

If $|V\rangle$ is a member of $\mathbf{V}^n(C)$, so is $\alpha|V\rangle$, for any complex number *a*. We require the inner product operation to commute with multiplication by a number, so $\langle W|(\alpha|V\rangle) = a\langle W|V\rangle$ The complex conjugate of the right hand side is $a^*\langle V|W\rangle$ For consistency, the bra corresponding to the ket $a|V\rangle$ must be $\langle V|a^*$.

It follows that if

$$|V\rangle = \sum v_i|i\rangle, |W\rangle = \sum W_i|i\rangle \, than \langle V|W\rangle = \sum v_i^* w_j \langle i|j\rangle.$$

Constructing an Orthonormal Basis: the Gram-Schmidt Process

To have something better resembling the standard dot product of ordinary three vectors, we need $\langle i|j\rangle = \delta_{ij}$, that is, we need to construct an *orthonormal basis* in the space. There is a straightforward procedure for doing this called the Gram-Schmidt process. We begin with a linearly independent set of basis vectors, $|1\rangle, |2\rangle, |3\rangle, \ldots$

We first normalize $|1\rangle$ by dividing it by its norm. Call the normalized vector $|1\rangle$. Now $|2\rangle$ cannot be parallel to $|1\rangle$, because the original basis was of linearly independent vectors, but $|2\rangle$ in general has a nonzero component parallel to $|1\rangle$ equal to $|I\rangle\langle I|2\rangle, |1\rangle\langle I|2\rangle$ since $|1\rangle$ is normalized.

Therefore, the vector $|2\rangle - |I\rangle\langle I|2\rangle$ is perpendicular to $|1\rangle$ as is easily verified. It is also easy to compute the norm of this vector, and divide by it to get $|II\rangle$ the second member of the orthonormal basis. Next, we take $|3\rangle$ and subtract off its components in the directions $|1\rangle$ and $|II\rangle$ and normalize the remainder, and so on.

In an n-dimensional space, having constructed an orthonormal basis with members $|i\rangle$, any vector $|V\rangle$ can be written as a column vector,

The corresponding bra is $\langle V| = \sum v_i{}^* \langle i|$, which we write as a row vector with the elements complex conjugated,

$$\langle V| = \left(v_1{}^*, v_2{}^*, v_n{}^*\right).$$

This operation, going from columns to rows and taking the complex conjugate, is called taking the *adjoint*, and can also be applied to matrices, as we shall see shortly.

$$|V\rangle = \sum V_i |i\rangle = \begin{pmatrix} v_1 \\ v_2 \\ \cdot \\ \cdot \\ v_n \end{pmatrix}, where |1\rangle \begin{pmatrix} 1 \\ 0 \\ \cdot \\ \cdot \\ 0 \end{pmatrix} and\ so\ on.$$

The reason for representing the bra as a row is that the inner product of two vectors is then given by standard matrix multiplication:

$$\langle V|W\rangle = (v_1, v_2, \ldots, v_n) \begin{pmatrix} w_1 \\ \cdot \\ \cdot \\ w_n \end{pmatrix}$$

(Of course, this only works with an orthonormal base.)

The Schwartz Inequality

The Schwartz inequality is the generalization to any inner product space of the result $\left|\vec{a}.\vec{a}\right|^2 \le \left|\vec{a}\right|^2 \left|\vec{b}\right|^2$ (or $\cos^2\theta \le 1$) for ordinary three-dimensional vectors. The equality sign in that result only holds when the vectors are parallel. To generalize to higher dimensions, one might just note that two vectors are in a two-dimensional subspace, but an illuminating way of understanding the inequality is to write the vector $\vec{a}$ as a sum of two components, one parallel to $\vec{b}$ and one perpendicular to. The component parallel $\vec{b}$ to is just $\vec{b}\left(\vec{a}\cdot\vec{b}\right)/\left|\vec{b}\right|^2$, so the component perpendicular to $\vec{b}$ is the vector $\vec{a}_\perp = \vec{a} - \vec{b}\left(\vec{a}\cdot\vec{b}\right)/\left|\vec{a}\right|^2$. Substituting this expression into $\vec{a}_\perp \cdot \vec{a}_\perp \ge 0$, the inequality follows.

This same point can be made in a general inner product space: if $|V\rangle, |W\rangle$ are two vectors, then

$$|Z\rangle = |V\rangle - \frac{|W\rangle,\langle W|V\rangle}{|W|^2}$$

is the component of $|V\rangle$ perpendicular to $|W\rangle$, as is easily checked by taking its inner product with $|W\rangle$.

$$\langle Z|Z\rangle \geq 0 \textit{ gives immediately } |\langle V|W\rangle|^2 \leq |V|^2 |W|^2$$

Linear Operators

A *linear operator* A takes any vector in a linear vector space to a vector in that space, $A|V\rangle = |V^1\rangle$, and satisfies

$$A(c_1|V_1\rangle + c_2|V_2\rangle) = c_1 A|V_1\rangle + c_2 A|V_2\rangle,$$

with c_1, c_2 arbitrary complex constants.

The *identity operator* I is (obviously!) defined by:

$$I|V\rangle = |V\rangle \textit{ for all } |V\rangle$$

For an n-dimensional vector space with an orthonormal basis $|1\rangle, \ldots, |n\rangle$, since any vector in the space can be expressed as a sum $|V\rangle = \sum v_i |i\rangle$, the linear operator is completely determined by its action on the basis vectors—this is all we need to know. It's easy to find an expression for the identity operator in terms of bras and kets.

Taking the inner product of both sides of the equation $|V\rangle = \sum v_i |i\rangle$, with the bra $\langle i|$ gives $\langle i|V\rangle = v_i$, so

$$|V\rangle = \sum v_i |i\rangle = \sum |i\rangle\langle i|V\rangle$$

Since this is true for any vector in the space, it follows that that the identity operator is just

$$I = \sum_1^n |i\rangle\langle i|$$

This is an important result that will reappear in many disguises. To analyse the action of a general linear operator A, we just need to know how it acts on each basis vector. Beginning with $A|1\rangle$, this must be some sum over the basis vectors, and since they are orthonormal, the component in the $|i\rangle$ direction must be just $\langle i|A|1\rangle$

That is,

$$A|1\rangle = \sum_1^n |i\rangle\langle i|A_{i1}|i\rangle = \sum_1^n A_{i1}|i\rangle \textit{ writing } \langle i|A|1\rangle = A_{i1}.$$

So if the linear operator A acting on $|V\rangle=\sum V_i|i\rangle$ gives $|V'\rangle\sum v_i{}'|i\rangle$,, that is, the $A|V\rangle=|V'\rangle$, linearity tells us that

$$\sum v_i^{'}|i\rangle=|V'\rangle=A|V\rangle=\sum v_jA|j\rangle=\sum_{i.j} v_j|i\rangle\langle i|A|i\rangle=\sum_{i.j} v_jA_{ij}|i\rangle$$

where in the fourth step we just inserted the identity operator. Since the $|i\rangle$'s are all orthogonal, the coefficient of a particular $|i\rangle$ on the left-hand side of the equation must be identical with the coefficient of the same $|i\rangle$ on the right-hand side. That is, $v_i^{'}=A_{ij}v_j$.

Therefore the operator A is simply equivalent to *matrix multiplication*:

$$\begin{pmatrix} v_1 \\ v_2 \\ \cdot \\ \cdot \\ v_n^{'} \end{pmatrix}=\begin{pmatrix} \langle 1|A|1\rangle & \langle 1|A|2\rangle & \cdot & \cdot & \langle 1|A|n\rangle \\ \langle 2|A|1\rangle & \langle 2|A|2\rangle & \cdot & \cdot & \cdot \\ \cdot & \cdot & \cdot & \cdot & \cdot \\ \cdot & \cdot & \cdot & \cdot & \cdot \\ \langle n|A|1\rangle & \langle n|A|2\rangle & \cdot & \cdot & \langle n|A|n\rangle \end{pmatrix}\begin{pmatrix} v_1 \\ v_2 \\ \cdot \\ \cdot \\ v_n^{'} \end{pmatrix}.$$

It is important to note that a linear operator applied successively to the members of an orthonormal basis might give a new set of vectors which *no longer span the entire space*. To give an example, the linear operator $|1\rangle\langle 1|$ applied to any vector in the space picks out the vector's component in the $|1\rangle$ direction. It is called a *projection operator*.

The operator $(|1\rangle\langle 1|+|2\rangle|2\rangle)$ projects a vector into its components in the subspace spanned by the vectors $|1\rangle$ and $|2\rangle$, and so on—if we extend the sum to be over the whole basis, we recover the identity operator.

The Adjoint Operator and Hermitian Matrices

As we've discussed, if a ket $|V\rangle$ in the n-dimensional space is written as a column vector with n (complex) components, the corresponding bra is a *row* vector having as elements the complex conjugates of the ket elements.

$$\langle W|V\rangle=\langle V|W\rangle^*$$

then follows automatically from standard matrix multiplication rules, and on multiplying $|V\rangle$ by a complex number a to get $a|V\rangle$ (meaning that each element in the column of numbers is multiplied by a) the corresponding bra goes to

$$\langle V|a^*=a^*\langle V|.$$

But suppose that instead of multiplying a ket by a number, we operate on it with a linear operator. What generates the parallel transformation among the bras? In other words, if $A|V\rangle = |V'\rangle$, can we find a corresponding operator that sends the bra $\langle V|$ to $\langle V'|$?

This must be another linear operator, because A is linear, that is, if under A $|V_1\rangle \to |V_1'\rangle, |V_2\rangle \to |V_2'\rangle$ and $|V_3\rangle = |V_1\rangle + |V_2\rangle$ then under A $\langle V_3|$ *is required to go to* $\langle V_3| = |V_1\rangle + |V_2'\rangle$. Consequently, under the parallel *bra* transformation we must have

$$\langle V_1| \to \langle V_1'|, \langle V_2| \to \langle V_2'| \; and \; \langle V_3| \to \langle V_3'|,$$

so the bra transformation is necessarily also linear. Recalling that the bra is an n-element *row* vector, the most general linear transformation sending it to another bra is an $n \times n$ matrix operating on the bra from the *right*.

This bra operator is called the *adjoint* of A, written $A^\dagger$. That is, the ket $A|V\rangle$ has corresponding bra $\langle V|A^\dagger$. In an orthonormal basis, using the notation $\langle Ai|$ to denote the bra $\langle i|A^\dagger$ corresponding to the ket $A|i\rangle = |Ai\rangle$, say,

$$\left(A^\dagger\right)_{ij} = \langle i|A^\dagger|j\rangle = \langle Ai|j\rangle = \langle j|Ai\rangle^* = A_{ji}{}^*.$$

So *the adjoint operator is the transpose complex conjugate*.

for a *product* of two operators (prove this!), $(AB)^\dagger = B^\dagger A^\dagger$.

An operator equal to its adjoint $A = A^\dagger$ is called *Hermitian*, one equal to minus its adjoint $A = A^\dagger$ is anti Hermitian (sometimes skew Hermitian). These operator types are in some ways generalizations of real and imaginary numbers. Any operator can be expressed as a sum of a Hermitian operator and an anti Hermitian operator, $A = \frac{1}{2}\left(A + A^\dagger\right) + \frac{1}{2}\left(A - A^\dagger\right)$

This notion naturally extends to *vectors and numbers*: the adjoint of a ket is the corresponding bra, the adjoint of a number is its complex conjugate.

This is useful to bear in mind when taking the adjoint of an operator which may be partially constructed of vectors and numbers, such as projection-type operators. The adjoint of a product of matrices, vectors and numbers is the product of the adjoints *in reverse order*. (Of course, for numbers the order doesn't matter.)

Unitary Operators

An operator is *unitary* if $U^{\dagger}U=1$ This implies first that U operating on any vector gives a vector having the same norm, since the new norm $\langle V|U^{\dagger}U|V\rangle=\langle V|V\rangle$ and furthermore inner products are preserved, $\langle W|U^{\dagger}U|V\rangle=\langle W|V\rangle$. The original orthonormal basis in the space must therefore under a unitary transformation go to another orthonormal basis.

Conversely, any transformation that takes one orthonormal basis into another one is a unitary transformation. To see this, suppose that a linear transformation A sends the members of the orthonormal basis $(|1\rangle_1,|2\rangle_1,...,|n\rangle_1)$ to the different orthonormal set $(|1\rangle_2,|2\rangle_2,...,|n\rangle_2),$, so $A|1\rangle_1=|1\rangle_2$, etc. Then the vector $|V\rangle=\sum v_i|i\rangle_1$ will go to $|V'\rangle=A|V\rangle=\sum v_i|i\rangle_2$, having the same norm, from $\langle V|V\rangle=\sum|v_i|^2$, and similarly $\langle W|V\rangle$ is invariant. A unitary operation amounts to a rotation (possibly combined with a reflection) in the space. Evidently, since $U^{\dagger}\ U=1,$ the adjoint $U^{\dagger}$ rotates the basis back—it is the inverse operation, and so $UU^{\dagger}=1,$ also, that is, U and commute.

Determinants

We review in this section the *determinant* of a matrix, a function closely related to the operator properties of the matrix.

Let's start with 2×2 matrices:

$$A=\begin{pmatrix} a_{11} & a_{12} \\ a_{21} & a_{22} \end{pmatrix}.$$

The determinant of this matrix is defined by:

$$\det A=|A|=a_{11}a_{22}-a_{12}a_{21}.$$

Writing the two rows of the matrix as vectors:

$$\vec{a}_1^R=(a_{11},a_{12})$$

$$\vec{a}_2^R=(a_{21},a_{22})$$

(R denotes row), $\det A=\vec{a}_1^R\times\vec{a}_2^R$ is just the *area* (with appropriate sign) of the parallelogram having the two row vectors as adjacent sides:

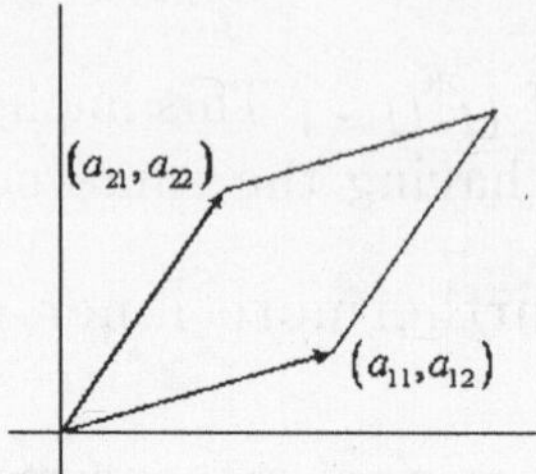

This is zero if the two vectors are parallel (linearly dependent) and is not changed by adding any multiple of $\vec{a}_2^R$ to $\vec{a}_1^R$ (because the new parallelogram has the same base and the same height as the original—check this by drawing).

Let's go on to the more interesting case of 3×3 matrices:

$$A = \begin{pmatrix} a_{11} & A_{12} & A_{13} \\ a_{21} & a_{22} & a_{23} \\ a_{31} & a_{32} & a_{33} \end{pmatrix}.$$

The determinant of A is defined as

$$\det A = \varepsilon_{ijk} a_{1i} a_{2j} a_{3k}$$

where $\varepsilon_{ijk} = 0$ if any two are equal, $+1$ if *ijk* = 123, 231 or 312 (that is to say, an even permutation of 123) and -1 if *ijk* is an odd permutation of 123. Repeated suffixes, of course, imply summation here.

Writing this out explicitly,

$$\det A = a_{11}a_{22}a_{33} + a_{21}a_{32}a_{13} + a_{31}a_{12}a_{23} - a_{11}a_{32}a_{23} - a_{21}a_{12}a_{33} - a_{31}a_{22}a_{13}.$$

Just as in two dimensions, it's worth looking at this expression in terms of vectors representing the rows of the matrix

$$\vec{a}_1^R = (a_{11}, a_{12}, a_{13})$$

$$\vec{a}_2^R = (a_{31}, a_{32}, a_{23})$$

$$\vec{a}_3^R = (a_{31}, a_{32}, a_{33})$$

so

$$A = \begin{pmatrix} \vec{a}_1^R \\ \vec{a}_2^R \\ \vec{a}_3^R \end{pmatrix}, \textit{and we see that } \det A = \left(\vec{a}_1^R \times \vec{a}_2^R \right) \cdot \vec{a}_3^R$$

This is the volume of the parallelopiped formed by the three vectors being adjacent sides (meeting at one corner, the origin). This parallelepiped volume will of course be zero if the three vectors lie in a plane, and it is not changed if a multiple of one of the vectors is added to another of the vectors. That is to say, ***the determinant of a matrix is not changed if a multiple of one row is added to another row.***

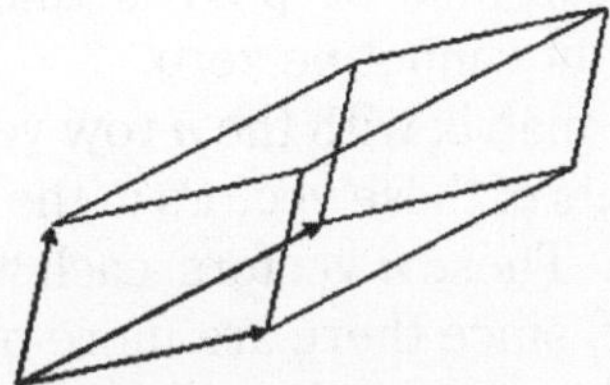

This is because the determinant is linear in the elements of a single row,

$$\det\begin{pmatrix} \vec{a}_1^R + \lambda\vec{a}_2^R \\ \vec{a}_2^R \\ \vec{a}_3^R \end{pmatrix} = \det\begin{pmatrix} \vec{a}_1^R \\ \vec{a}_2^R \\ \vec{a}_3^R \end{pmatrix} + \lambda\det\begin{pmatrix} \vec{a}_2^R \\ \vec{a}_2^R \\ \vec{a}_3^R \end{pmatrix}$$

and the last term is zero because two rows are identical—so the triple vector product vanishes. A more general way of stating this, applicable to larger determinants, is that for a determinant with two identical rows, the symmetry of the two rows, together with the ***antisymmetry*** ϵ_{ijk}, of ensures that the terms in the sum all cancel in pairs.

Since the determinant is not altered by adding some multiple of one row to another, if the rows are linearly dependent, one row could be made identically zero by adding the right multiples of the other rows. Since every term in the expression for the determinant has one element from each row, the determinant would then be identically zero. For the three-dimensional case, the linear dependence of the rows means the corresponding vectors lie in a plane, and the parallelepiped is flat.

The algebraic argument generalizes easily to n×n determinants: they are *identically zero if the rows are linearly dependent*.

The generalization from $3\times 3 to n\times n$ determinants is that $\det A = \varepsilon_{ijk} a_{1i} a_{2j} a_{3k}$ becomes:

$$\det A = \varepsilon_{ijk\ldots\gamma}\, a_{1i} a_{2j} a_{3k} \ldots a_{np}$$

where *ijk...p* is summed over all permutations of 132...*n*, and the symbol is zero if any two of its suffixes are equal, +1 for an even permutation and -1 for an odd permutation. (*Note*: any permutation can be written as a product of swaps of neighbours. Such a representation is in general not unique, but for a given permutation, all such representations will have either an odd number of elements or an even number.) An important theorem is that for a product of two matrices A, B the determinant of the product is the product of the determinants, $det\, AB = det\, A \times det\, B$. This can be verified by brute force for 2×2 matrices, and a proof in the general case can be found in any book on mathematical physics. It can also be proved that *if* the rows are linearly independent, the determinant cannot be zero.

(*Here's a proof*: take an matrix with the n row vectors linearly independent. Now consider the components of those vectors in the $n-1$ dimensional subspace perpendicular to (1, 0, ...,0). These n vectors, each with only $n-1$ components, must be linearly dependent, since there are more of them than the dimension of the space. So we can take some combination of the rows below the first row and subtract it from the first row to leave the first row (a, 0, 0, ...,0), and a cannot be zero since we have a matrix with n linearly independent rows.

We can then subtract multiples of this first row from the other rows to get a determinant having zeros in the first column below the first row. Now look at the $n-1$ by $n-1$ determinant to be multiplied by a.

Its rows must be linearly independent since those of the original matrix were. Now proceed by induction.)

To return to three dimensions, it is clear from the form of

$$\det A = a_{11}a_{22}a_{33} + a_{21}a_{32}a_{13} - a_{31}a_{12}a_{23} - a_{11}a_{32}a_{23} - a_{21}a_{12}a_{33} - a_{31}a_{22}a_{13}$$

that we could equally have taken the *columns* of A as three vectors,

$$A = \vec{a}_1^{\,C}, \vec{a}_2^{\,C}, \vec{a}_3^{\,C}$$

in an obvious notation,

$$\det A = \left(\vec{a}_1^{\,C} \times \vec{a}_2^{\,C}\right)\vec{a}_3^{\,C},$$

and linear dependence among the *columns* will also ensure the vanishing of the determinant—so, in fact, linear dependence of the columns ensures linear dependence of the rows.

This, too, generalizes to: in the definition of determinant

$$\det A = \varepsilon_{ijk\ldots p}a_{1i}a_{2j}a_{3k}\ldots a_{np},$$

the row suffix is fixed and the column suffix goes over all permissible permutations, with the appropriate sign—but the same terms would be generated by having the *column* suffixes kept in numerical order and allowing the row suffix to undergo the permutations.

An Aside: Reciprocal Lattice Vectors

It is perhaps worth mentioning how the inverse of a 3×3 matrix operator can be understood in terms of vectors. For a set of linearly independent vectors $(\vec{a}_1, \vec{a}_2, \vec{a}_3)$, a reciprocal set $(\vec{b}_1, \vec{b}_2, \vec{b}_3)$ can be defined by

$$\vec{b}_1 = \frac{\vec{a}_2 \times \vec{a}_3}{\vec{a}_1 \times \vec{a}_2 \cdot \vec{a}_3}$$

and the obvious cyclic definitions for the other two reciprocal vectors. We see immediately that

$$\vec{a}_i \cdot \vec{b}_j = \delta_{ij}$$

from which it follows that the inverse matrix to

$$A = \begin{pmatrix} \vec{a}_1^R \\ \vec{a}_2^R \\ \vec{a}_3^R \end{pmatrix} \text{is } B = \begin{pmatrix} \vec{b}_1^C & \vec{b}_2^C & \vec{b}_3^C \end{pmatrix}.$$

(These reciprocal vectors are important in x-ray crystallography, for example. If a crystalline lattice has certain atoms at positions

$$n_1\vec{a}_1 + n_2\vec{a}_2 + n_3\vec{a}_3,$$

where n_1, n_2, n_3 are integers, the reciprocal vectors are the set of normals to possible planes of the atoms, and these planes of atoms are the important elements in the diffractive x-ray scattering.)

Eigenkets and Eigenvalues

If an operator A operating on a ket $|V\rangle$ gives a multiple of the same ket,

$$A|V\rangle = \lambda|V\rangle$$

then $|V\rangle$ is said to be an *eigenket* (or, just as often, *eigenvector*, or *eigenstate*!) of A with *eigenvalue.*

Eigenkets and eigenvalues are of central importance in quantum mechanics: dynamical variables are operators, a physical measurement of a dynamical variable yields an eigenvalue of the operator, and forces the system into an eigenket. In this section, we shall show how to find the eigenvalues and corresponding eigenkets for an operator A. We'll use the notation $A|a_i\rangle = a_i|a_i\rangle$ for the set of eigenkets $|a_i\rangle$ with corresponding eigenvalues a_i. (Obviously, in the eigenvalue equation here the suffix i not summed over.) The first step in solving $A|V\rangle = \lambda|V\rangle$ is to find the allowed eigenvalues a_i.

Writing the equation in matrix form:

$$\begin{pmatrix} A_{11}-\lambda & A_{12} & \cdot & \cdot & A_{1n} \\ A_{21} & A_{22}-\lambda & \cdot & \cdot & \cdot \\ \cdot & \cdot & \cdot & \cdot & \cdot \\ \cdot & \cdot & \cdot & \cdot & \cdot \\ A_{n1} & \cdot & \cdot & \cdot & A_{nm}-\lambda \end{pmatrix} \begin{pmatrix} v_1 \\ v_2 \\ \cdot \\ \cdot \\ v_n \end{pmatrix} = 0$$

This equation is actually telling us that the *columns* of the matrix $A-\lambda I$ are linearly dependent! To see this, write the matrix as a row vector each element of which is one of its columns, and the equation becomes

$$\left(\overrightarrow{M}_1^C, \overrightarrow{M}_2^C, \ldots, \overrightarrow{M}_n^C\right)\begin{pmatrix} v_1 \\ \cdot \\ \cdot \\ \cdot \\ v_n \end{pmatrix} = 0$$

which is to say

$$v_1\overrightarrow{M}_1^C + v_2\overrightarrow{M}_2^C = \ldots + v_n\overrightarrow{M}_n^C = 0,$$

the columns of the matrix are indeed a *linearly dependent* set. We know that means the determinant of the matrix $A-\lambda I$ is zero,

$$\begin{vmatrix} A_{11}-\lambda & A_{12} & \cdot & \cdot & A_{1n} \\ A_{21} & A_{22}-\lambda & \cdot & \cdot & \cdot \\ \cdot & \cdot & \cdot & \cdot & \cdot \\ \cdot & \cdot & \cdot & \cdot & \cdot \\ A_{n1} & \cdot & \cdot & \cdot & A_{nn}-\lambda \end{vmatrix} = 0$$

Evaluating the determinant using $\det A = e_{jk\ldots p}a_{1i}a_{2j}a_{3k}\ldots a_{np}$ gives an nth order polynomial in sometimes called the *characteristic polynomial*. Any polynomial can be written in terms of its roots:

$$C(\lambda - a_1)(\lambda - a_2)\ldots(\lambda - a_n) = 0$$

where the a_i's, the roots of the polynomial, and C is an overall constant, which from inspection of the determinant we can see to be $(-1)^n$. (It's the coefficient of λ^n.) The polynomial roots (which we don't yet know) are in fact the eigenvalues.

For example, putting $\lambda = a_1$ in the matrix, $\det(A - a_1 I) = 0$ which means that $(A - a_1 I)|V\rangle = 0$ has a nontrivial solution $|V\rangle$, and this is our eigenvector $|a_1\rangle$. Notice that the diagonal term in the determinant $(A_{11}-\lambda)(A_{22}-\lambda)\ldots(A_{nn}-\lambda)$ generates the leading two orders in the polynomial $(-1)^n\left(\lambda^n - (A_{11}+\ldots+A_{nn})\lambda^{n-1}\right)$, (and some lower order terms too). Equating the coefficient of λ^{n-1} here with that in $(-1)^n(\lambda - a_1)(\lambda - a_2)\ldots(\lambda - a_n)$,

$$\sum_{i-1}^{n} a_i = \sum_{i-1}^{n} A_{ij} = TrA.$$

Putting $\lambda = 0$ in both the determinantal and the polynomial representations (in other words, equating the λ -independent terms),

$$\prod_{i-1}^{n} a_i = \det A.$$

So we can learn something about the eigenvalues directly from the determinant. But if the space is bigger than 2×2 this is not enough to solve the problem.

We must solve the polynomial equation $\det(A - \lambda I) = 0$ to find the set of eigenvalues, then use them to calculate the corresponding eigenvectors. This is done one at a time.

Labeling the first eigenvalue found as a_1, the corresponding equation for the components v_i of the eigenvector $|a_1\rangle$ is

$$\begin{pmatrix} A_{11} - a_1 & A_{12} & \cdot & \cdot & A_{1n} \\ A_{21} & A_{22-a_1} & \cdot & \cdot & \cdot \\ \cdot & \cdot & \cdot & \cdot & \cdot \\ \cdot & \cdot & \cdot & \cdot & \cdot \\ A_{n1} & \cdot & \cdot & \cdot & A_{nm} - a_1 \end{pmatrix} \begin{pmatrix} v_1 \\ v_2 \\ \cdot \\ \cdot \\ v_n \end{pmatrix} = 0$$

This looks like n equations for the n numbers v_i, but it isn't: remember the rows are linearly dependent, so there are only $n - 1$ independent equations. However, that's enough to determine the ratios of the vector components v_1, ..., v_n, then finally the eigenvector is normalized.

Eigenvalues and Eigenstates of Hermitian Matrices

For a *Hermitian* matrix, it is easy to establish that the eigenvalues are always *real*. Taking A to be hermitian, $A = A^\dagger$, and labeling the eigenkets by the eigenvalue, that is,

$$A|a_1\rangle = a_1|a_1\rangle$$

the inner product with the bra $\langle a_1|$ gives $\langle a_1|A|a_1\rangle = a_1\langle a_1|a_1\rangle$ But the inner product of the *adjoint* equation (remembering $A = A^1$)

$$\langle a_1|A = a_1 * \langle a_1|$$

with $|a_1\rangle$ gives $\langle a_1|A|a_1\rangle = a_1 * \langle a_1|a_1\rangle$, so $a_1 = a_1*$, and all the eigenvalues must be real. Furthermore, the eigenkets belonging to *different* eigenvalues are *orthogonal*, if

$$A|a_1\rangle = a_1|a_1\rangle$$
$$A|a_2\rangle = a_2|a_2\rangle$$

taking the adjoint of the first equation and taking its inner product with $|a_2\rangle$,

$$\langle a_1|A|a_2\rangle = a_1\langle a_1|a_2\rangle = a_2\langle a_1|a_2\rangle$$

so $\langle a_1|a_2\rangle = 0$ unless the eigenvalues are equal. If they *are* equal, then any linear combination of them is also an eigenvector with the same eigenvalue, and the Gram Schmidt procedure can be used to construct an orthonormal basis in this subspace as part of a full orthonormal basis.

A subspace spanned by eigenvectors having the same eigenvalue is termed *degenerate*.

Diagonalizing a Hermitian Matrix

Diagonalizing a matrix means transforming to an orthonormal basis in which the only nonzero matrix elements are on the diagonal, $i = j$.

In fact, it is evident from the discussion at the end of the previous section that a hermitian matrix is diagonal in the orthonormal basis of its set of eigenvectors: $|a_1\rangle, |a_2\rangle, \ldots, |a_n\rangle$, since

$$\langle a_i|A|a_j\rangle = \langle a_i|a_j|a_j\rangle = a_j\langle a_i|a_j\rangle = a_j\delta_{ij}.$$

However, we are assuming that we know the matrix elements of A in some other orthonormal basis, so to diagonalize A we need to rotate from the initial orthonormal basis to one made up of the eigenkets of A.

$$|1\rangle = \begin{pmatrix}1\\0\\0\\\vdots\\0\end{pmatrix}, |2\rangle = \begin{pmatrix}0\\1\\0\\\vdots\\0\end{pmatrix}, |i\rangle = \begin{pmatrix}0\\\vdots\\1\\\vdots\\0\end{pmatrix} \cdots \left(1\,in\,1^{th}\ place\,down\right), |n\rangle = \begin{pmatrix}0\\0\\0\\\vdots\\1\end{pmatrix}$$

Denoting the initial orthonormal basis in the standard fashion.

The elements of the matrix are $A_{ij} = \langle i|A|j\rangle$.

A transformation from one orthonormal basis to another is a *unitary* transformation, as discussed above, so we write it

$$|V\rangle \to |V'\rangle = U|V\rangle.$$

Under this transformation, the matrix element

$$\langle W|A|V\rangle \to \langle W'|A|V'\rangle = \langle W|U^{\dagger} AU|V\rangle.$$

So we can find the appropriate transformation matrix U by requiring that $U^{\dagger} AU$ be diagonal with respect to the *original* set of basis vectors.

(Transforming the operator in this way, leaving the vector space alone, is equivalent to rotating the vector space and leaving the operator alone. Of course, in a system with more than one operator, the same transformation would have to be applied to all the operators).

In fact, the unitary matrix U we need is just composed of the normalized eigenkets of the operator A,

$$U = \left(|a_1\rangle, |a_2\rangle, \ldots, |a_n\rangle\right).$$

It is easy to see that this must be true:

$$AU = \left(a_1|a_1\rangle, a_2|a_2\rangle, \ldots, a_n|a_n\rangle\right)$$

so

$$\left(U^{\uparrow}AU\right)_{ij} = \langle a_i|a_j|a_j\rangle = \delta_{ij}a_j, \textit{a diagonal matrix.}$$

(The repeated suffixes here are of course *not* summed over.)

To actually find the eigenkets, then, we need to go through the routine outlined in the previous section, that is, first find the eigen*values* by finding the zeros of the determinantal equation, $|A - \lambda 1| = 0$, then enter the roots a_1, $a_2, \ldots a_n$ one at a time into the matrix equation

$$\begin{pmatrix} A_{11} - a_i & A_{12} & \cdot & \cdot & \cdot \\ A_{21} & A_{22} - a_i & \cdot & \cdot & \cdot \\ \cdot & \cdot & \cdot & \cdot & \cdot \\ \cdot & \cdot & \cdot & \cdot & \cdot \\ A_{n1} & \cdot & \cdot & \cdot & A_{nm} - a_i \end{pmatrix} \begin{pmatrix} v_1 \\ v_2 \\ \cdot \\ \cdot \\ v_n \end{pmatrix} = 0$$

to give n - 1 linear equations for the n unknown components of the eigenket $|a_i\rangle$.

Then

$$U^{\uparrow}AU = \begin{pmatrix} a_1 & 0 & 0 & \cdot & 0 \\ 0 & a_2 & 0 & \cdot & 0 \\ 0 & 0 & a_3 & \cdot & 0 \\ \cdot & \cdot & \cdot & \cdot & \cdot \\ 0 & \cdot & \cdot & \cdot & A_n \end{pmatrix}$$

If some of the eigenvalues are the same, the Gram Schmidt procedure may be needed to generate an orthogonal set, as mentioned earlier.

Functions of Matrices

The same unitary operator U that diagonalizes an Hermitian matrix A will also diagonalize A^2, because

$$U^{-1}A^2U = U^{-1}AAU = U^{-1}AUU^{-1}AU$$

so

$$U^{\dagger}A^2U = \begin{pmatrix} a_1^2 & 0 & 0 & \cdot & 0 \\ 0 & a_2^2 & 0 & \cdot & 0 \\ 0 & 0 & a_3^2 & \cdot & 0 \\ \cdot & \cdot & \cdot & \cdot & \cdot \\ 0 & \cdot & \cdot & \cdot & a_n^2 \end{pmatrix}$$

Evidently, this same process works for any power of A, and formally for any function of A expressible as a power series, but of course convergence properties need to be considered, and this becomes trickier on going from finite matrices to operators on infinite spaces.

Commuting Hermitian Matrices

From the above, the set of powers of an Hermitian matrix all commute with each other, and have a common set of eigenvectors (but not the same eigenvalues, obviously). In fact it is not difficult to show that any two Hermitian matrices that commute with each other have the same set of eigenvectors (after possible Gram Schmidt rearrangements in degenerate subspaces).

If two n×n Hermitian matrices A, B commute, that is, $AB = BA$, and A has a *nondegenerate* set of eigenvectors $A|a_i\rangle = a_i|a_i\rangle$, then $AB|a_i\rangle = BA|a_i\rangle = Ba_i|a_i\rangle = a_iB|a_i\rangle$ that is, $B|a_i\rangle$ is an eigenvector of A with eigenvalue a_i. Since A is nondegenerate $B|a_i\rangle$, must be some multiple of $|a_i\rangle$, and we conclude that A, B have the same set of eigenvectors.

Now suppose A *is* degenerate, and consider the $m \times m$ subspace S_α spanned by the eigenvectors $|a_i,1\rangle, |a_i,2\rangle...,$ of A having eigenvalue a_i. Applying the argument in the paragraph above, $B|a_i,1\rangle, B|a_i,2\rangle,....$ must also lie in this subspace. Therefore, if we transform B with the same unitary transformation that diagonalized A, B will not in general be diagonal in the subspace S_α, but it will be what is termed *block diagonal*, in that if B operates on any vector in S_α it gives a vector in S_α.

B can be written as two diagonal blocks: one $m \times m$, one $(n-m)\times(n-m)$with zeroes outside these diagonal blocks, for example, for $m = 2$, $n = 5$:

$$\begin{pmatrix} b_{11} & b_{12} & 0 & 0 & 0 \\ b_{21} & b_{22} & 0 & 0 & 0 \\ 0 & 0 & b_{33} & b_{34} & b_{35} \\ 0 & 0 & b_{43} & b_{44} & b_{45} \\ 0 & 0 & b_{53} & b_{54} & b_{55} \end{pmatrix}.$$

And, in fact, if there is only *one* degenerate eigenvalue that second block will only have nonzero terms on the diagonal:

$$\begin{pmatrix} b_{11} & b_{12} & 0 & 0 & 0 \\ b_{21} & b_{22} & 0 & 0 & 0 \\ 0 & 0 & b_3 & 0 & 0 \\ 0 & 0 & 0 & b_4 & 0 \\ 0 & 0 & 0 & 0 & b_5 \end{pmatrix}.$$

B therefore operates on two subspaces, one m-dimensional, one $(n\text{-}m)$-dimensional, *independently*—a vector entirely in one subspace stays there.

This means we can complete the diagonalization of B with a unitary operator that *only* operates on the m × m block S_α. Such an operator will also affect the eigenvectors of A, but that doesn't matter, because all vectors in this subspace are eigenvectors of A with the *same* eigenvalue, so as far as A is concerned, we can choose any orthonormal basis we like—the basis vectors will still be eigenvectors

This establishes that any two *commuting* Hermitian matrices can be diagonalized at the same time. Obviously, this can never be true of *noncommuting* matrices, since all diagonal matrices commute.

Diagonalizing a Unitary Matrix

As previously stated, a unitary matrix is an operator that rotates an orthonormal basis into another orthonormal basis. It is clear that in three dimensions the axis through the origin about which the basis is rotated is an eigenvector of this transformation.

It's less clear what the other two eigenvectors might be—or, equivalently, what are the eigenvectors corresponding to a two dimensional rotation of basis in a plane? The way to find out is to write down the matrix and diagonalize it.

The matrix

$$U(\theta) = \begin{pmatrix} \cos\theta & \sin\theta \\ -\sin\theta & \cos\theta \end{pmatrix}$$

Note that the determinant is equal to unity. The eigenvalues are given by solving

$$\begin{vmatrix} \cos-\lambda & \sin\theta \\ -\sin\theta & \cos-\lambda \end{vmatrix} = 0 \, to \, give \, \lambda^{\pm ie}.$$

The corresponding eigenvectors satisfy

$$\begin{pmatrix} \cos\theta & \sin\theta \\ -\sin\theta & \cos\theta \end{pmatrix}\begin{pmatrix} u_1^{\pm} \\ u_2^{\pm} \end{pmatrix} = e^{\pm i\theta}\begin{pmatrix} u_1^{\pm} \\ u_2^{\pm} \end{pmatrix}$$

The eigenvectors, normalized, are:

$$\begin{pmatrix} u_1^{\pm} \\ u_2^{\pm} \end{pmatrix} = \frac{1}{\sqrt{2}}\begin{pmatrix} 1 \\ \pm i \end{pmatrix}.$$

Note that, in contrast to a Hermitian matrix, the eigenvalues of a unitary matrix do not have to be real. In fact, from $U^{\uparrow} U = 1,$, sandwiched between the bra and ket of an eigenvector, we see that any eigenvalue of a unitary matrix must have unit modulus—it's a complex number on the unit circle. With hindsight, we should have realised that one eigenvalue of a two-dimensional rotation had to be $e^{i\theta}$, the product of two two-dimensional rotations is given be adding the angles of rotation, and a rotation through π changes all signs, so has eigenvalue -1.

Note that the eigenvector itself is independent of the angle of rotation—the rotations all commute, so they must have common eigenvectors. Successive rotation operators applied to the plus eigenvector add their angles, when applied to the minus eigenvector, all angles are subtracted.

5

Statistical Mechanics and Thermodynamics

STATISTICAL MECHANICS

The kinetic theory of matter attempts to explain the empirical regularities occurring in the macroscopic properties of material objects in terms of the microscopic behaviour of their atomic and molecular constituents as described by Newton's laws of mechanics.

The early studies in the revival of this theory which occurred in the second half of the nineteenth century had only limited explanatory power since they assumed that:

- These constituents all had the same velocity; and
- The influence of intermolecular forces and impacts were negligible. The development of the subject took a major leap forward when Maxwell allowed intermolecular collisions to play a role and demonstrated that their effect was to produce a statistical distribution of molecular velocities in which all velocities would occur with a known probability. This is given by Maxwell's famous distribution law:

$$f(x) = \frac{1}{a\sqrt{\pi}} e^{-x^2/a2}$$

where a is a constant and x is the component of the velocity in the x-direction.

Of course, the statistical element was introduced purely as a matter of convenience in order to overcome the practical problems in handling enormously large numbers of atoms and molecules. The latter themselves were regarded as traversing distinct, continuous space-time trajectories and as obeying Newton's laws.

The original proof of the above law was widely regarded as unsatisfactory and subsequent attempts to derive it more rigourously can be divided into three types:

1. If a Maxwellian distribution is already established then it can be shown that conservation of energy implies that further collisions will leave

it unchanged. Hence this distribution is the only one which is stable. This was the approach taken by Maxwell himself.

2. One may define a quantity, H, which depends on the velocity distribution and then show that the effect of molecular collisions is always to decrease H, unless the distribution is Maxwellian, in which case H remains constant. This constitutes the essence of Boltzmann's 'H-Theorem' approach.
3. One may regard the atomic and molecular velocities as random quantities and obtain the best possible estimate of their distribution, subject to the constraints of fixed energy and number of atoms or molecules, using probability theory. Maxwell's derivation may be improved by calculating all the possible ways of dividing the energy among the atoms or molecules, subject to the above constraints. This was the thinking which subsequently developed into the 'Combinatorial Approach' (and which lies behind the permutation argument above).

Approaches II and III rest on very different foundations: the H-Theorem Approach is explicitly based on a consideration of molecular trajectories, whereas the Combinatorial Approach eschewed such detailed consideration and was simply concerned with the distribution of molecules over energy states. Historically, the Combinatorial Approach played an absolutely fundamental role in the development of quantum theory.

Maxwell's work had an enormous impact upon Boltzmann, who was seeking a mechanical explanation of the apparent irreversibility of natural processes as expressed by the Second Law of Thermodynamics. Put rather crudely, this states that it is impossible for heat to flow from a colder to a hotter body. The law was expressed by Clausius in 1865 in terms of his newly introduced entropy function S, where,

$$dQ \leq T.\,dS$$

where dQ is the heat change, T is the temperature change and dS is the change in entropy. The equality holds for reversible processes, of course. The task facing Boltzmann was then two-fold: first, to demonstrate the existence of an entropy function satisfying Clausius's definition and secondly to show that this function could only increase in an irreversible process. After reading Maxwell's 1867 paper, Boltzmann realised that the key lay with the above distribution function and in 1871 he was able to demonstrate the existence of an entropy function in purely mechanical terms and lay down a procedure for finding it.

The following year he took the next step and analysed the behaviour of his entropy function in irreversible processes by considering the way in which the velocity distribution changed with time due to intermolecular collisions. This work can be understood as both an attempt to complete the statistical mechanical reduction of the Second Law, and as showing that the effect of intermolecular collisions on a gas in a non-equilibrium state would be to drive it to equilibrium as described by Maxwell's law. Boltzmann was able to show that the Maxwellian distribution represents the equilibrium state and obtained an explicit formula for the rate of change of f, $\frac{\partial f}{\partial t}$, on the basis of an "exact consideration of the collision process" between two molecules of a spatially homogenous, low-density gas with no external forces present. With this formula in hand, he was able to show that f *always* tends towards the Maxwellian form. To do this he introduced an auxiliary quantity E (which later came to be called H), defined by,

$$E = \int_0^\infty f(xt)\left\{\log\left[f(x,t\right]t / \sqrt{x} - 1\right\} dx .$$

By considering the symmetrical character of the collision and the possibility of inverse collisions, and assuming, as Maxwell had before him, that the velocities of the two molecules before they collide are statistically independent, Boltzmann demonstrated that E could only decrease with time; that is,

$$\frac{dE}{dz} \leq 0 .$$

With H substituted for E, this corresponds to Boltzmann's H-Theorem. E cannot decrease to infinity, of course, and so the distribution function must approach a form for which E has a minimum value and its derivative vanishes. At this value the function has, and can only have, the Maxwellian form.

Thus $-E$ increases in the irreversible approach to equilibrium and hence behaves like the entropy function. Furthermore, $-E$ was actually proportional to the entropy in the equilibrium state. This implies, as Boltzmann himself indicated, an entirely new approach to proving the Second Law, one that could deal with the increase in entropy in irreversible processes as well as with its existence as an equilibrium state function. In this way, the H-Theorem effectively extended the definition of entropy to non-equilibrium situations and completed the kinetic reduction of the Second Law of Thermodynamics.

Now, although Maxwell's distribution function introduced a certain statistical 'coarse-graining' into the description, the development of the H-Theorem was still based upon consideration of binary collisions between molecules traversing distinct continuous trajectories. Thus, despite the probabilistic elements, the underlying basis was still, of course, Newtonian and deterministic. The tensions this caused within Boltzmann's edifice are well known.

In particular, his conclusion, that the E/H-function always decreased, was disputed by Maxwell, Tait and Thompson and, independently and famously, by Loschmidt, who noted that in any system, "the entire course of events will be retraced if at some instant the velocities of all its parts are reversed". If entropy is a specifiable function of the positions and velocities of the particles of a system, and if that function increases during some particular motion of the system, then reversing the direction of time in the equations of motion will specify a trajectory through which the entropy must decrease. For every possible motion that leads towards equilibrium, there is another, equally possible, that leads away and is therefore incompatible with the Second Law. Loschmidt concluded that if the kinetic theory were true, then the Second Law could not hold universally and thus Boltzmann's proof of the *H*-Theorem could not be correct.

Boltzmann's 1877 response to the Loschmidt 'paradox' is illuminating. He basically admitted that one could not, in fact, prove that entropy increased 'with absolute necessity' and that, according to probability theory, even the most improbable non-uniform distribution is still not absolutely impossible. However, he then claimed that the existence of such improbable entropy-decreasing situations did not contradict the fact that for the overwhelming majority of initial states the entropy could be counted on to increase and that the improbabilities associated with the former case were, for all practical purposes, impossibilities. Boltzmann now focused on the probabilistic aspect of the *H*-Theorem and argued that it followed from this that the number of states leading to a uniform, equilibrium distribution after a certain time must be much larger than the number leading to a non-uniform distribution, since, he claimed, there are infinitely many more uniform states than non-uniform ones. No justification was actually given for this last claim, although Boltzmann intimated where one might be found:

> One could even calculate the possibilities of the various states from the ratios of the number of ways in which their distributions could be achieved, which could perhaps lead to an interesting method of calculating thermal equilibrium.

This 'interesting method' was subsequently elaborated later that same year in one of the most significant papers of classical statistical mechanics, entitled 'Probabilistic Foundations of Heat Theory'.

Prior to this Boltzmann had, in a series of papers from 1868 and 1871, re-derived and extended Maxwell's results and, significantly, in 1868, sketched an alternative derivation that was free from any assumptions regarding intermolecular collisions. Thus he considered the distribution of a fixed amount of energy over a finite number of molecules in such a way that all combinations were equally probable. By regarding the energy as divided into small but finite packets, Boltzmann could treat this as a problem in combinatorial analysis. In this manner he obtained a complicated expression which reduced to Maxwell's law in the limit of an infinite number of molecules and infinitesimal energy elements. This marks the beginnings of his Combinatorial Approach. In the 1877 work, Boltzmann drew on this earlier work and presented a new and radical alternative to the *H*-Theorem approach. In line with his general philosophical attitude towards physical theories this 'Combinatorial Approach' was elaborated through a succession of models of increasing complexity and closer approximation to the physical situation.

The first such model was a highly simplified and explicitly fictional discrete energy model in which he considered a collection of molecules whose individual energies were restricted to a finite, discrete set, with the total energy held fixed. If ω_k is the number of such molecules with energy $k\varepsilon$, then the set ω_0, α_1., ε_p is sufficient to define a particular macro-state (*Zustandverteilung*) of the gas. Boltzmann then noted that such a macro-state could be achieved in many different ways, each of which he called a 'complexion'. In general, if a complexion was specified by a set of numbers, each fixing the energy $k\,\varepsilon$ of the *i*th molecule, then, he wrote, a second complexion belonging to the same macro-state would be achieved by any permutation of the two molecules *i* and *j* which have different energies. Thus a permutation of particles between different energy states gave rise to a different macro-state of the system as a whole.

POWER LAW DISTRIBUTION IN STATISTICAL MECHANICS

From the notion of statistical mechanics, it is known that distributions dominate the nature phenomenon and the world structures. Some of them are discovered and well interpreted; some of them are invisible to us or beyond our understanding. Barabási and Albert in 1999 suggest a power law distribution which was found in a number of complex networks. They are called scale free networks. Such kind of power law distribution can describe a variety of systems both in the nature, society and man-made high-tech visual world, such as the World Wide Web, which is an enormous visual network of human intelligence. In this network, the nodes are the documents and the edges are the hyperlinks that point from one document to another. It is discovered that the degree distribution of web pages follows a power law distribution. Both probability that a document has k outgoing hyperlinks, P_{our}, (k) and the distribution of incoming edges, p_{in} (k) have power-law tails.

$$p_{our}(k) \sim k^{-\gamma_{our}} * \text{ and } p_{in}(k) \sim k^{-\gamma_{in}}$$

This can be interpreted as the more links one has today, the more links it will have tomorrow. However, this is only on the surface level of the experiment understanding. The deep physics meaning are not revealed.

We try to understand such phenomenon from different perspective by treating networks as thermodynamics system. At first glance, it sounds rude because network is not highly thermalised. However, further reflection tells that it is actually its intrinsic properties that we have ignored before. Take social networks for example. As members of a community and citizens of a country, we never stop doing our effort to keep our society stable against the "second law of thermodynamics". Otherwise, the social network will break up. This gives us a hint that, in most cases, the "uncertainty" of a stable network will not change over time. To interpret it in physics is that the mean entropy of a network is a constant. On this assumption, we will in the following sections show how it works to demonstrate the power law distribution.

INFORMATION THEORY AND MAXIMUM-ENTROPY ESTIMATES

Information theory provides a constructive criterion for setting up probability distributions on the basis of partial knowledge, and leads to a type of statistical inference which is called the maximum entropy estimate. Where the Shannon entropy are usually used:

$$H[P(x)] = -k\sum_i P_i \ln P_i(x)$$

It is the least biased estimate possible on the give information. Suppose,

$$\langle f(x)\rangle = \sum_i P_i f(x_i)$$

The quantity x is capable of assuming the discrete values x_i (i = 1, 2..., n). We are not given the corresponding probabilities p_i; all we know is the expectation value of function $f(x)$; and the normalisation condition,

$$\sum_i P_i = 1$$

From the merely facts equations above, we want to find the probability assignment which avoids bias, while agreeing with whatever information is given. Therefore, in making inferences on the basis of partial information we must use that probability distribution which has maximum entropy subject whatever is know. The entropy is defined by Shannon, which reads,

$$H[P(x)] = -k\sum_i P_i \ln P_i(x)$$

Lagrangian multipliers λ, μ are introduced,

$$F = -\sum_i P_i \ln P_i(x) - \mu(\sum_i P_i f(x_i) - \langle f(x)\rangle) - (\lambda - 1)(\sum_i P_i - 1)$$

$$\frac{\partial F}{\partial P_i} = -[\ln P_i(x) + 1] - \mu f(x_i) - (\lambda - 1) = 0$$

from equation (above) we could obtain the result,

$$P_i = e^{-\lambda - \mu f(x_i)}$$

the constants λ, μ are determined by substituting into equation. The result can be written in the form,

$$\langle f(x)\rangle = -\frac{\partial}{\partial \mu}\ln Z(\mu)$$

$$\lambda = \ln Z(\mu)$$

$$Z(\mu) = \sum_i e^{-\mu f(x_i)}$$

where

it is called the partition function.

The Derivation of Power Law Distribution

We consider the mean entropy of system as a constant. In the language of information theory, it means the expectation value of entropy $S(x)$ is known, namely,

$$\langle S(x) \rangle = \sum_i P_i[\Omega_i(x)] S_i(x)$$

P_i is the probability of finding a system in one of the states corresponding to the ith element (or group). where the entropy of each element in the system is represented by the Boltzmann entropy,

$$S_i(x) = k_B \ln[\Omega_i(x)]$$

where the Ω_I is number of states of the ith element (or group). And the normalisation condition reads,

$$\sum_i P_i = 1$$

The Shannon entropy is,

$$H[P(x)] = -k \sum_i P_i[\Omega_i(x)] \ln P_i[\Omega_i(x)]$$

maximize above equation subject to the constraints,

$$F = -\sum_i P_i[\Omega_i(x)] \ln P_i[\Omega_i(x)] - \mu\left(\sum_i k_B P_i[\Omega_i(x)] \ln[\Omega_i(x)] - \langle S \rangle\right) - (\lambda - 1)\left(\sum_i P_i - 1\right)$$

$$\frac{\partial F}{\partial P_i(\Omega_i(x))} = -\sum_i \ln P_i[\Omega_i] - \mu \sum_i k_b \ln[\Omega_i(x)] - \lambda = 0$$

therefore,

$$P_i[\Omega_i(x)] = \exp(\lambda - \mu k_B \ln[\Omega_i(x)]) = A\Omega_i^{-\alpha}$$

where $A = e^{\lambda}$; use constraint, A could be determined,

$$A = \frac{1}{\sum_i \Omega_i^{-\alpha}} = \frac{1}{Z}$$

hence that,

$$P_i = \frac{\Omega_i^{-\alpha}}{\sum_i \Omega_i^{-\alpha}}$$

The power law we expect.

Application on Networks

Networks are likely to be interpreted as graphs with nodes linked by edges. Simple behaviour of each node (identical or not) can give a complex collective behaviours in a network. However, there are some important quantities that would generalise the topology of a network. The first one is the degree (the number of edges of a node) distribution. It describes how links distributed among different nodes. Here, we can treat the number of state of the ith element (Ω_i) as the degree of the ith node, because the more connections one node have the more opportunity it can interact the other nodes, namely,

$$\Omega_i \sim n_i$$

where n_i is the degree of the ith node. Then substitute it into equation:

$$P_i = \frac{n_i^{-\alpha}}{\sum_i n_i^{-\alpha}} = An_i^{-\alpha}$$

This give us very clear physics meaning: if a network is dynamically stable (or temporarily say stable), the degree distribution would follow the power law. Remember, dynamically stable is very important here. If new nodes are inserted randomly into the network continuously, the power law will break up gradually.

The second one is the aggregation. The aggregation in networks is termed as clustering. Usually, the clusters in a network can be treated as sub-networks, which have various amounts of nodes and are relatively isolated to the other notes of their supplementary network. Take the telecommunication network for example.

Local telecom network, such as a network of city, can be treated as a subsystem of the state network. Traffic in such networks are main local calls. However, the toughest thing to deal with is how we count the number of state of a sub-network, if we want to take advantage. To think about it, we'd better trace back, where we used the opportunity interpretation for a single node. Can we use the same inference? The answer is no but a little bit similar. Suppose a sub-group with M nodes.

The maximum links the network could have is $M(M-1)/2$, but it is not the number of states of the sub-network because it only relates to the number of nodes in the network and has nothing to do with the other properties of the network. It seems we have to discover a new quantity to serve our need. However, predecessors of network topology have already solved the problem. They suggest another quantity named average path length which is the average of the shortest distance between any nodes. Therefore the number of states could be easily defined as,

$$\Omega_i = C_{L_i}^{l}$$

Where L is total path length of ith network and l is the average path length; usually in an effective network $L \sim M^{\beta}$

$$0 < \beta \leq 2$$

Then,

$$\Omega_i \cong C^l_{M^{\beta}} = \frac{M^{\beta}(M^{\beta}-1)\cdots(M^{\beta}-l+1)}{l!} \approx \frac{M^{l\beta}}{l!}$$

where $M >> l$; substitute it into equation,

$$P_i = BM^{-\beta\alpha}$$

equation (above) shows very clear physics meaning that the size distribution of sub-network follow the power law.

STATISTICAL THERMODYNAMICS

Let us briefly review the material which we have covered so far in this course. We started off by studying the mathematics of probability. We then used probabilistic reasoning to analyse the dynamics of many particle systems, a subject area known as statistical mechanics. Next, we explored the physics of heat and work, the study of which is termed thermodynamics.

The final step in our investigation is to combine statistical mechanics with thermodynamics: in other words, to investigate heat and work via statistical arguments. This discipline is called statistical thermodynamics, and forms the central subject matter of this course.

THERMAL INTERACTION BETWEEN MACROSYSTEMS

Let us begin our investigation of statistical thermodynamics by examining a purely thermal interaction between two macroscopic systems, A and A', from a microscopic point of view. Suppose that the energies of these two systems are E and E', respectively. The external parameters are held fixed, so that A and A' cannot do work on one another.

However, we assume that the systems are free to exchange heat energy (*i.e.*, they are in thermal contact). It is convenient to divide the energy scale into small subdivisions of width δE. The number of microstates of A consistent with a macrostate in which the energy lies in the range E to $E + \delta E$ is denoted $\Omega(E)$. Likewise, the number of microstates of consistent with a macrostate in which the energy lies between E' and $E' + \delta E$ is denoted $\Omega'(E')$. The combined system $A^{(0)} = A + A'$ is assumed to be isolated (*i.e.*, it neither does work on nor exchanges heat with its surroundings). It follows from the first law of thermodynamics that the total energy $E^{(0)}$ is constant. When speaking of thermal contact between two distinct systems, we usually assume that the mutual interaction is sufficiently weak for the energies to be additive. Thus,

$$E + E \simeq E^{(0)} = \text{constant}$$

Of course, in the limit of zero interaction the energies are strictly additive. However, a small residual interaction is always required to enable the two systems to exchange heat energy and, thereby, eventually reach thermal equilibrium. In fact, if the interaction between A and A' is too strong for the energies to be additive then it makes little sense to consider each system in isolation, since the presence of one system clearly strongly perturbs the other, and vice versa. In this case, the smallest system which can realistically be examined in isolation is $A^{(0)}$.

According to Equation if the energy of A lies in the range E to $E + \delta E$ then the energy of A' must lie between $E^{(0)} - E - \delta E$ and $E^{(0)} - E$. Thus, the number of microstates accessible to each system is given by $\Omega'(E)$ and $(E^{(0)} - E)$, respectively. Since every possible state of A can be combined with every possible state of A' to form a distinct microstate, the total number of distinct states accessible to $A^{(0)}$ when the energy of A lies in the range E to $E + \delta E$ is $\Omega^{(0)} = \Omega(E)\, \Omega'(E^{(0)} - E)$.

Consider an ensemble of pairs of thermally interacting systems, A and A', which are left undisturbed for many relaxation times so that they can attain thermal equilibrium. The principle of equal a priori probabilities is applicable to this situation.

According to this principle, the probability of occurrence of a given macrostate is proportional to the number of accessible microstates, since all microstates are equally likely. Thus, the probability that the system A has an energy lying in the range E to $E + \delta E$can be written

$$P(E) = C\, \Omega(E)\, \Omega'\, (E^{(0)} - E),$$

where C is a constant which is independent of E.

The typical variation of the number of accessible states with energy is of the form,

$$\Omega \propto E^f,$$

where f is the number of degrees of freedom. For a macroscopic system f is an exceedingly large number. It follows that the probability $P(E)$in Equation is the product of an extremely rapidly increasing function of E and an extremely rapidly decreasing function of E. Hence, we would expect the probability to exhibit a very pronounced maximum at some particular value of the energy.

Let us Taylor expand the logarithm of $P(E)$ in the vicinity of its maximum value, which is assumed to occur at $E = \tilde{E}$. We expand the relatively slowly varying logarithm, rather than the function itself, because the latter varies so rapidly with the energy that the radius of convergence of its Taylor expansion is too small for this expansion to be of any practical use. The expansion of $\ln \Omega(E)$ yields

$$\ln \Omega(E) = \ln \Omega(\tilde{E}) + \beta(\tilde{E})\eta - \frac{1}{2}\lambda(\tilde{E})\eta^2 + \cdots,$$

where,

$$\eta = E - \tilde{E},$$

$$\frac{\partial \ln \Omega}{\partial E},$$

$$\frac{\partial^2 \ln \Omega}{\partial E^2} = -\frac{\partial \beta}{\partial E}.$$

Now, since $E' = E^{(0)} - E$, we have

$$E' - \tilde{E}' = -(E - \tilde{E}) = -\eta.$$

It follows that,

$$\ln \Omega'(E') = \ln \Omega'(\tilde{E}') + \beta'(\tilde{E}')(-\eta) - \frac{1}{2}\lambda'(\tilde{E}')(-\eta)^2 + \cdots,$$

where β' and λ' are defined in an analogous manner to the parameters β and λ. Equations can be combined to give, At the maximum of ln $[\Omega(E)\Omega'(E')]$ the linear term in the Taylor expansion must vanish, so

$$\beta(\tilde{E}) = \beta(\tilde{E}'),$$

which enables us to determine $\tilde{E}$. It follows that

$$\ln P(E) = \ln P(\tilde{E}) - \frac{1}{2}\lambda_0 \eta^2,$$

or,

$$P(E) = P(\tilde{E}) \exp\left[-\frac{1}{2}\lambda_0 (E - \tilde{E})^2\right],$$

where,

$$\lambda_0 = \lambda(\tilde{E}) + \lambda'(\tilde{E}').$$

Now, the parameter λ_0 must be positive, otherwise the probability $P(E)$ does not exhibit a pronounced maximum value: *i.e.*, the combined system $A^{(0)}$ does not possess a well-defined equilibrium state as, physically, we know it must.

It is clear that $\lambda(\tilde{E})$ must also be positive, since we could always choose for A' a system with a negligible contribution to λ_0, in which case the constraint $\lambda_0 > 0$ would effectively correspond to $\lambda(\tilde{E}) > 0$. [A similar argument can be used to show that $\lambda'(\tilde{E}')$ must be positive.] The same conclusion also follows from the estimate $\Omega \propto E^f$, which implies that

$$\lambda(\tilde{E}) \sim \frac{f}{\tilde{E}^2} > 0$$

According to Equation the probability distribution function $P(E)$ is a Gaussian. This is hardly surprising, since the central limit theorem ensures that the probability distribution for any macroscopic variable, such as $\tilde{E}$, is Gaussian in nature.

It follows that the mean value of E corresponds to the situation of maximum probability (*i.e.*, the peak of the Gaussian curve), so that,

$$\bar{E} = \tilde{E}.$$

The standard deviation of the distribution is,

$$\Delta^* E = \lambda_0^{-1/2} \sim \frac{\bar{E}}{\sqrt{f}},$$

where use has been made of Equation (assuming that system makes the dominant contribution to λ_0). It follows that the fractional width of the probability distribution function is given by,

$$\frac{\Delta^* E}{\bar{E}} \sim \frac{1}{\sqrt{f}}.$$

Hence, if A contains 1 mole of particles then $f \sim N_A \simeq 10^{24}$ and $\Delta^* E/\bar{E} \sim 10^{-12}$. Clearly, the probability distribution for E has an exceedingly sharp maximum. Experimental measurements of this energy will almost always yield the mean value, and the underlying statistical nature of the distribution may not be apparent.

TEMPERATURE

Suppose that the systems A and A' are initially thermally isolated from one another, with respective energies E_i and E_i'. (Since the energy of an isolated system cannot fluctuate, we do not have to bother with mean energies here.)

If the two systems are subsequently placed in thermal contact, so that they are free to exchange heat energy, then, in general, the resulting state is an extremely improbable one [*i.e.*, $P(E_i)$is much less than the peak probability]. The configuration will, therefore, tend to change in time until the two systems attain final mean energies $\bar{E}_f$ and $\bar{E}_f'$ which are such that

$$\beta_f = \beta_f',$$

where $\beta_f \equiv \beta(\bar{E}_f)$ and $\beta_f' \equiv \beta'(\bar{E}_f')$. This corresponds to the state of maximum probability. In the special case where the initial energies, E_i and E_i', lie very close to the final mean energies, $\bar{E}_f$ and $\bar{E}_f'$, respectively, there is no change in the two systems when they are brought into thermal contact, since the initial state already corresponds to a state of maximum probability.

It follows from energy conservation that,

$$\overline{E}_f + \overline{E}'_f = E_i + E'_i.$$

The mean energy change in each system is simply the net heat absorbed, so that,

$$Q \equiv \overline{E}_f - E_i,$$

$$Q' \equiv \overline{E}'_f - E'_i.$$

The conservation of energy then reduces to,

$$Q + Q' = 0:$$

i.e., the heat given off by one system is equal to the heat absorbed by the other (in our notation absorbed heat is positive and emitted heat is negative).

It is clear that if the systems A and A' are suddenly brought into thermal contact then they will only exchange heat and evolve towards a new equilibrium state if the final state is more probable than the initial one. In other words, if

$$P(\overline{E}_f) > (E_i),$$

or,

$$\ln P(\overline{E}_f) > \ln P(E_i),$$

since the logarithm is a monotonic function. The above inequality can be written,

$$\ln \Omega(\overline{E}_f) + \ln \Omega'(\overline{E}'_i) > \ln \Omega(E_i) + \ln \Omega'(E'_i),$$

with the aid of Equation. Taylor expansion to first order yields,

$$\frac{\partial \ln \Omega(E_i)}{\partial E}(\overline{E}_f - E_i) + \frac{\partial \ln \Omega'(E'_l)}{\partial E'}(\overline{E}'_f - E'_i) > 0,$$

which finally gives,

$$(\beta_i - \beta'_i)\, Q > 0,$$

where $\beta_i \equiv \beta(E_i)$, $\beta'_i \equiv \beta'(E'_i)$, and use has been made of Equation. It is clear, from the above, that the parameter β, defined,

$$\beta = \frac{\partial \ln \Omega}{\partial E},$$

has the following properties:

If two systems separately in equilibrium have the same value of β then the systems will remain in equilibrium when brought into thermal contact with one another. If two systems separately in equilibrium have different values of β then the systems will not remain in equilibrium when brought into thermal contact with one another. Instead, the system with the higher value of β will absorb heat from the other system until the two β values are the same. Incidentally, a partial derivative is used in Equation because in a purely thermal interaction the external parameters of the system are held constant whilst the energy changes.

Let us define the dimensionless parameter T, such that,

$$\frac{1}{kT} \equiv \beta \equiv \frac{\partial \ln \Omega}{\partial E},$$

where k is a positive constant having the dimensions of energy. The parameter T is termed the thermodynamic temperature, and controls heat flow in much the same manner as a conventional temperature. Thus, if two isolated systems in equilibrium possess the same thermodynamic temperature then they will remain in equilibrium when brought into thermal contact.

However, if the two systems have different thermodynamic temperatures then heat will flow from the system with the higher temperature (*i.e.*, the "hotter" system) to the system with the lower temperature until the temperatures of the two systems are the same. In addition, suppose that we have three systems A, B, and C.

We know that if A and B remain in equilibrium when brought into thermal contact then their temperatures are the same, so that $T_A = T_B$. Similarly, if B and C remain in equilibrium when brought into thermal contact, then $T_B = T_C$. But, we can then conclude that $T_A = T_C$, so systems A and C will also remain in equilibrium when brought into thermal contact. Thus, we arrive at the following statement, which is sometimes called the zeroth law of thermodynamics: If two systems are separately in thermal equilibrium with a third system then they must also be in thermal equilibrium with one another. The thermodynamic temperature of a macroscopic body, as defined in Equation depends only on the rate of change of the number of accessible microstates with the total energy. Thus, it is possible to define a thermodynamic temperature for systems with radically different microscopic structures (*e.g.*, matter and radiation).

The thermodynamic, or absolute, scale of temperature is measured in degrees kelvin. The parameter k is chosen to make this temperature scale accord as much as possible with more conventional temperature scales.

The choice,

$$k = 1.381 \times 10^{-23} \text{ joules/kelvin},$$

ensures that there are 100 degrees kelvin between the freezing and boiling points of water at atmospheric pressure (the two temperatures are 273.15 and 373.15 degrees kelvin, respectively). The above number is known as the Boltzmann constant. In fact, the Boltzmann constant is fixed by international convention so as to make the triple point of water (*i.e.*, the unique temperature

at which the three phases of water co-exist in thermal equilibrium) exactly 273.16°K.
Note that the zero of the thermodynamic scale, the so called absolute zero of temperature, does not correspond to the freezing point of water, but to some far more physically significant temperature which.

The familiar $\Omega \propto E^f$ scaling for translational degrees of freedom yields,

$$kT \sim \frac{\bar{E}}{f}$$

using Equation so kT is a rough measure of the mean energy associated with each degree of freedom in the system. In fact, for a classical system (*i.e.*, one in which quantum effects are unimportant) it is possible to show that the mean energy,

$$(1/2)kT$$

associated with each degree of freedom is exactly. This result, which is known as the equipartition theorem.

The absolute temperature T is usually positive, since $\Omega(E)$ is ordinarily a very rapidly increasing function of energy. In fact, this is the case for all conventional systems where the kinetic energy of the particles is taken into account, because there is no upper bound on the possible energy of the system, and $\Omega(E)$ consequently increases roughly like E^f.

It is, however, possible to envisage a situation in which we ignore the translational degrees of freedom of a system, and concentrate only on its spin degrees of freedom. In this case, there is an upper bound to the possible energy of the system (*i.e.*, all spins lined up anti-parallel to an applied magnetic field). Consequently, the total number of states available to the system is finite. In this situation, the density of spin states $\Omega_{spin}(E)$ first increases with increasing energy, as in conventional systems, but then reaches a maximum and decreases again. Thus, it is possible to get absolute spin temperatures which are negative, as well as positive.

In Lavoisier's calorific theory, the basic mechanism which forces heat to flow from hot to cold bodies is the supposed mutual repulsion of the constituent particles of calorific fluid. In statistical mechanics, the explanation is far less contrived. Heat flow occurs because statistical systems tend to evolve towards their most probable states, subject to the imposed physical constraints.

When two bodies at different temperatures are suddenly placed in thermal contact, the initial state corresponds to a spectacularly improbable state of the overall system. For systems containing of order 1 mole of particles, the only reasonably probable final equilibrium states are such that the two bodies differ in temperature by less than 1 part in 10^{12}. The evolution of the system towards these final states (*i.e.*, towards thermal equilibrium) is effectively driven by probability.

MECHANICAL INTERACTION BETWEEN MACROSYSTEMS

Let us now examine a purely mechanical interaction between macrostates, where one or more of the external parameters is modified, but there is no exchange of heat energy. Consider, for the sake of simplicity, a situation where only one external parameter x of the system is free to vary. In general, the number of microstates accessible to the system when the overall energy lies between E and $E + \delta E$ depends on the particular value of x, so we can write $\Omega \equiv \Omega(E, x)$. When x is changed by the amount dx, the energy $E_r(x)$ of a given microstate r changes by $(\partial E_r/\partial x)dx$. The number of states $\sigma(E, x)$ whose energy is changed from a value less than E to a value greater than E when the parameter changes from x to $x + dx$ is given by the number of microstates per unit energy range multiplied by the average shift in energy of the microstates.

Hence,

$$\sigma(E,x) = \frac{\Omega(E,x)}{\delta E}\frac{\overline{\partial E_r}}{\partial x}dx,$$

where the mean value of $\partial E_r/\partial x$is taken over all accessible microstates (*i.e.*, all states where the energy lies between E and $E + \delta E$ and the external parameter takes the value). The above equation can also be written

$$s(E, x) = \frac{\Omega(E,x)}{\delta E}\overline{X}$$

where

$$\overline{X}(E,x) = -\frac{\overline{\partial E_r}}{\partial x}$$

is the mean generalized force conjugate to the external, parameter x.

Consider the total number of microstates between E and $E + \delta E$. When the external parameter changes from to $x + dx$, the number of states in this energy range changes by $(\partial \Omega/\partial x)\, dx$. This change is due to the difference between the number of states which enter the range because their energy is changed from a value less than E to one greater than E and the number which leave because their energy is changed from a value less than $E + \delta E$ to one greater than $E + \delta E$. In symbols,

$$\frac{\partial \Omega(E,x)}{\partial x}dx = \sigma(E) - \sigma(E+\delta E) \simeq -\frac{\partial \sigma}{\partial E}\delta E,$$

which yields

$$\frac{\partial \Omega}{\partial x} = \frac{\partial(\Omega \overline{X})}{\partial E},$$

where use has been made of Equation Dividing both sides by Ω gives,

$$\frac{\partial \ln \Omega}{\partial x} = \frac{\partial \ln \Omega}{\partial E}\bar{X} + \frac{\partial \bar{X}}{\partial E}.$$

However, according to the usual estimate $\Omega \propto E^f$, the first term on the right-hand side is of order $(f/\bar{E})\bar{X}$, whereas the second term is only of order. Clearly, for a macroscopic system with many degrees of freedom, the second term is utterly negligible, so we have

$$\frac{\partial \ln \Omega}{\partial x} = \frac{\partial \ln \Omega}{\partial E}\bar{X} = \beta\bar{X}.$$

When there are several external parameters $x_1..., x_n$ so that $\Omega \equiv \Omega(E, x_1..., x_n)$, the above derivation is valid for each parameter taken in isolation.

Thus,

$$\frac{\partial \ln \Omega}{\partial x_\alpha} = \beta\bar{X}_\alpha,$$

where $\bar{X}_\alpha$ is the mean generalized force conjugate to the parameter x_α.

General Interaction between Macrosystems

Consider two systems, A and A', which can interact by exchanging heat energy and doing work on one another. Let the system A' have energy E' and adjustable external parameters $x_1...x_n$. Likewise, let the system $A^{(0)} = A + A'$ have energy and adjustable external parameters $x'_1...x'_n$. The combined system $A^{(0)} = A + A'$ is assumed to be isolated. It follows from the first law of thermodynamics that

$$E + E' = E^{(0)} = \text{constant}.$$

Thus, the energy E' of system A' is determined once the energy E of system A is given, and vice versa. In fact, E' could be regarded as a function of E. Furthermore, if the two systems can interact mechanically then, in general, the parameters x' are some function of the parameters x. As a simple example, if the two systems are separated by a movable partition in an enclosure of fixed volume $V^{(0)}$, then,

$$V + V' = V^{(0)} = \text{constant},$$

where V and V' are the volumes of systems A and A', respectively. The total number of microstates accessible to $A^{(0)}$ is clearly a function of E and the parameters x_α (where runs from 1 to n), so $\Omega^{(0)} \equiv \Omega^{(0)}(E, x_1...x_n)$. We have already demonstrated that $\Omega^{(0)}$ exhibits a very pronounced maximum at one particular value of the energy $E = \tilde{E}$ when E is varied but the external parameters are held constant. This behaviour comes about because of the very strong,

$$\Omega \propto E^f,$$

increase in the number of accessible microstates of A(or A') with energy. The number of accessible microstates exhibits a similar strong increase with the volume, which is a typical external parameter, so that

$$\Omega \propto V^f.$$

It follows that the variation of $\Omega^{(0)}$ with a typical parameter x_α, when all the other parameters and the energy are held constant, also exhibits a very sharp maximum at some particular value $x_\alpha = \tilde{x}_\alpha$. The equilibrium situation corresponds to the configuration of maximum probability, in which virtually all systems $A^{(0)}$ in the ensemble have values of E and x_α very close to $\tilde{E}$ and $\tilde{x}_\alpha$. The mean values of these quantities are thus given by $\bar{E} = \tilde{E}$ and $\bar{x}_\alpha = \tilde{x}_\alpha$.

Consider a quasi-static process in which the system A is brought from an equilibrium state described by $\bar{E}$ and $\bar{x}_\alpha$ to an infinitesimally different equilibrium state described by and. Let us calculate the resultant change in the number of microstates accessible to A. Since $\Omega \equiv \Omega(E, x_1..., x_n)$, the change in ln Ω follows from standard mathematics:

$$d\ln\Omega = \frac{\partial \ln\Omega}{\partial E} d\bar{E} + \sum_{\alpha=1}^{n} \frac{\partial \ln\Omega}{\partial x_\alpha} d\bar{x}_\alpha.$$

However, we have previously demonstrated that,

$$\beta = \frac{\partial \ln\Omega}{\partial E}, \beta\bar{X}_\alpha = \frac{\partial \ln\Omega}{\partial x_\alpha}$$

$$d\ln\Omega = \beta\left(d\bar{E} + \sum_\alpha \bar{X}_\alpha d\bar{x}_\alpha\right).$$

Note that the temperature β parameter and the mean conjugate forces $\bar{X}_\alpha$ are only well-defined for equilibrium states. This is why we are only considering quasi-static changes in which the two systems are always arbitrarily close to equilibrium.

Let us rewrite Equation in terms of the thermodynamic temperature, T using the relation $\beta \equiv 1/kT$. We obtain

$$dS = \left(d\bar{E} + \sum_\alpha \bar{X}_\alpha d\bar{x}_\alpha\right)/T,$$

where,

$$S = k \ln \Omega.$$

Equation is a differential relation which enables us to calculate the quantity S as a function of the mean energy $\bar{E}$ and the mean external parameters $\bar{x}_\alpha$, assuming that we can calculate the temperature T and mean conjugate forces $\bar{X}_\alpha$ for each equilibrium state. The function $S(\bar{E}, \bar{x}_\alpha)$ is termed the entropy of system A. The word entropy is derived from the Greek en+trepien, which means "in change." The reason for this etymology will become apparent presently. It can be seen from Equation that the entropy is merely a parameterization of the number of accessible microstates.

Hence, according to statistical mechanics, is essentially a measure of the relative probability of a state characterized by values of the mean energy and mean external parameters $\bar{E}$ and $\bar{x}_\alpha$, respectively.

According to Equation the net amount of work performed during a quasi-static change is given by

$$đW = \sum_\alpha \bar{X}_\alpha d\bar{x}_\alpha.$$

It follows from Equation that,

$$dS \frac{đW + đW}{T} = \frac{đQ}{T}.$$

Thus, the thermodynamic temperature T is the integrating factor for the first law of thermodynamics,

$$đQ = d\bar{E} + đW,$$

which converts the inexact differential $đQ$ into the exact differential dS. It follows that the entropy difference between any two macrostates and i can f be written

$$S_f - S_i = \int_i^f dS = \int_i^f \frac{đQ}{T},$$

where the integral is evaluated for any process through which the system is brought quasi-statically via a sequence of near-equilibrium configurations from its initial to its final macrostate.

The process has to be quasi-static because the temperature T, which appears in the integrand, is only well-defined for an equilibrium state. Since the left-hand side of the above equation only depends on the initial and final states, it follows that the integral on the right-hand side is independent of the particular sequence of quasi-static changes used to get from to f. Thus,

$$\int_i^f đQ/T$$

is independent of the process (provided that it is quasi-static). All of the concepts which we have encountered up to now in this course, such as temperature, heat, energy, volume, pressure, etc., have been fairly familiar to us from other branches of Physics. However, entropy, which turns out to be of crucial importance in thermodynamics, is something quite new. Let us consider the following questions. What does the entropy of a system actually signify? What use is the concept of entropy?

ENTROPY

Consider an isolated system whose energy is known to lie in a narrow range. Let Ω be the number of accessible microstates. According to the principle of equal a priori probabilities, the system is equally likely to be found in any one of these states when it is in thermal equilibrium. The accessible states are just that set of microstates which are consistent with the macroscopic constraints imposed on the system.

These constraints can usually be quantified by specifying the values of some parameters $y_1..., y_n$ which characterize the macrostate. Note that these parameters are not necessarily external: *e.g.*, we could specify either the volume (an external parameter) or the mean pressure (the mean force conjugate to the volume). The number of accessible states is clearly a function of the chosen parameters, so we can write $\Omega \equiv \Omega\ (y_1..., y_n)$for the number of microstates consistent with a macrostate in which the general parameter y_α lies in the range y_α to $y_\alpha + dy_\alpha$.

Suppose that we start from a system in thermal equilibrium. According to statistical mechanics, each of the Ω_i, say, accessible states are equally likely. Let us now remove, or relax, some of the constraints imposed on the system. Clearly, all of the microstates formally accessible to the system are still accessible, but many additional states will, in general, become accessible.

Thus, removing or relaxing constraints can only have the effect of increasing, or possibly leaving unchanged, the number of microstates accessible to the system. If the final number of accessible states is Ω_f, then we can write,

$$\Omega_f \geq \Omega_i.$$

Immediately after the constraints are relaxed, the systems in the ensemble are not in any of the microstates from which they were previously excluded. So the systems only occupy a fraction,

$$P_i = \frac{\Omega_i}{\Omega_f}$$

of the W_f states now accessible to them. This is clearly not a equilibrium situation. Indeed, if $W_f >> W_i$ then the configuration in which the systems are only distributed over the original W_i states is an extremely unlikely one.

In fact, its probability of occurrence is given by Equation According to the theorem, *H* the ensemble will evolve in time until a more probable final state is reached in which the systems are evenly distributed over the W_f available states. As a simple example, consider a system consisting of a box divided into two regions of equal volume.

Suppose that, initially, one region is filled with gas and the other is empty. The constraint imposed on the system is, thus, that the coordinates of all of the molecules must lie within the filled region.

In other words, the volume accessible to the system $V = V_i$ is, where V_i is half the volume of the box. The constraints imposed on the system can be relaxed by removing the partition and allowing gas to flow into both regions. The volume accessible to the gas is now

$$V = V_f = 2V_i.$$

Immediately after the partition is removed, the system is in an extremely improbable state that at constant energy the variation of the number of accessible states of an ideal gas with the volume is

$$\Omega \propto V^N,$$

where N is the number of particles. Thus, the probability of observing the state immediately after the partition is removed in an ensemble of equilibrium systems with volume $V = V_f$ is

$$P_i = \frac{\Omega_i}{\Omega_f} = \left(\frac{V_i}{V_f}\right)^N = \left(\frac{1}{2}\right)^N.$$

If the box contains of order 1 mole of molecules then $N \sim 10^{24}$ and this probability is fantastically small:

$$P_i \sim \exp(-10^{24}).$$

Clearly, the system will evolve towards a more probable state. This can also be phrased in terms of the parameters $y_1 \ldots y_n$ of the system. Suppose that a constraint is removed. For instance, one of the parameters, y, say, which originally had the value $y = y_i$, is now allowed to vary. According to statistical mechanics, all states accessible to the system are equally likely. So, the probability $P(y)$of finding the system in equilibrium with the parameter in the range to is just proportional y to $y + \delta y$ the number of microstates in this interval: *i.e.*,

$$P(y) \propto \Omega(y).$$

Usually, $\Omega(y)$ has a very pronounced maximum at some particular value $\tilde{y}$. This means that practically all systems in the final equilibrium ensemble have values of y close to $\tilde{y}$. Thus, if $y_i \neq \tilde{y}$ initially then the parameter y will change until it attains a final value close to $\tilde{y}$, where Ω is maximum.

If some of the constraints of an isolated system are removed then the parameters of the system tend to readjust themselves in such a way that

$$\Omega(y_1 \ldots y_n) \to \text{maximum}.$$

Suppose that the final equilibrium state has been reached, so that the systems in the ensemble are uniformly distributed over the Ω_f accessible final states. If the original constraints are reimposed then the systems in the ensemble still occupy these Ω_f states with equal probability.

Thus, if $\Omega_f > \Omega_i$, simply restoring the constraints does not restore the initial situation. Once the systems are randomly distributed over the Ω_f states they cannot be expected to spontaneously move out of some of these states and occupy a more restricted class of states merely in response to the reimposition of a constraint. The initial condition can also not be restored by removing further constraints. This could only lead to even more states becoming accessible to the system. Suppose that some process occurs in which an isolated system goes from some initial configuration to some final configuration. If the final configuration is such that the imposition or removal of constraints cannot by itself restore the initial condition then the process is deemed irreversible. On the other hand, if it is such that the imposition or removal of constraints can restore the initial condition then the process is deemed reversible.

From what we have already said, an irreversible process is clearly one in which the removal of constraints leads to a situation where $\Omega_f > \Omega_i$. A reversible process corresponds to the special case where the removal of constraints does not change the number of accessible states, so that $\Omega_f = \Omega_i$. In this situation, the systems remain distributed with equal probability over these states irrespective of whether the constraints are imposed or not.

Our microscopic definition of irreversibility is in accordance with the macroscopic. Recall that on a macroscopic level an irreversible process is one which "looks unphysical" when viewed in reverse. On a microscopic level it is clearly plausible that a system should spontaneously evolve from an improbable to a probable configuration in response to the relaxation of some constraint.

However, it is quite clearly implausible that a system should ever spontaneously evolve from a probable to an improbable configuration. Let us consider our example again. If a gas is initially restricted to one half of a box, via a partition, then the flow of gas from one side of the box to the other when the partition is removed is an irreversible process.

This process is irreversible on a microscopic level because the initial configuration cannot be recovered by simply replacing the partition. It is irreversible on a macroscopic level because it is obviously unphysical for the molecules of a gas to spontaneously distribute themselves in such a manner that they only occupy half of the available volume.

It is actually possible to quantify irreversibility. In other words, in addition to stating that a given process is irreversible, we can also give some indication of how irreversible it is. The parameter which measures irreversibility is just the number of accessible states Ω.

Thus, if Ω for an isolated system spontaneously increases then the process is irreversible, the degree of irreversibility being proportional to the amount of the increase. If Ω stays the same then the process is reversible. Of course, it is unphysical for Ω to ever spontaneously decrease. In symbols, we can write

$$\Omega_f - \Omega_i \equiv \Delta\Omega \geq 0,$$

for any physical process operating on an isolated system. In practice, Ω itself is a rather unwieldy parameter with which to measure irreversibility. For instance, in the previous example, where an ideal gas doubles in volume (at constant energy) due to the removal of a partition, the fractional increase in Ω is,

$$\frac{\Omega_f}{\Omega_i} \simeq 10^{2v \times 10^{23}},$$

where is the number of moles. This is an extremely large number! It is far more convenient to measure irreversibility in terms of ln Ω. If Equation is true then it is certainly also true that,

$$\ln \Omega_f - \ln \Omega_i \equiv \Delta \ln \Omega \geq 0$$

for any physical process operating on an isolated system. The increase in when an ideal gas doubles in volume (at constant energy) is

$$\ln \Omega_f - \ln \Omega_i = \nu N_A \ln 2,$$

where $N_A = 6 \times 10^{23}$. This is a far more manageable number! Since we usually deal with particles by the mole in laboratory physics, it makes sense to pre-multiply our measure of irreversibility by a number of order $1/N_A$. For historical reasons, the number which is generally used for this purpose is the Boltzmann constant k, which can be written,

$$k = \frac{R}{N_A} \text{ joules/kelvin},$$

where,

$$\text{R} = 8.3143 \text{ joules/kelvin/mole}$$

is the ideal gas constant which appears in the well-known equation of state for an ideal gas, $PV = v\,RT$. Thus, the final form for our measure of irreversibility is

$$S = k \ln \Omega$$

This quantity is termed "entropy", and is measured in joules per degree kelvin. The increase in entropy when an ideal gas doubles in volume (at constant energy) is

$$S_f - S_i = v\,R \ln 2,$$

which is order unity for laboratory scale systems (*i.e.*, those containing about one mole of particles). The essential irreversibility of macroscopic phenomena can be summed up as follows:

$$S_f - S_i = \Delta S \geq 0,$$

for a process acting on an isolated system. Thus, the entropy of an isolated system tends to increase with time and can never decrease. This proposition is known as the second law of thermodynamics.

One way of thinking of the number of accessible states Ω is that it is a measure of the disorder associated with a macrostate. For a system exhibiting a high degree of order we would expect a strong correlation between the motions of the individual particles. For instance, in a fluid there might be a strong tendency for the particles to move in one particular direction, giving rise to an ordered flow of the system in that direction. On the other hand, for a system exhibiting a low degree of order we expect far less correlation between the motions of individual particles.

It follows that, all other things being equal, an ordered system is more constrained than a disordered system, since the former is excluded from microstates in which there is not a strong correlation between individual particle motions, whereas the latter is not. Another way of saying this is that an ordered system has less accessible microstates than a corresponding disordered system.

Thus, entropy is effectively a measure of the disorder in a system (the disorder increases with $\mathcal{S}$). With this interpretation, the second law of thermodynamics reduces to the statement that isolated systems tend to become more disordered with time, and can never become more ordered.

Note that the second law of thermodynamics only applies to isolated systems. The entropy of a non-isolated system can decrease. For instance, if a gas expands (at constant energy) to twice its initial volume after the removal of a partition, we can subsequently recompress the gas to its original volume. The energy of the gas will increase because of the work done on it during compression, but if we absorb some heat from the gas then we can restore it to its initial state. Clearly, in restoring the gas to its original state, we have restored its original entropy.

This appears to violate the second law of thermodynamics because the entropy should have increased in what is obviously an irreversible process (just try to make a gas spontaneously occupy half of its original volume!). However, if we consider a new system consisting of the gas plus the compression and heat absorption machinery, then it is still true that the entropy of this system (which is assumed to be isolated) must increase in time.

Thus, the entropy of the gas is only kept the same at the expense of increasing the entropy of the rest of the system, and the total entropy is increased. If we consider the system of everything in the Universe, which is certainly an isolated system since there is nothing outside it with which it could interact, then the second law of thermodynamics becomes:

The disorder of the Universe tends to increase with time and can never decrease. An irreversible process is clearly one which increases the disorder of the Universe, whereas a reversible process neither increases nor decreases disorder. This definition is in accordance with our previous definition of an irreversible process as one which "does not look right" when viewed backwards.

One easy way of viewing macroscopic events in reverse is to film them, and then play the film backwards through a projector. There is a famous passage in the novel "Slaughterhouse 5," by Kurt Vonnegut, in which the hero, Billy

Pilgrim, views a propaganda film of an American World War II bombing raid on a German city in reverse. This is what the film appeared to show: "American planes, full of holes and wounded men and corpses took off backwards from an airfield in England. Over France, a few German fighter planes flew at them backwards, sucked bullets and shell fragments from some of the planes and crewmen. They did the same for wrecked American bombers on the ground, and those planes flew up backwards and joined the formation.

The formation flew backwards over a German city that was in flames. The bombers opened their bomb bay doors, exerted a miraculous magnetism which shrunk the fires, gathered them into cylindrical steel containers, and lifted the containers into the bellies of the planes. The containers were stored neatly in racks. The Germans had miraculous devices of their own, which were long steel tubes. They used them to suck more fragments from the crewmen and planes. But there were still a few wounded Americans, though, and some of the bombers were in bad repair. Over France, though, German fighters came up again, made everything and everybody as good as new."

Vonnegut's point, I suppose, is that the morality of actions is inverted when you view them in reverse. What is there about this passage which strikes us as surreal and fantastic? What is there that immediately tells us that the events shown in the film could never happen in reality? It is not so much that the planes appear to fly backwards and the bombs appear to fall upwards. After all, given a little ingenuity and a sufficiently good pilot, it is probably possible to fly a plane backwards. Likewise, if we were to throw a bomb up in the air with just the right velocity we could, in principle, fix it so that the velocity of the bomb matched that of a passing bomber when their paths intersected. Certainly, if you had never seen a plane before it would not be obvious which way around it was supposed to fly.

However, certain events are depicted in the film, "miraculous" events in Vonnegut's words, which would immediately strike us as the wrong way around even if we had never seen them before.

For instance, the film might show thousands of white hot bits of shrapnel approach each other from all directions at great velocity, compressing an explosive gas in the process, which slows them down such that when they meet they fit together exactly to form a metal cylinder enclosing the gases and moving upwards at great velocity. What strikes us as completely implausible about this event is the spontaneous transition from the disordered motion of the gases and metal fragments to the ordered upward motion of the bomb.

Properties of Entropy

Entropy, as we have defined it, has some dependence on the resolution δE to which the energy of macrostates is measured. Recall that $\Omega(E)$ is the number of accessible microstates with energy in the range E to $E + \delta E$. Suppose that we choose a new resolution $\delta^* E$ and define a new density of states $\Omega^* E$

which is the number of states with energy in the range E to $E + \delta^* E$. It can easily be seen that,

$$\Omega^*(E) = \frac{\delta^* E}{\delta E}\,\Omega(E).$$

It follows that the new entropy $S^* = k \ln \Omega^*$ is related to the previous entropy $S^* = k \ln \Omega$ via

$$S^* = S + k \ln \frac{\delta^* E}{\delta E}.$$

Now, our usual estimate that $\Omega \sim E^f$ gives $S \sim kf$, where f is the number of degrees of freedom. It follows that even if $\delta^* E$ were to differ from δE by of order f (*i.e.*, twenty four orders of magnitude), which is virtually inconceivable, the second term on the right-hand side of the above equation is still only of order $k \ln f$, which is utterly negligible compared to kf. It follows that,

$$S^* = S$$

to an excellent approximation, so our definition of entropy is completely insensitive to the resolution to which we measure energy (or any other macroscopic parameter).

Note that, like the temperature, the entropy of a macrostate is only well-defined if the macrostate is in equilibrium.

The crucial point is that it only makes sense to talk about the number of accessible states if the systems in the ensemble are given sufficient time to thoroughly explore all of the possible microstates consistent with the known macroscopic constraints.

In other words, we can only be sure that a given microstate is inaccessible when the systems in the ensemble have had ample opportunity to move into it, and yet have not done so. Note that for an equilibrium state, the entropy is just as well-defined as more familiar quantities such as the temperature and the mean pressure.

Consider, again, two systems A and A' which are in thermal contact but can do no work on one another. Let E and E' be the energies of the two systems, and $\Omega'(E)$ and $\Omega'(E')$ the respective densities of states. Furthermore, let $E^{(0)}$ be the conserved energy of the system as a whole and the $\Omega^{(0)}$ corresponding density of states. We have from Equation that

$$\Omega^{(0)}(E) = \Omega(E)\,\Omega'(E'),$$

where $E' = E^{(0)} - E$. In other words, the number of states accessible to the whole system is the product of the numbers of states accessible to each subsystem, since every microstate of A can be combined with every microstate of A' to form a distinct microstate of the whole system that in equilibrium the mean energy of A takes the value $\bar{E} = \tilde{E}$ for which $\Omega^{(0)}(E)$ is maximum, and the temperatures of A and A' are equal.

The distribution of E around the mean value is of order

$$\Delta^* E = \tilde{E}/\sqrt{f},$$

where f is the number of degrees of freedom. It follows that the total number of accessible microstates is approximately the number of states which lie within $\Delta^* E$ of $\tilde{E}$. Thus,

$$\Omega_{tot}^{(0)} \simeq \frac{\Omega^{(0)}(\tilde{E})}{\delta E} \Delta^* E.$$

The entropy of the whole system is given by

$$S^{(0)} = k \ln \Omega_{tot}^{(0)} = k \ln \Omega^{(0)}(\tilde{E}) + k \ln \frac{\Delta^* E}{\delta E}.$$

According to our usual estimate, $\Omega \sim E^f$, the first term on the right-hand side is of order kf whereas the second term is of order $k \ln(\tilde{E}/\sqrt{f}\,\delta E)$. Any reasonable choice for the energy subdivision δE should be greater than $\tilde{E}/f$, otherwise there would be less than one microstate per subdivision. It follows that the second term is less than or of order $k \ln f$, which is utterly negligible compared to kf.

Thus,

$$S^{(0)} = k \ln \Omega^{(0)}(\tilde{E}) = k\ln[\Omega(\tilde{E})\Omega(\tilde{E}')] = k \ln \Omega(\tilde{E}) + k \ln \Omega'(\tilde{E}')$$

to an excellent approximation, giving,

$$S^{(0)} = S(\tilde{E}) + S'(\tilde{E}').$$

It can be seen that the probability distribution for $\Omega^{(0)}(E)$is so strongly peaked around its maximum value that, for the purpose of calculating the entropy, the total number of states is equal to the maximum number of states [*i.e.*, $\Omega_{tot}^{(0)} \sim \Omega^{(0)}(\tilde{E})$]. One consequence of this is that the entropy has the simple additive property Equation Thus, the total entropy of two thermally interacting systems in equilibrium is the sum of the entropies of each system in isolation.

Uses of Entropy

We have defined a new function called entropy, denoted S, which parameterizes the amount of disorder in a macroscopic system. The entropy of an equilibrium macrostate is related to the number of accessible microstates Ω via,

$$S = k \ln \Omega.$$

On a macroscopic level, the increase in entropy due to a quasi-static change in which an infinitesimal amount of heat $đQ$ is absorbed by the system is given by,

$$dS = \frac{đQ}{T},$$

where T is the absolute temperature of the system. The second law of thermodynamics states that the entropy of an isolated system can never spontaneously decrease. Let us now briefly examine some consequences of these results.

Consider two bodies, A and A', which are in thermal contact but can do no work on one another. We know what is supposed to happen here. Heat flows from the hotter to the colder of the two bodies until their temperatures are the same. Consider a quasi-static exchange of heat between the two bodies. According to the first law of thermodynamics, if an infinitesimal amount of heat $đQ$ is absorbed by A then infinitesimal heat $đQ' = -đQ$ is absorbed by A'. The increase in the entropy of system A is $dS = đQ/T$ and the corresponding increase in the entropy of A' is $dS = đQ/T'$. Here, T and T' are the temperatures of the two systems, respectively.

Note that $đQ$ is assumed to the sufficiently small that the heat transfer does not substantially modify the temperatures of either system. The change in entropy of the whole system is,

$$dS^{(0)} = dS + dS' = \left(\frac{1}{T} - \frac{1}{T'}\right) đQ,$$

This change must be positive or zero, according to the second law of thermodynamics, so $dS^{(0)} \geq 0$. It follows that $đQ$ is positive (*i.e.*, heat flows from A' to A) when $T' > T$, and vice versa. The spontaneous flow of heat only ceases when $T = T'$. Thus, the direction of spontaneous heat flow is a consequence of the second law of thermodynamics.

Note that the spontaneous flow of heat between bodies at different temperatures is always an irreversible process which increases the entropy, or disorder, of the Universe.

Consider, now, the slightly more complicated situation in which the two systems can exchange heat and also do work on one another via a movable partition. Suppose that the total volume is invariant, so that

$$V^{(0)} = V + V' \text{ constant,}$$

where V and V' are the volumes of A and A', respectively. Consider a quasi-static change in which system A absorbs an infinitesimal amount of heat $đQ$ and its volume simul-taneously increases by an infinitesimal amount dV. The infinitesimal amount of work done by system A is $đW = \bar{p}dV$, where $\bar{p}$ is the mean pressure of A. According to the first law of thermodynamics,

$$đQ = dE + đW = dE + \bar{p}dV,$$

where dE is the change in the internal energy of A. Since $dS = đQ/T$, the increase in entropy of system A is written

$$dS = \frac{dE + \bar{p}dV}{T}.$$

Likewise, the increase in entropy of system A' is given by

$$dS' = \frac{dE' + \bar{p}' dV'}{T'}.$$

According to Equation,

$$\frac{1}{T} = \left(\frac{\partial S}{\partial E}\right)_V,$$

$$\frac{\bar{p}}{T} = \left(\frac{\partial S}{\partial V}\right)_E,$$

where the subscripts are to remind us what is held constant in the partial derivatives. We can write a similar pair of equations for the system A'. The overall system is assumed to be isolated, so conservation of energy gives $dE + dE' = 0$. Furthermore, Equation implies that $dV + dV' = 0$. It follows that the total change in entropy is given by,

$$dS^{(0)} = dS + dS' = \left(\frac{1}{T} - \frac{1}{T'}\right)dE + \left(\frac{\bar{p}}{T} - \frac{\bar{p}'}{T'}\right)dV.$$

The equilibrium state is the most probable state. According to statistical mechanics, this is equivalent to the state with the largest number of accessible microstates. Finally, Equation implies that this is the maximum entropy state. The system can never spontaneously leave a maximum entropy state, since this would imply a spontaneous reduction in entropy, which is forbidden by the second law of thermodynamics.

A maximum or minimum entropy state must satisfy $dS^{(0)} = 0$ for arbitrary small variations of the energy and external parameters.

It follows from Equation that,

$$T = T'$$

$$\bar{p} = \bar{p}',$$

for such a state. This corresponds to a maximum entropy state (*i.e.*, an equilibrium state) provided

$$\left(\frac{\partial^2 S}{\partial E^2}\right)_V < 0, \quad \left(\frac{\partial^2 S}{\partial V^2}\right)_E < 0,$$

with a similar pair of inequalities for system $A¢$.

The usual estimate $\Omega \propto E^f V^f$, giving

$$S = kf \ln E + kf \ln V + ...,$$

ensures that the above inequalities are satisfied in conventional macroscopic systems. In the maximum entropy state the systems A and A' have equal temperatures (*i.e.*, they are in thermal equilibrium) and equal pressures (*i.e.*, they are in mechanical equilibrium). The second law of thermodynamics implies that the two interacting systems will evolve towards this state, and will then remain in it indefinitely (if left undisturbed).

ENTROPY AND QUANTUM MECHANICS

The entropy of a system is defined in terms of the number Ω of accessible microstates consistent with an overall energy in the range E to $E + \delta E$via,

$$S = k \ln \Omega.$$

We have already demonstrated that this definition is utterly insensitive to the resolution δE to which the macroscopic energy is measured. In classical mechanics, if a system possesses f degrees of freedom then phase-space is conventionally subdivided into cells of arbitrarily chosen volume h_0^f. The number of accessible microstates is equivalent to the number of these cells in the volume of phase-space consistent with an overall energy of the system lying in the range E to $E + \delta E$. Thus,

$$\Omega = \frac{1}{h_0{}^f} \int \cdots \int dq_1 \cdots dq_f dp_1 \cdots dp_f,$$

giving,

$$S = k \ln\left(\int \cdots \int dq_1 \cdots dq_f dp_1 \cdots dp_f \right) - kf \ln h_0.$$

Thus, in classical mechanics the entropy is undetermined to an arbitrary additive constant which depends on the size of the cells in phase-space. In fact, S increases as the cell size decreases. The second law of thermodynamics is only concerned with changes in entropy, and is, therefore, unaffected by an additive constant.

Likewise, macroscopic thermodynamical quantities, such as the temperature and pressure, which can be expressed as partial derivatives of the entropy with respect to various macroscopic parameters are unaffected by such a constant. So, in classical mechanics the entropy is rather like a gravitational potential: it is undetermined to an additive constant, but this does not affect any physical laws.

The non-unique value of the entropy comes about because there is no limit to the precision to which the state of a classical system can be specified.

In other words, the cell size h_0 can be made arbitrarily small, which corresponds to specifying the particle coordinates and momenta to arbitrary accuracy. However, in quantum mechanics the uncertainty principle sets a definite limit to how accurately the particle coordinates and momenta can be specified. In general,

$$\delta q_i \; \delta p_i \geq h_i$$

where p_i is the momentum conjugate to the generalized coordinate q_i, and δq_i, δp_i are the uncertainties in these quantities, respectively. In fact, in quantum mechanics the number of accessible quantum states with the overall energy in the range E to $E + \delta E$ is completely determined.

This implies that, in reality, the entropy of a system has a unique and unambiguous value. Quantum mechanics can often be "mocked up" in classical mechanics by setting the cell size in phase-space equal to Planck's constant, so that $h_0 = h$.

This automatically enforces the most restrictive form of the uncertainty principle, $\delta q_i \delta p_i = h$. In many systems, the substitution $h_0 \to h$ in Equation gives the same, unique value for as that obtained from a full quantum mechanical calculation. Consider a simple quantum mechanical system consisting of N non-interacting spinless particles of mass m confined in a cubic box of dimension L. The energy levels of the th particle are given by,

$$e_i = \frac{\hbar^2 \pi^2}{2mL^2}\left(n_{il}^2 + n_{i2}^2 + n_{i3}^2\right),$$

where n_{i1}, n_{i2}, and n_{i3} are three (positive) quantum numbers. The overall energy of the system is the sum of the energies of the individual particles, so that for a general state τ,

$$E_r = \sum_{i=1}^{N} e_i.$$

The overall state of the system is completely specified by $3N$ quantum numbers, so the number of degrees of freedom is $f = 3N$. The classical limit corresponds to the situation where all of the quantum numbers are much greater than unity. In this limit, the number of accessible states varies with energy according to our usual estimate $\Omega \propto E^f$.

The lowest possible energy state of the system, the so-called ground-state, corresponds to the situation where all quantum numbers take their lowest possible value, unity. Thus, the ground-state energy E_0 is given by

$$E_0 = \frac{f \hbar^2 \pi^2}{2mL^2}.$$

There is only one accessible microstate at the ground-state energy (*i.e.*, that where all quantum numbers are unity), so by our usual definition of entropy

$$S(E_0) = k \ln 1 = 0.$$

In other words, there is no disorder in the system when all the particles are in their ground-states. Clearly, as the energy approaches the ground-state energy, the number of accessible states becomes far less than the usual classical estimate E^f. This is true for all quantum mechanical systems. In general, the number of microstates varies roughly like

$$\Omega(E) \sim 1 + C(E - E_0)^f,$$

where C is a positive constant. According to Equation the temperature varies approximately like

$$T \sim \frac{E - E_0}{kf},$$

provided $\Omega >> 1$. Thus, as the absolute temperature of a system approaches zero, the internal energy approaches a limiting value E_0(the quantum mechanical ground-state energy), and the entropy approaches the limiting value zero. This proposition is known as the third law of thermodynamics.

At low temperatures, great care must be taken to ensure that equilibrium thermodynamical arguments are applicable, since the rate of attaining equilibrium may be very slow. Another difficulty arises when dealing with a system in which the atoms possess nuclear spins.

Typically, when such a system is brought to a very low temperature the entropy associated with the degrees of freedom not involving nuclear spins becomes negligible. Nevertheless, the number of microstates Ω_s corresponding to the possible nuclear spin orientations may be very large. Indeed, it may be just as large as the number of states at room temperature.

The reason for this is that nuclear magnetic moments are extremely small, and, therefore, have extremely weak mutual interactions. Thus, it only takes a tiny amount of heat energy in the system to completely randomize the spin orientations. Typically, a temperature as small as 10^{-3} degrees kelvin above absolute zero is sufficient to randomize the spins.

Suppose that the system consists of N atoms of spin 1/2. Each spin can have two possible orientations. If there is enough residual heat energy in the system to randomize the spins then each orientation is equally likely. If follows that there are $\Omega_s = 2^N$ accessible spin states.

The entropy associated with these states is $S_0 = k \ln \Omega_s = v R \ln 2$. Below some critical temperature, T_0, the interaction between the nuclear spins becomes significant, and the system settles down in some unique quantum mechanical ground-state (*e.g.*, with all spins aligned). In this situation, $S \to 0$,in accordance with the third law of thermodynamics. However, for temperatures which are small, but not small enough to "freeze out" the nuclear spin degrees of freedom, the entropy approaches a limiting value S_0 which depends only on the kinds of atomic nuclei in the system.

This limiting value is independent of the spatial arrangement of the atoms, or the interactions between them. Thus, for most practical purposes the third law of thermodynamics can be written,

$$\text{as } T \to 0_+, S \to S_0,$$

where 0_+ denotes a temperature which is very close to absolute zero, but still much larger than. This modification of the third law is useful because it can be applied at temperatures which are not prohibitively low.

STATISTICAL THERMODYNAMICS AND TEMPERATURE

We deal with the simplest possible model of gas molecules in a closed container. All molecules have the same mass. The walls of the container are rigid and therefore do nothing to change the energy of any impacting molecule reflected into a different direction.

The walls, however, are not precisely smooth but with enough irregularities that the molecule is reflected randomly and not specularly. The molecules are modelled as a massive point with no internal structure and thus not only is momentum conserved in a collision between them but so is energy; the scattering is elastic. In setting up the equations for the exchange of energy and momentum between two molecules the reader will find it best to work in a centre of mass coordinate system.

There will be a scattering distribution function that will give the probability of turning one of the molecules into a specified new direction after scattering. It is not necessary to know this function for the equilibrium case but only to invoke the same axioms that support the Third Law of Thermodynamics, that is, the non-cyclic nature of equilibrium already used in our discussion of Onsager's equations.

Thus our model is that of an ideal gas. We can arrive at the result of an equilibrium distribution as follows. Molecules will gain energy only as a result of an unusual sequence of collisions imparting energy in the same direction. There will be very few molecules with large momentum. If the overall system has no momentum most molecules must have small momentum. Not surprisingly therefore, the distribution is a multi-dimensional Gaussian, the well-known bell curve, symmetric in positive and negative directions of momentum. That is the probability distribution in momentum space is,

$$p(P) \sim^{-\frac{\frac{1}{2}mc^2}{hT}}$$

where $p = mv = m(v_x, v_y, v_z)$ and speed is given by,

$$c = \left|\sqrt{v_x^2 + v_y^2 + v_z^2}\right|.$$

[c is a positive scalar speed but the v_x, etc., are the signed vector elements of the velocity.] Then the molecule's energy is $E = \frac{1}{2}mc^2$ to be compared with the mean energy at the equilibrium temperature. For convenience, put $\varepsilon = E/kT$. Here k is Boltzmann's constant, the universal gas constant per molecule.

Table below compares this distribution with that of photons satisfying Bose-Einstein statistics and their counter-part, Fermi-Dirac statistics.

Table. Momentum Probabilities by Statistics

Maxwell-Boltzmann (molecule distribution)	**Bose-Einstein (photon distribution)**	**Fermi-Dirac (free conducting electron distribution)**
$\frac{1}{Ae^{\varepsilon}}$	$\frac{1}{Ae^{\varepsilon}-1}$	$\frac{1}{Ae^{\varepsilon}+1}$

Scaling factor A; the additional term in the denominator indicates the consequences of the discrete nature of quantum theory. Bose-Einstein: $A > 0$.. Fermi-Dirac: $-A > 0$.

To find the mean or expected energy we have the mean pressure in our six-sided unit volume box, $\bar{P}$ to be in the ideal gas model as

$$PV = \bar{P} = kT\ .$$

The probable change of momentum on impact in the *x,y,z* planes is 2p(P) and the frequency $\frac{1}{m}p(p)$, The probable pressure is

$$\frac{2}{6m}p(p^2) = \frac{2}{3}p(\frac{1}{2}mc^2)$$

and hence $\bar{\varepsilon} = \frac{3}{2}$, the mean value. I leave to the Reader the more difficult problem of the median value; it could always be done numerically.

THE GAUSSIAN

The Gaussian distribution can be developed from the binomial distribution. Suppose in one dimension there have been N collisions (tending to infinity) with equal probability of imparting the average momentum change, left and right. Then the probability of seeing an absolute momentum to the left proportional to n is given by the binomial expansion,

$$p_n = \frac{N!}{n!(N-n)!}\left(\frac{1}{2}\right)^N$$

Write this symmetrically around $N/2$ as $m=N/2-n$ and use Stirling's approximation: $m! \approx \sqrt{2\pi m}m^m e^m$ to obtain the momentum distribution as e^{-m^2}.

Normalisation is needed for these distributions to ensure the probabilities sum to 1 representing certainty of some result. To guide my Reader through some standard but less usual integrals, consider a generalisation of the factorial to fractional and negative numbers,

$$m! = \int_0^{\infty} s^m e^{-s} ds\ .$$

Justify this by integrating by parts yielding $m! = m(m-1)!$ and in particular that $1! = 0! = 1$. This last result may not be new to you if you have used it in combinatorial theory. What can you say of –1! and –2!? Putting $s = c^2$ each of our three Cartesian integrals over the Gaussian becomes $(-\frac{1}{2})!$. But our spherically symmetrical argument is readily integrated over the solid angle and thus we have

$$[(-\frac{1}{2})!]^3 = 2\pi(\frac{1}{2})! = \pi(-\frac{1}{2})!$$

We have $\left(-\frac{1}{2}\right)! = \sqrt{\pi}$ and $\left(\frac{1}{2}\right)! = \frac{1}{2}\sqrt{\pi}$ and the normalisation of the momentum probability is $\pi^{-\frac{3}{2}}$ In like manner the integral for the Boltzmann distribution in energy is $\frac{1}{2}\left(\frac{1}{2}\right)!$ and the normalising factor is $4/\sqrt{\pi}$. The Reader may deal with the speed distribution in like fashion. Welcome in disguise to Euler's gamma function. Can you sketch m! in the real plane?

This momentum space has an element of volume $mdv_x mdv_y mdv_z$. Combined with normal space we speak of a six-dimensional phase space. The normalisation of energy in this equation involves the system equilibrium temperature T and Boltzmann's constant k. The term kT can be thought of as $\frac{2}{3}$ the average thermal energy per molecule, as we shall see.

To find the probability as a function of other arguments we can use the connection that $p(p)dp = p(\varepsilon)d\varepsilon = p(c)dc$, noting that spherical symmetry in momentum space gives a $dp = 4\pi c^2 dc$. Then within a normalisation we have the results given in Table below.

Table. Boltzmann Distributions in Momentum, Speed and Energy

In momentum	In speed	In energy
$p(p) \sim e^{-c^2}$	$p(c) \sim c^2 e^{-c^2}$	$p(\varepsilon) \sim \varepsilon^{\frac{1}{2}} e^{-\varepsilon}$

$$\varepsilon = E/kT = \frac{1}{2}mc^2/kT$$

Speeds normalised to *r.m.s.* values: $\sqrt{<c^2>}$

The Boltzmann distribution has an asymptote towards large energy that falls off as exp(–*E*/*kT*) that is the origin of the Arrhenius term. That is at any energy *E* that represents a barrier in metastable equilibrium the area of the curve to the right shows the proportion of molecules having sufficient energy to cross the barrier. If the temperature is decreased then correspondingly the number of molecules of sufficient energy is decreased and will slow the rate of crossing of the barrier and hence the rate of reaction.

Examples abound; we have already spoken of the effect in a refrigerator slowing down the rate of decomposition of food and another example arises in combustion engineering where the fuel reaction can be accelerated as the temperature rises with particular interest to problems of 'knock' and auto-ignition. Given the molecular mass in unified atomic mass units, the reciprocal of Avogadro's number $u = 1/No$ the reader can calculate the *r.m.s.* speed as a function of temperature. If the rest mass of the hydrogen molecule is 2.016 u, what is this speed at 25 °C?

The Gaussian distribution of velocities in equilibrium can be derived directly using the conservation of momentum, conservation of energy and the principle of detailed balance.

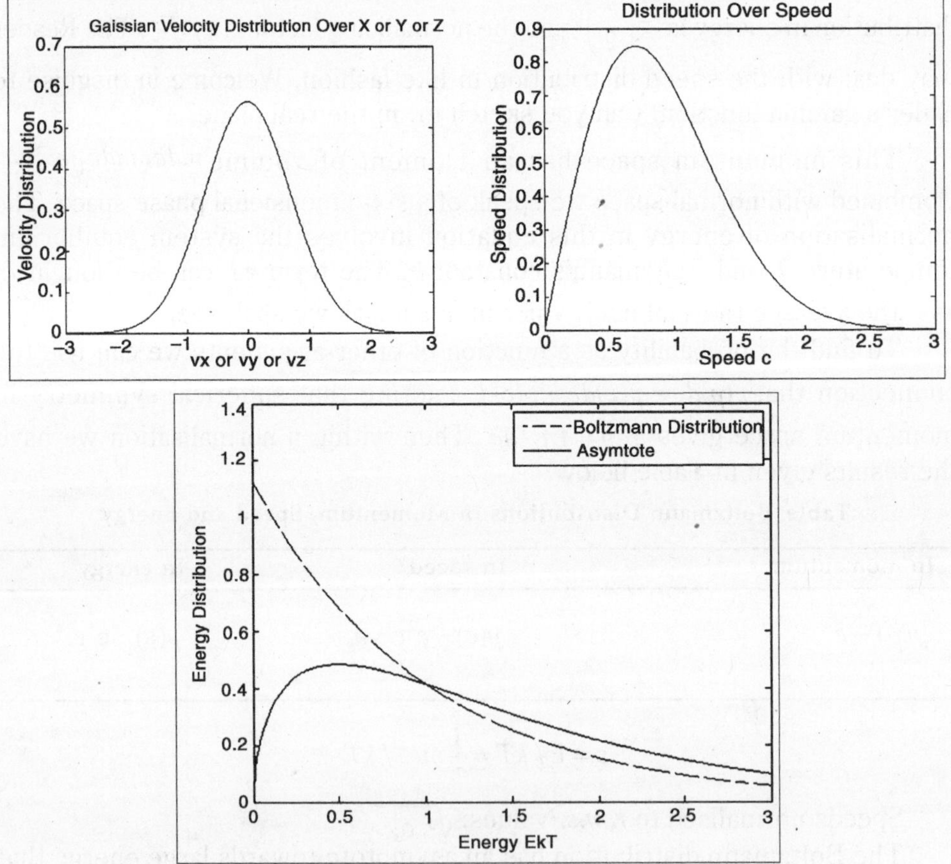

Fig. Boltzmann's Distribution in Momentum, Speed and Energy. Speeds Normalized to r.m.s. Values. My thanks to Jon Heffer who Drew the Figures.

Reader: Is the distribution in velocity as speed or momentum? You may like to see Pippard's account of Josephson circuit noise in terms of statistical thermodynamics in his text book. And rubber, regarded as long polymer chains with many links; would you expect all to be open and the chain extended?

AUTO-IGNITION AND THE ARRHENIUS RATE FACTOR

To illustrate the use of the Arrhenius rate factor in chemical kinetics, as an offshoot from Boltzmann's distribution, we consider the problem of auto-ignition or premature conflagration in fuel-oxidant mixtures, one cause of 'knock' or pinking in carburetted internal combustion engines. Very similar arguments apply to the speed of propagation of fuels, spontaneous combustion of hayricks and to food technology with its rule of thumb, that a 10 kelvin drop in refrigerator temperature halves food spoilage rates.

A mixture of fuel and oxidant may lie quiescent in a metastable state before a spark, say, initiates a runaway chain reaction releasing the energy of combustion. We explain this in terms of a metastable energy barrier due to the need to tear the molecules of fuel and oxygen apart before their radicals combine to release energy. This barrier ΔE_b can be calculated or with more difficulty measured per mole of fuel. We can call the corresponding temperature the activation temperature $T_a = \Delta E_b / R$ in molar units. At the activation temperature many of the molecules will have enough energy to breach the metastable barrier. If the spark has enough energy to initiate many molecules, then we can have a runaway chain reaction. [The length and possible extinction of such chains is studied in genealogy and nuclear reactor theory.]

In the absence of a spark it may happen within the temperature distribution there are enough molecules with above average energy not only to breach the local barrier but also to pass on enough energy to continue the reaction, raising the local temperature until combustion is fully initiated and finally all the fuel is consumed. This obviously depends upon the richness of the fuel, the combustion enthalpy and the molar heat capacity as well as the starting temperature. The insight of Arrhenius was to put forward a rate term in the equation $Ae^{-\frac{1}{\theta}}$ where A is a rate constant to be determined and the non-dimensional temperature is,

$$\theta = \frac{RT}{\Delta E_b}.$$

The exponential comes from the asymptotic form of the Boltzmann distribution at high enough energies being proportional to the proportion of molecules having more than the barrier energy and thus being capable of initiating the burn. The lower limit for the right-hand asymptote to be realistic is the turning point at $\theta = 1$ and we should therefore impose $\theta \leq 1$. We note that at this limit, $\theta = 1$, we have $T = \Delta E_b / R = T_a$, the activation temperature where nearly half the molecules have enough energy to breach the metastable barrier.

In a standard notation we define

$$C = \frac{1}{\theta_0}\frac{\Delta H^0}{c_p} p_0 \frac{A}{MW_{ox}} Y_{fu,0} Y_{ox,0}$$

and a non-dimensional time $\tau = -Ct$.

Then the rate equation is,

$$\frac{d\theta}{d\tau} = -e^{-\frac{1}{\theta}}$$

That is, we measure a non-dimensional ignition time backwards from some yet-to-be-agreed moment when we can say normal combustion that might be set off by a spark is taking place. A very suitable definition of the start of normal ignition and the end of the auto-ignition delay time is when the temperature reaches the activation temperature $\theta = 1$.

Such a definition is widely accepted in a range of studies. It can be given another interpretation in the context of the present model that makes no allowance for fuel consumption or burn-up. That is, after an infinite time the temperature would be infinite but the rate of combustion would reach an asymptotic slope of 1. At $\theta = 1$ however the rate is *1/ e*. The e-folding is typical of time constants in electrical circuits.

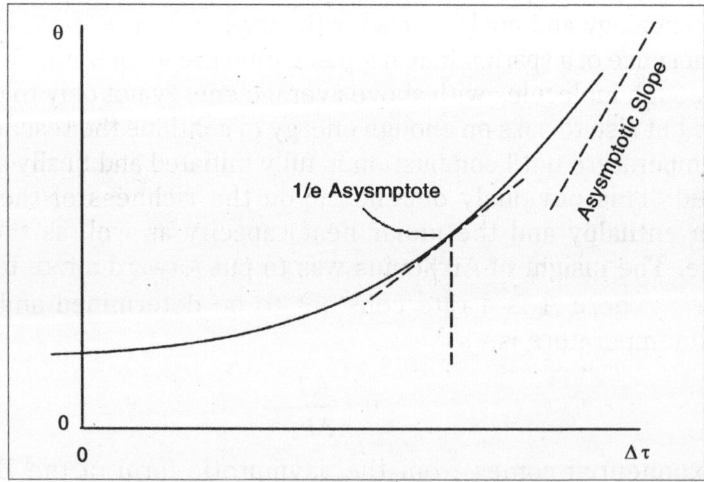

Fig. Auto-Ignition Delay Times Definition

This is not readily integrated analytically perhaps and a standard approximation in the literature is to expand around the activation temperature where $\theta = 1$, retaining the terms up to θ only. The argument is that $\theta : 1$; too high and ignition has surely started while too low and the delay time is so long as to make the result and the model unrealistic. We return to these approximations after deriving the exact solution.

Fortunately the equation can be integrated as a power series and we have,

$$d\tau = -\left[\frac{1}{0!} + \frac{1}{1!\theta} + \frac{1}{2!\theta^2} + \ldots\right] d\theta$$

with $\tau(\theta) = -[\theta + \log\theta - fn(\theta)] + C$.

The reader interested in combustion engineering might take data to see what auto-ignition temperatures might be and auto-ignition times from say 25 °C. A weakness of the model is the disregard of fuel consumption or burn-up. How might this be incorporated?

The figure shows our definition of 'take-off' and the table some non-dimensional auto-ignition delay times. Reader: How many terms before the series converges?

An elaboration is to divide the original rate equation by θ^n where the exponent depends on the order of the chemical rate reaction and is typically $n = \frac{1}{2}$, having the effect of shortening the delay time. Replace A with $\theta^n A_a$ (evaluating A_a at the activation temperature $\theta = 1$) and modify τ accordingly.

$$\frac{d\tau}{d\theta} = \left(\frac{1}{\theta}\right)^{\frac{1}{2}}\left[1 + \left(\frac{1}{\theta}\right) + \frac{1}{2!}\left(\frac{1}{\theta}\right)^2 \ldots\right]$$

and

$$\tau = -\frac{1}{2}\left(\frac{1}{\theta}\right)^{\frac{1}{2}} - \frac{3}{2}\left(\frac{1}{\theta}\right)^{\frac{3}{2}} - \frac{5}{2!2}\left(\frac{1}{\theta}\right)\frac{5}{2} + \ldots c$$

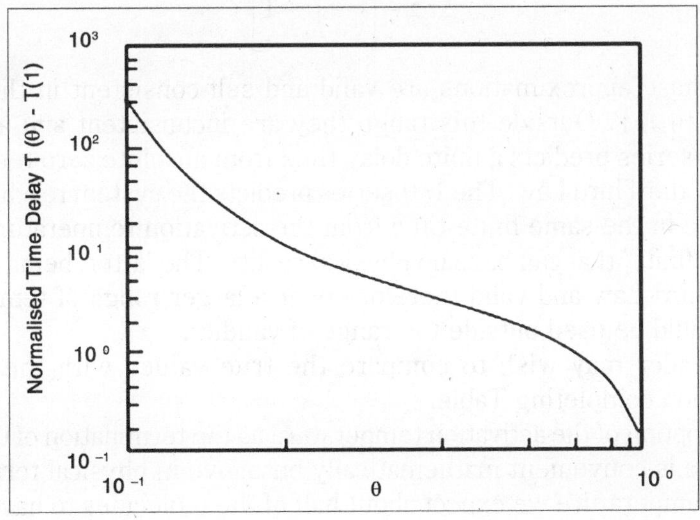

Fig. Auto-Ignition Delay Times Versus Non-Dimensional Temperature

Table. Some Delay Time

Model	$\tau @ \theta = 1$	$\Delta\tau_{delay}(0.5)$	$\Delta\tau_{delay}(0.3)$	$\Delta\tau_{delay}(0.2)$	$\Delta\tau_{delay}(10.1)$
$n = 0$	-0.40_1	2.08_3	3.85_1	11.3_3	$290._4$
$\frac{2}{\theta}$-series approximation	?	?	?	?	?
$n = \frac{1}{3}$	-1.53_7	?	?	?	?

Return to the question of an approximate solution of eqn 10.3. The exact series solution is slowly converging so that the first few terms are insufficient. The power series solution about the e-point, the activation temperature $\theta = 1$ is to be proffered. Even so we have two choices: to expand as a series in $\theta - 1$ or $1/\theta - 1$. The latter is proffered in the literature.

Since,

$$\frac{d(1/\theta)}{d\tau} = -\frac{d\theta}{d\tau} / \theta^2$$

the first derivative $\frac{d\tau}{d\theta}$ or $\frac{d\tau}{d1/\theta}$ evaluated at $\theta = 1 \to e$ or $-e$ while the second derivative also $\to -e$ and $+e$ respectively. Hence the expansion to second order is parabolic and

$$\Delta\tau \cong \frac{e}{2}\left(1 - \theta^2\right)$$

or,

$$\Delta\tau \cong \frac{e}{2}\left[\left(\frac{1}{\theta}\right)^2 - 1\right]$$

Both these approximations are valid and self-consistent in the range of applicability $\theta \to 1$. Outside this range they are inconsistent and anomalous. Thus the θ-series predicts a finite delay time from absolute zero temperature, ruled out by the Third Law. The $1/\theta$-series predicts the system reaching infinite temperature in the same finite time from the activation temperature, a 'finite escape to infinity' that can have no physical reality. The latter, being consistent with the Third Law and valid therefore over a larger range of temperatures. Neither should be used outside the range of validity.

The reader may wish to compare the true values with the two-term approximation completing Table.

Our adoption of the activation temperature as the termination of the ignition waiting time is convenient mathematically but naïve in physical terms. At the activation temperature we expect about half of the molecules to have a kinetic energy higher than the activation energy. We might more realistically say that ignition had commenced when 0.5 per cent of the bonds were broken, a factor of about 4.5 lower in temperature therefore. In addition, combustion is rarely a simple first-order reaction but involves weaker bonds. The table below shows some values for nominal activation temperatures. The values should be interpreted in the light of these remarks.

Table. Bond Enthalpies and Activation Temperatures

Radical	$H-H$	$O-O$	$H-O$	$C-H$	CH_3-H	$OC-O$
Bond enthalpy* $MJ/kmol$	432	493	97	112	430	116
Activation temperature kK	52	56	12	13.5	52	14

THE IDEAL GAS

To do the ideal gas properly we need to know the quantum states of particles in a box. We learned this in PC210 last semester, so we will be able to tackle it properly. However with much less work we can at least discover how the entropy depends on volume at fixed particle number and energy.

Consider an isolated system of N atoms in a box of volume *V*. Imagine the box subdivided into many tiny cells of volume ΔV, so that there are $V/\Delta V$ cells in all (this number should be much greater than *N*). Now each atom can be in any cell, so there are $V/\Delta V$ microstates for each atom, and $(V/\Delta V)^N$ microstates for the gas as a whole. Thus

$$S = Nk_B \ln\left(\frac{V}{\Delta V}\right)$$

and

$$\frac{P}{T} = \left(\frac{\partial S}{\partial V}\right)_{E,N}$$

$$\Rightarrow \quad P = \frac{Nk_B T}{V}$$

$$\Rightarrow \quad PV = Nk_B T$$

A problem with this expression for the entropy is that it depends on the size ΔV of the imaginary cells into which we subdivided our box. This is clearly unsatisfactory (though at least entropy changes are independent of it), but classical physics can't do any better.

STATISTICAL THERMODYNAMICS OF IDEAL GASES

The statistical thermodynamics of ideal gases using a rather ad hoc combination of classical and quantum mechanics. In fact, we employed classical mechanics to deal with the translational degrees of freedom of the constituent particles, and quantum mechanics to deal with the non-translational degrees of freedom.

Let us now discuss ideal gases from a purely quantum mechanical standpoint. It turns out that this approach is necessary to deal with either low temperature or high density gases. Furthermore, it also allows us to investigate completely non-classical "gases," such as photons or the conduction electrons in a metal.

Symmetry Requirements in Quantum Mechanics

Consider a gas consisting of N identical, non-interacting, structureless particles enclosed within a container of volume V. Let Q_i denote collectively all the coordinates of the ith particle: i.e., the three Cartesian coordinates which determine its spatial position, as well as the spin coordinate which determines its internal state.

Let s_i be an index labeling the possible quantum states of the th particle: i.e., each possible value of s_i corresponds to a specification of the three momentum components of the particle, as well as the direction of its spin orientation. According to quantum mechanics, the overall state of the system when the th particle is in state s_i, etc., is completely determined by the complex wave-function

$$\Psi_{s_1,\ldots,s_N}(Q_1, Q_2, \ldots, Q_N).$$

In particular, the probability of an observation of the system finding the ith particle with coordinates in the range Q_i to $Q_i + dQ_i$, etc., is simply

$$|\Psi_{s_1,\ldots,s_N}(Q_1, Q_2, \ldots, Q_N)|^2 \, dQ_1 dQ_2 \ldots dQ_N.$$

One of the fundamental postulates of quantum mechanics is the essential indistinguishability of particles of the same species. What this means, in practice, is that we cannot label particles of the same species: i.e., a proton is just a proton—we cannot meaningfully talk of proton number 1 and proton number 2, etc. Note that no such constraint arises in classical mechanics. Thus, in classical mechanics particles of the same species are regarded as being distinguishable, and can, therefore, be labelled. Of course, the quantum mechanical approach is the correct one.

Suppose that we interchange the ith and jth particles: i.e.,

$$Q_i \leftrightarrow Q_j$$

$$s_i \leftrightarrow s_j.$$

If the particles are truly indistinguishable then nothing has changed: i.e., we have a particle in quantum state s and a particle in quantum state s_j both before and after the particles are swapped. Thus, the probability of observing the system in a given state also cannot have changed: i.e.,

$$|\Psi(Q_i \cdots Q_j \cdots)|^2 = |\Psi(\cdots Q_j \cdots Q_i \cdots)|^2 .$$

Here, we have omitted the subscripts $s_1, \ldots, s_N$ for the sake of clarity. Note that we cannot conclude that the wave-function Ψ is unaffected when the particles are swapped, because Ψ cannot be observed experimentally. Only the probability density $|\Psi|^2$ is observable. Equation implies that

$$\Psi(\cdots Q_i \cdots Q_j \cdots) = A\Psi(\cdots Q_i \cdots Q_j \cdots),$$

where A is a complex constant of modulus unity: i.e., $|A|^2 = 1$.

Suppose that we interchange the ith and jth particles a second time. Swapping the ith and jth particles twice leaves the system completely unchanged: i.e., it is equivalent to doing nothing to the system. Thus, the wave-functions before and after this process must be identical. It follows from previous equation that

$$A^2 = 1$$

Of course, the only solutions to the above equation are $A = \pm 1$.

We conclude, from the above discussion, that the wave-function Ψ is either completely symmetric under the interchange of particles, or it is completely anti-symmetric. In other words, either

$$\Psi(\cdots Q_i \cdots Q_j \cdots) = +\Psi(\cdots Q_j \cdots Q_i \cdots),$$

or

$$\Psi(\cdots Q_i \cdots Q_j \cdots) = -\Psi(\cdots Q_j \cdots Q_i \cdots).$$

In 1940 the Nobel prize winning physicist Wolfgang Pauli demonstrated, via arguments involving relativistic invariance, that the wave-function associated with a collection of identical integer-spin (i.e., spin 0, 1, 2, etc.) particles satisfies previous equation whereas the wave-function associated with a collection of identical half-integer-spin (i.e., spin 1/2,3/2,5/2, etc.) particles satisfies previous equation. The former type of particles are known as bosons [after the Indian physicist S. N. Bose, who first put forward earlier equation on empirical grounds]. The latter type of particles are called fermions (after the Italian physicists Enrico Fermi, who first studied the properties of fermion gases). Common examples of bosons are photons and He^4 atoms. Common examples of fermions are protons, neutrons, and electrons.

Consider a gas made up of identical bosons. Equation implies that the interchange of any two particles does not lead to a new state of the system. Bosons must, therefore, be considered as genuinely indistinguishable when enumerating the different possible states of the gas. Note that previous equation imposes no restriction on how many particles can occupy a given single-particle quantum state s. Consider a gas made up of identical fermions. Equation implies that the interchange of any two particles does not lead to a new physical state of the system (since $|\Psi|^2$ is invariant). Hence, fermions must also be considered genuinely indistinguishable when enumerating the different possible states of the gas.

Consider the special case where particles i and j lie in the same quantum state. In this case, the act of swapping the two particles is equivalent to leaving the system unchanged, so

$$\Psi(\cdots Q_i \cdots Q_j \cdots) = \Psi(\cdots Q_j \cdots Q_i \cdots).$$

However, previous equation is also applicable, since the two particles are fermions. The only way in which these equations and can be reconciled is if

$$\Psi = 0$$

wherever particles i and j lie in the same quantum state. This is another way of saying that it is impossible for any two particles in a gas of fermions to lie in the same single-particle quantum state. This proposition is known as the Pauli exclusion principle, since it was first proposed by W. Pauli in 1924 on empirical grounds.

Consider, for the sake of comparison, a gas made up of identical classical particles. In this case, the particles must be considered distinguishable when enumerating the different possible states of the gas. Furthermore, there are no constraints on how many particles can occupy a given quantum state.

According to the above discussion, there are three different sets of rules which can be used to enumerate the states of a gas made up of identical particles. For a boson gas, the particles must be treated as being indistinguishable, and there is no limit to how many particles can occupy a given quantum state. This set of rules is called Bose-Einstein statistics, after S.N. Bose and A. Einstein, who first developed them. For a fermion gas, the particles must be treated as being indistinguishable, and there can never be more than one particle in any given quantum state. This set of rules is called Fermi-Dirac statistics, after E. Fermi and P.A.M. Dirac, who first developed them. Finally, for a classical gas, the particles must be treated as being distinguishable, and there is no limit to how many particles can occupy a given quantum state.

This set of rules is called Maxwell-Boltzmann statistics, after J.C. Maxwell and L. Boltzmann, who first developed them.

An Illustrative Example

Consider a very simple gas made up of two identical particles. Suppose that each particle can be in one of three possible quantum states, $s = 1, 2, 3$. Let us enumerate the possible states of the whole gas according to Maxwell-

Boltzmann, Bose-Einstein, and Fermi-Dirac statistics, respectively. For the case of Maxwell-Boltzmann (MB) statistics, the two particles are considered to be distinguishable. Let us denote them *A* and *B*. Furthermore, any number of particles can occupy the same quantum state.

Table. Two Particles Distributed Amongst three States According to Maxwell-Boltzmann Statistics.

1	2	3
AB	...	...
...	*AB*	...
...	...	*AB*
A	*B*	...
B	*A*	...
A	...	*B*
B	...	*A*
...	*A*	*B*
...	*B*	*A*

There are Clearly 9 Distinct States: For the case of Bose-Einstein (BE) statistics, the two particles are considered to be indistinguishable.

Let us denote them both as *A*. Furthermore, any number of particles can occupy the same quantum state. The possible different states of the gas are shown in Table.

Table. Two Particles Distributed Amongst three States According to Bose-Einstein Statistics.

1	2	3
AA	...	...
...	*AA*	...
...	...	*AA*
A	*A*	...
A	...	*A*
...	*A*	*A*

There are clearly 6 distinct states: Finally, for the case of Fermi-Dirac (FD) statistics, the two particles are considered to be indistinguishable. Let us again denote them both as *A*. Furthermore, no more than one particle can occupy a given quantum state.

Table.Two Particles Distributed Amongst three States According to Fermi-Dirac Statistics.

1	2	3
A	*A*	...
A	...	*A*
...	*A*	*A*

There are clearly only 3 distinct states: It follows, from the above example, that Fermi-Dirac statistics are more restrictive (i.e., there are less possible states of the system) than Bose-Einstein statistics, which are, in turn, more restrictive than Maxwell-Boltzmann statistics. Let

$$\xi \equiv \frac{\text{probability that the two particles are found in the same state}}{\text{probability that the two particles are found in different state}}.$$

For the case under investigation,

$$\xi_{MB} = 1/2,$$

$$\xi_{BE} = 1,$$

$$\xi_{FD} = 0.$$

We conclude that in Bose-Einstein statistics there is a greater relative tendency for particles to cluster in the same state than in classical statistics.

On the other hand, in Fermi-Dirac statistics there is less tendency for particles to cluster in the same state than in classical statistics.

Formulation of the Statistical Problem

Consider a gas consisting of N identical non-interacting particles occupying volume V and in thermal equilibrium at temperature T. Let us label the possible quantum states of a single particle by τ (or s). Let the energy of a particle in state τ be denoted $\in_\tau$.

Let the number of particles in state r be written n_τ. Finally, let us label the possible quantum states of the whole gas by R.

The particles are assumed to be non-interacting, so the total energy of the gas in state R, where there are n_τ particles in quantum state τ, etc., is simply

$$E_R = \sum_\tau n_\tau \in_\tau,$$

where the sum extends over all possible quantum states τ. Furthermore, since the total number of particles in the gas is known to be N, we must have

$$N = \sum_\tau n_\tau.$$

In order to calculate the thermodynamic properties of the gas (i.e., its internal energy or its entropy), it is necessary to calculate its partition function,

$$N = \sum_R e^{-\beta E_R} = \sum_R e^{-\beta(n_1 c_1 + n_2 c_2 + \cdots)}.$$

Here, the sum is over all possible states R of the whole gas: i.e., over all the various possible values of the numbers $n_1, n_2, \cdots$.

Now, $\exp[-\beta(n_1 \in_1 + n_2 \in_2 + \cdots)]$ is the relative probability of finding the gas in a particular state in which there are n_1 particles in state 1, n_2 particles in state 2, etc. Thus, the mean number of particles in quantum state s can be written

$$\bar{n}_s = \frac{\Sigma_R n_s \exp[-\beta(n_1 \epsilon_1 + n_2 \epsilon_2 + \cdots)]}{\Sigma_R \exp[-\beta(n_1 \epsilon_1 + n_2 \epsilon_2 + \cdots)]}.$$

A comparison of Eqs. yields the result

$$\bar{n}_s = \frac{1}{\beta} \frac{\partial \ln Z}{\partial \epsilon_s}.$$

Here, $\beta \equiv 1/kT$.

FERMI-DIRAC STATISTICS

Let us, first of all, consider Fermi-Dirac statistics. According to prvious equation the average number of particles in quantum state S can be written

$$\bar{n}_s = \frac{\sum_{n_s} n_s e^{-\beta n_s c_s} \sum^{(s)}_{n_1, n_2, \ldots} e^{-\beta(n_1 c_1 + n_2 c_2 + \cdots)}}{\sum_{n_s} e^{-\beta n_s c_s} \sum^{(s)}_{n_1, n_2, \ldots} e^{-\beta(n_1 c_1 + n_2 c_2 + \cdots)}}.$$

Here, we have rearranged the order of summation, using the multiplicative properties of the exponential function. Note that the first sums in the numerator and denominator only involve n_s, whereas the last sums omit the particular state s from consideration (this is indicated by the superscript s on the summation symbol). Of course, the sums in the above expression range over all values of the numbers $n_1, n_2, \ldots$ such that $n_\tau = 0$ and 1 for each τ, subject to the overall constraint that

$$\sum_\tau n_\tau = N.$$

Let us introduce the function

$$Z_s(N) = \sum^{(s)}_{n_1, n_2, \cdots} e^{-\beta(n_1 c_1 + n_2 c_2 + \cdots)},$$

which is defined as the partition function for N particles distributed over all quantum states, excluding state s, according to Fermi-Dirac statistics. By explicitly performing the sum over $n_s = 0$ and 1, the expression reduces to

$$\bar{n}_s = \frac{0 + e^{-\beta c_s} Z_s(N-1)}{Z_s(N) + e^{-\beta c_s} Z_s(N-1)},$$

which yields

$$\bar{n}_s = \frac{1}{[Z_s(N)/Z_s(N-1)]\, e^{-\beta c_s} + 1}$$

In order to make further progress, we must somehow relate $Z_s(N-1)$ to $Z_s(N)$. Suppose that $\Delta N << N$. It follows that $\ln Z_s(N - \Delta N)$ can be Taylor expanded to give

$$\ln Z_s(N-\Delta N) \simeq \ln Z_s(N) - \frac{\partial \ln Z_s}{\partial N}\Delta N = \ln Z_s(N) - \alpha_s\,\Delta N,$$

where

$$\alpha_s \equiv \frac{\partial \ln Z_s}{\partial N}.$$

As always, we Taylor expand the slowly varying function ln $Z_s(N)$, rather than the rapidly varying function $Z_s(N)$, because the radius of convergence of the latter Taylor series is too small for the series to be of any practical use. Equation can be rearranged to give

$$Z_s(N-\Delta N) = Z_s(N)e^{-\alpha_s \Delta N}$$

Now, since $Z_s(N)$ is a sum over very many different quantum states, we would not expect the logarithm of this function to be sensitive to which particular state s is excluded from consideration. Let us, therefore, introduce the approximation that α_s is independent of s, so that we can write

$$\alpha_s \simeq \alpha$$

for all s. It follows that the derivative can be expressed approximately in terms of the derivative of the full partition function $Z(N)$ (in which the N particles are distributed over all quantum states). In fact,

$$\alpha \simeq \frac{\partial \ln Z}{\partial N}.$$

Making use of above equation with $\Delta N = 1$, plus the approximation the expression reduces to

$$\bar{n}_s = \frac{1}{e^{\alpha+\beta\,c_s}+1}.$$

This is called the Fermi-Dirac distribution. The parameter α is determined by the constraint that $\Sigma_r \bar{n}_r = N$: i.e.,

$$\sum_{\tau} \frac{1}{e^{\alpha+\beta\,c_r}+1} = N.$$

Note that $\bar{n}_s \to 0$ if ϵ_s becomes sufficiently large. On the other hand, since the denominator in previous equation can never become less than unity, no matter how small ϵ_s becomes, it follows that $\bar{n}_s \leq 1$. Thus,

$$0 \leq \bar{n}_s \leq 1,$$

in accordance with the Pauli exclusion principle.

Equations can be integrated to give

$$\ln Z = \alpha N + \sum_\tau \ln(1 + e^{-\alpha - \beta c_\tau}),$$

where use has been made of previous equation.

PHOTON STATISTICS

Up to now, we have assumed that the number of particles N contained in a given system is a fixed number. This is a reasonable assumption if the particles possess non-zero mass, since we are not generally considering relativistic systems in this course.

However, this assumption breaks down for the case of photons, which are zero-mass bosons. In fact, photons enclosed in a container of volume V, maintained at temperature T, can readily be absorbed or emitted by the walls. Thus, for the special case of a gas of photons there is no requirement which limits the total number of particles.

It follows, from the above discussion, that photons obey a simplified form of Bose-Einstein statistics in which there is an unspecified total number of particles. This type of statistics is called photon statistics.

Consider the expression. For the case of photons, the numbers $n_1, n_2 \ldots$ assume all values $n_\tau = 0, 1, 2, \ldots$ for each τ, without any further restriction. It follows that the sums $\Sigma^{(s)}$ in the numerator and denominator are identical and, therefore, cancel. Hence, previous equation reduces to

$$\bar{n}_s = \frac{\Sigma_{n_s} n_s e^{-\beta n_s c_s}}{\Sigma_{n_s} e^{-\beta n_s c_s}}.$$

However, the above expression can be rewritten

$$\bar{n}_s = -\frac{1}{\beta}\frac{\partial}{\partial \epsilon_s}\left(\ln \sum_{n_s} e^{-\beta n_s c_s}\right).$$

Now, the sum on the right-hand side of the above equation is an infinite geometric series, which can easily be evaluated. In fact,

$$\sum_{n_s=0}^{\infty} e^{-\beta n_s c_s} = 1 + e^{-\beta c_s} + e^{-2\beta c_s} + \cdots = \frac{1}{1 - e^{-\beta c_s}}.$$

Thus, earlier equation gives

$$\bar{n}_s = \frac{1}{\beta}\frac{\partial}{\partial \epsilon_s}\ln(1 - e^{-\beta c_s}) = \frac{e^{-\beta c_s}}{1 - e^{-\beta c_s}},$$

or

$$\bar{n}_s = \frac{1}{e^{\beta c_s} - 1}.$$

This is known as the Planck distribution, after the German physicist Max Planck who first proposed it in 1900 on purely empirical grounds.

Equation can be integrated to give

$$\ln Z = -\sum_{\tau} \ln(1 - e^{-\beta c_r}),$$

where use has been made of previous equation.

BOSE-EINSTEIN STATISTICS

Let us now consider Bose-Einstein statistics. The particles in the system are assumed to be massive, so the total number of particles N is a fixed number.

Consider the expression. For the case of massive bosons, the numbers n_1, n_2,. . . assume all values $n_\tau = 0, 1, 2,...$for each r, subject to the constraint that $\Sigma_\tau n_\tau = N$. Performing explicitly the sum over n_s, this expression reduces to

$$\bar{n}_s = \frac{0 + e^{-\beta c_s} Z_s(N-1) + 2e^{-2\beta c_s} Z_s(N-2) + \cdots}{Z_s(N) + e^{-\beta c_s} Z_s(N-1) + e^{-2\beta c_s} Z_s(N-2) + \cdots},$$

where $Z_s(N)$is the partition function for N particles distributed over all quantum states, excluding state s, according to Bose-Einstein statistics. Using earlier equation, and the approximation the above equation reduces to

$$\bar{n}_s = \frac{\Sigma_s n_s e^{-n_s(\alpha + \beta c_s)}}{\Sigma_s e^{-n_s(\alpha+\beta c_s)}}.$$

Note that this expression is identical to except that $\beta \in_s$ is replaced by $\alpha + \beta \in_s$. Hence, an analogous calculation to that outlined in the previous subsection yields

$$\bar{n}_s = \frac{1}{e^{\alpha + \beta c_s} - 1}.$$

This is called the Bose-Einstein distribution. Note that $\bar{n}_s$ can become very large in this distribution. The parameter α is again determined by the constraint on the total number of particles: i.e.,

$$\sum_{\tau} \frac{1}{e^{\alpha + \beta c_\tau} - 1} = N.$$

Equations can be integrated to give

$$\ln Z = \alpha N - \sum_{\tau} \ln(1 - e^{\alpha - \beta c_\tau}),$$

where use has been made of previous equation.

Note that photon statistics correspond to the special case of Bose-Einstein statistics in which the parameter α takes the value zero, and the constraint does not apply.

MAXWELL-BOLTZMANN STATISTICS

For the purpose of comparison, it is instructive to consider the purely classical case of Maxwell-Boltzmann statistics. The partition function is written

$$Z = \sum_{R} e^{-\beta(a_1 c_1 + n_2 c_2 + \cdots)},$$

where the sum is over all distinct states R of the gas, and the particles are treated as distinguishable. For given values of $n_1, n_2 \ldots$ there are

$$\frac{N!}{n_1! n_2! \cdots}$$

possible ways in which N distinguishable particles can be put into individual quantum states such that there are n_1 particles in state 1, n_2 particles in state 2, etc. Each of these possible arrangements corresponds to a distinct state for the whole gas. Hence, previous equation can be written

$$Z = \sum_{n_1! n_2! \ldots} \frac{N!}{n_1! n_2! \cdots} e^{-\beta(n_1 c_1 + n_2 c_2 + \cdots)},$$

where the sum is over all values of $n_\tau = 0, 1, 2, \ldots$ for each r, subject to the constraint that

$$\sum_{\tau} n_\tau = N.$$

Now, earlier equation can be written

$$Z = \sum_{n_1, n_2 \ldots} \frac{N!}{n_1! n_2!} (e^{-\beta c_1})^{n_1} (e^{-\beta c_2})^{n_2} \cdots,$$

which, by virtue of given equation is just the result of expanding a polynomial. In fact,

$$Z = (e^{-\beta c_1} + e^{-\beta c_2} + \cdots)^N,$$

or

$$\ln Z = N \ln\left(\sum_{\tau} e^{-\beta c_\tau}\right).$$

Note that the argument of the logarithm is simply the partition function for a single particle.

Equations can be combined to give

$$\overline{n}_s = N \frac{e^{-\beta c_s}}{\Sigma_r e^{-\beta c_r}}.$$

This is known as the Maxwell-Boltzmann distribution. It is, of course, just the result obtained by applying the Boltzmann distribution to a single particle

QUANTUM STATISTICS IN THE CLASSICAL LIMIT

The preceding analysis regarding the quantum statistics of ideal gases is summarized in the following statements. The mean number of particles occupying quantum state s is given by

$$\bar{n}_s = \frac{1}{e^{\alpha+\beta c_s} \pm 1},$$

where the upper sign corresponds to Fermi-Dirac statistics and the lower sign corresponds to Bose-Einstein statistics. The parameter α is determined via

$$\sum_{\tau} \bar{n}_s = \sum_{\tau} \frac{1}{e^{\alpha+\beta c_s} \pm 1} = N.$$

Finally, the partition function of the gas is given by

$$\ln Z = \alpha N \pm \sum_{\tau} \ln(1 \pm e^{-\alpha-\beta c_r}).$$

Let us investigate the magnitude of α in some important limiting cases. Consider, first of all, the case of a gas at a given temperature when its concentration is made sufficiently low: i.e., when N is made sufficiently small. The relation can only be satisfied if each term in the sum over states is made sufficiently small; i.e., if $\bar{n}_\tau << 1$ or $\exp(\alpha + \beta \in_\tau) >> 1$ for all states τ.

Consider, next, the case of a gas made up of a fixed number of particles when its temperature is made sufficiently large: i.e., when β is made sufficiently small.

In the sum in previous equation the terms of appreciable magnitude are those for which $\beta\in_\tau << \alpha$. Thus, it follows that as $\beta \to 0$ an increasing number of terms with large values of $\in_\tau$ contribute substantially to this sum. In order to prevent the sum from exceeding N, the parameter α must become large enough that each term is made sufficiently small: i.e., it is again necessary that $\bar{n}_\tau << 1$ or $\exp(\alpha + \beta \in_\tau) >>$ for all states τ.

The above discussion suggests that if the concentration of an ideal gas is made sufficiently low, or the temperature is made sufficiently high, then α must become so large that

$$e^{\alpha + \beta c_\tau} >> 1$$

for all τ. Equivalently, this means that the number of particles occupying each quantum state must become so small that

$$\bar{n}_\tau << 1$$

for all τ. It is conventional to refer to the limit of sufficiently low concentration, or sufficiently high temperature, in which Eqs. are satisfied, as the classical limit. According to Eqs. both the Fermi-Dirac and Bose-Einstein distributions reduce to

$$\bar{n}_s = e^{-\alpha-\beta c_s}$$

in the classical limit, whereas the constraint yields

$$\sum_{\tau} e^{-\alpha-\beta c_\tau} = N.$$

The above expressions can be combined to give

$$\bar{n}_s = N \frac{e^{-\beta c_s}}{\sum_{\tau} e^{-\beta c_\tau}}.$$

It follows that in the classical limit of sufficiently low density, or sufficiently high temperature, the quantum distribution functions, whether Fermi-Dirac or Bose-Einstein, reduce to the Maxwell-Boltzmann distribution. It is easily demonstrated that the physical criterion for the validity of the classical approximation is that the mean separation between particles should be much greater than their mean de Broglie wavelengths.

Let us now consider the behaviour of the partition function in the classical limit. We can expand the logarithm to give

$$\ln Z = \alpha N \pm \sum_{\tau} \left(\pm e^{-\alpha-\beta c_\tau}\right) = \alpha N + N.$$

However, according to given equation

$$\alpha = -\ln N + \ln\left(\sum_{\tau} e^{-\beta c_\tau}\right).$$

It follows that

$$\ln Z = -N \ln N + N + N \ln\left(\sum_{\tau} e^{-\beta c_\tau}\right).$$

Note that this does not equal the partition function Z_{MB} computed in earlier equation from Maxwell-Boltzmann statistics: i.e.,

$$\ln Z_{MB} = N \ln\left(\sum_{\tau} e^{-\beta c_\tau}\right).$$

In fact,

$$\ln Z = \ln Z_{MB} - \ln N!,$$

or

$$Z = \frac{Z_{MB}}{N!},$$

where use has been made of Stirling's approximation ($N! \simeq N \ln N - N$), since N is large. Here, the factor $N!$ simply corresponds to the number of different permutations of the N particles: permutations which are physically meaningless when the particles are identical. Recall, that we had to introduce precisely this factor, in an ad hoc fashion in order to avoid the non-physical consequences of the Gibb's paradox. Clearly, there is no Gibb's paradox when an ideal gas is treated properly via quantum mechanics.

In the classical limit, a full quantum mechanical analysis of an ideal gas reproduces the results obtained , except that the arbitrary parameter h_0 is replaced by Planck's constant $h = 6.61 \times 10^{-34}$ Js.

A gas in the classical limit, where the typical de Broglie wavelength of the constituent particles is much smaller than the typical inter-particle spacing, is said to be non-degenerate. In the opposite limit, where the concentration and temperature are such that the typical de Broglie wavelength becomes comparable with the typical inter-particle spacing, and the actual Fermi-Dirac or Bose-Einstein distributions must be employed, the gas is said to be degenerate.

THE PLANCK RADIATION LAW

Let us now consider the application of statistical thermodynamics to electromagnetic radiation. According to Maxwell's theory, an electromagnetic wave is a coupled self-sustaining oscillation of electric and magnetic fields which propagates though a vacuum at the speed of light, $c = 3 \times 10^8$ ms^{-1}. The electric component of the wave can be written

$$E = E_0 \exp[i(k.r - \omega t],$$

where E_0 is a constant, k is the wave-vector which determines the wavelength and direction of propagation of the wave, and ω is the frequency. The dispersion relation

$$\omega = kc$$

ensures that the wave propagates at the speed of light. Note that this dispersion relation is very similar to that of sound waves in solids. Electromagnetic waves always propagate in the direction perpendicular to the coupled electric and magnetic fields (i.e., electromagnetic waves are transverse waves).
This means that $k.E_0 = 0$. Thus, once k is specified, there are only two possible independent directions for the electric field. These correspond to the two independent polarizations of electromagnetic waves.

Consider an enclosure whose walls are maintained at fixed temperature T. What is the nature of the steady-state electromagnetic radiation inside the enclosure? Suppose that the enclosure is a parallelepiped with sides of lengths L_x, L_y, and L_z.

Alternatively, suppose that the radiation field inside the enclosure is periodic in the x–, y–, and z–directions, with periodicity lengths L_x, L_y, and L_z, respectively. As long as the smallest of these lengths, L, say, is much greater than the longest wavelength of interest in the problem, $\lambda = 2\pi/k$, then these assumptions should not significantly affect the nature of the radiation inside the enclosure.

We find, just as in our earlier discussion of sound waves, that the periodicity constraints ensure that there are only a discrete set of allowed wave-vectors (i.e., a discrete set of allowed modes of oscillation of the electromagnetic field inside the enclosure). Let $\rho(k)\, d^3k$ be the number of allowed modes per unit volume with wave-vectors in the range k to $k + dk$. We know, by analogy with previous equation that

$$\rho(k)d^3k = \frac{d^3k}{(2\pi)^3}.$$

The number of modes per unit volume for which the magnitude of the wave-vector lies in the range k to $k + dk$ is just the density of modes, $\rho(k)$, multiplied by the "volume" in k-space of the spherical shell lying between radii k and $k + dk$. Thus,

$$\rho_k(k)dk = \frac{4\pi k^2 dk}{(2\pi)^3} = \frac{k^2}{2\pi^2}dk.$$

Finally, the number of modes per unit volume whose frequencies lie between ω and $\omega + d\omega$ is, by given equation),

$$\sigma(\omega)d\omega = 2\frac{\omega^2 dk}{2\pi^2 c^3}d\omega.$$

Here, the additional factor 2 is to take account of the two independent polarizations of the electromagnetic field for a given wave-vector k.

Let us consider the situation classically. By analogy with sound waves, we can treat each allowable mode of oscillation of the electromagnetic field as an independent harmonic oscillator. According to the equipartition theorem, each mode possesses a mean energy kT in thermal equilibrium at temperature T. In fact, (1/2) kT resides with the oscillating electric field, and another (1/2)kT with the oscillating magnetic field. Thus, the classical energy density of electromagnetic radiation (i.e., the energy per unit volume associated with modes whose frequencies lie in the range ω to $\omega + d\omega$) is

$$\overline{u}(\omega)d\omega = kT\sigma(\omega)d\omega\frac{kT}{\pi^2 c^3}\omega^2 d\omega.$$

This result is known as the Rayleigh-Jeans radiation law, after Lord Rayleigh and James Jeans who first proposed it in the late nineteenth century.

According to Debye theory, the energy density of sound waves in a solid is analogous to the Rayleigh-Jeans law, with one very important difference. In Debye theory there is a cut-off frequency (the Debye frequency) above which no modes exist. This cut-off comes about because of the discrete nature of solids (i.e., because solids are made up of atoms instead of being continuous). It is, of course, impossible to have sound waves whose wavelengths are much less than the inter-atomic spacing.

On the other hand, electromagnetic waves propagate through a vacuum, which possesses no discrete structure. It follows that there is no cut-off frequency for electromagnetic waves, and so the Rayleigh-Jeans law holds for all frequencies. This immediately poses a severe problem. The total classical energy density of electromagnetic radiation is given by

$$U = \int_0^\infty \overline{u}(\omega)d\omega = \frac{kT}{\pi^2 C^3}\int_0^\infty \omega^2 d\omega.$$

This is an integral which obviously does not converge. Thus, according to classical physics, the total energy density of electromagnetic radiation inside an enclosed cavity is infinite! This is clearly an absurd result, and was recognized as such in the latter half of the nineteenth century. In fact, this prediction is known as the ultra-violet catastrophe, because the Rayleigh-Jeans law usually starts to diverge badly from experimental observations (by over-estimating the amount of radiation) in the ultra-violet region of the spectrum.

So, how do we obtain a sensible answer? Well, as usual, quantum mechanics comes to our rescue. According to quantum mechanics, each allowable mode of oscillation of the electromagnetic field corresponds to a photon state with energy and momentum

$$\in = \hbar\omega,$$
$$P = \hbar k,$$

respectively. Incidentally, it follows from previous equation that

$$\in = pc,$$

which implies that photons are massless particles which move at the speed of light. According to the Planck distribution, the mean number of photons occupying a photon state of frequency ω is

$$n(\omega) = \frac{1}{e^{\beta\hbar\omega} - 1}.$$

Hence, the mean energy of such a state is given by

$$\overline{\in}(\omega) = \hbar\omega\, n(\omega) = \frac{\hbar\omega}{e^{\beta\hbar\omega} - 1}.$$

Note that low frequency states (i.e., $\hbar\omega << kT$) behave classically: i.e.,

$$\overline{\in} \simeq kT.$$

On the other hand, high frequency states (i.e., $\hbar\omega >> kT$) are completely "frozen out": i.e.,

$$\bar{\epsilon} << kT.$$

The reason for this is simply that it is very difficult for a thermal fluctuation to create a photon with an energy greatly in excess of kT, since kT is the characteristic energy associated with such fluctuations.

According to the above discussion, the true energy density of electromagnetic radiation inside an enclosed cavity is written

$$\bar{u}d\omega = \epsilon(\omega)\sigma(\omega)d\omega,$$

giving

$$\bar{u}d\omega = \frac{\hbar}{\pi^2 C^3}\frac{\omega^3 d\omega}{\exp(\beta\hbar\omega)-1}.$$

This is famous result is known as the Planck radiation law. The Planck law approximates to the classical Rayleigh-Jeans law for $\hbar\omega << kT$, peaks at about $\hbar\omega \simeq 3kT$, and falls off exponentially for $\hbar\omega >> kT$. The exponential fall off at high frequencies ensures that the total energy density remains finite.

BLACK-BODY RADIATION

Suppose that we were to make a small hole in the wall of our enclosure, and observe the emitted radiation. A small hole is the best approximation in Physics to a black-body, which is defined as an object which absorbs, and, therefore, emits, radiation perfectly at all wavelengths. What is the power radiated by the hole? Well, the power density inside the enclosure can be written

$$u(\omega)d\omega = \hbar\omega\, n(\omega)d\omega,$$

where $n(\omega)$ is the mean number of photons per unit volume whose frequencies lie in the range ω to $\omega + d\omega$. The radiation field inside the enclosure is isotropic (we are assuming that the hole is sufficiently small that it does not distort the field). It follows that the mean number of photons per unit volume whose frequencies lie in the specified range, and whose directions of propagation make an angle in the range θ to $\theta + d\theta$ with the normal to the hole, is

$$u(\omega,\theta)d\omega\, d\theta = \frac{1}{2}n(\omega)\, d\omega \sin\theta\, d\theta,$$

where $\sin\theta$ is proportional to the solid angle in the specified range of directions, and

$$\int_0^{\pi} n(\omega,\theta)d\omega\, d\theta = n(\omega)\, d\omega.$$

Photons travel at the velocity of light, so the power per unit area escaping from the hole in the frequency range ω to $\omega + d\omega$ is

$$P(\omega)d\omega \int_0^{\pi/2} c\cos\theta\hbar\omega\, n(\omega,\theta)\, d\omega\, d\theta,$$

where c cos θ is the component of the photon velocity in the direction of the hole. This gives

$$P(\omega)\, d\omega = c\overline{u}(\omega)\, d\omega \frac{1}{2}\int_0^{\pi/2} \cos\theta \sin\theta\, d\theta = \frac{c}{4}\overline{u}(\omega)\, d\omega,$$

so

$$P(\omega)\, d\omega = \frac{\hbar}{4\pi^2 c^2} \frac{\omega^3 d\omega}{\exp(\beta\hbar\omega) - 1}$$

is the power per unit area radiated by a black-body in the frequency range ω to ω + $d\omega$. A black-body is very much an idealization. The power spectra of real radiating bodies can deviate quite substantially from black-body spectra. Nevertheless, we can make some useful predictions using this model.

The black-body power spectrum peaks when $\hbar\omega \simeq 3kT$. This means that the peak radiation frequency scales linearly with the temperature of the body. In other words, hot bodies tend to radiate at higher frequencies than cold bodies.

This result (in particular, the linear scaling) is known as Wien's displacement law.

It allows us to estimate the surface temperatures of stars from their colours (surprisingly enough, stars are fairly good black-bodies).

Some stellar temperatures determined by this method (in fact, the whole emission spectrum is fitted to a black-body spectrum). It can be seen that the apparent colours (which correspond quite well to the colours of the peak radiation) scan the whole visible spectrum, from red to blue, as the stellar surface temperatures gradually rise.

Table. Physical Properties of Some Well-known Stars

Name	*Constellation*	*Spectral Type*	*Surf. Temp. (°K)*	*Colour*
Antares	Scorpio	M	3300	Very Red
Aldebaran	Taurus	K	3800	Reddish
Sun		G	5770	Yellow
Procyon	Canis Minor	F	6570	Yellowish
Sirius	Canis Major	A	9250	White
Rigel	Orion	B	11,200	Bluish White

Probably the most famous black-body spectrum is cosmological in origin. Just after the "big bang" the Universe was essentially a "fireball," with the energy associated with radiation completely dominating that associated with matter. The early Universe was also pretty well described by equilibrium statistical thermodynamics, which means that the radiation had a black-body spectrum. As the Universe expanded, the radiation was gradually Doppler shifted to ever larger wavelengths (in other words, the radiation did work against the expansion of the Universe, and, thereby, lost energy), but its spectrum remained invariant.

Nowadays, this primordial radiation is detectable as a faint microwave background which pervades the whole universe. The microwave background was discovered accidentally by Penzias and Wilson in 1961. Until recently, it was difficult to measure the full spectrum with any degree of precision, because of strong microwave absorption and scattering by the Earth's atmosphere.

However, all of this changed when the COBE satellite was launched in 1989. It took precisely nine minutes to measure the perfect black-body spectrum reproduced. This data can be fitted to a black-body curve of characteristic temperature 2.735°K. In a very real sense, this can be regarded as the "temperature of the Universe."

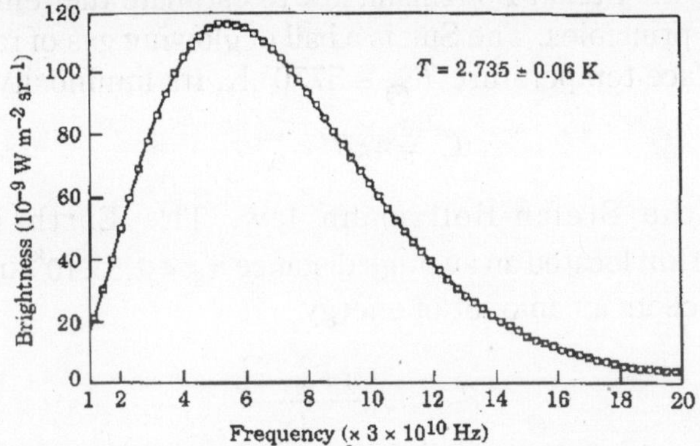

Fig. Cosmic Background Radiation Spectrum Measured by the Far Infrared Absolute Spectrometer (FIRAS) aboard the Cosmic Background Explorer Satellite (COBE).

THE STEFAN-BOLTZMANN LAW

The total power radiated per unit area by a black-body at all frequencies is given by

$$P_{\text{tot}}(T) = \int_0^\infty P(\omega)\, d\omega = \frac{\hbar}{4\pi^2 c^2} \int_0^\infty \frac{\omega^3 d\omega}{\exp(\hbar\omega/kT) - 1},$$

or

$$P_{\text{tot}}(T) = \frac{k^4}{4\pi^2 c^2 \hbar^3} \int_0^\infty \frac{\eta^3 d\eta}{\exp\eta - 1},$$

where $\eta = \hbar\omega/kT$. The above integral can easily be looked up in standard mathematical tables. In fact,

$$\int_0^\infty \frac{\eta^3 d\eta}{\exp\eta - 1} = \frac{\pi^4}{15}.$$

Thus, the total power radiated per unit area by a black-body is

$$P_{\rm tot}(T) = \frac{\pi^2}{60}\frac{k^4}{c^2\hbar^3}T^4 = \sigma T^4.$$

This T^4 dependence of the radiated power is called the Stefan-Boltzmann law, after Josef Stefan, who first obtained it experimentally, and Ludwig Boltzmann, who first derived it theoretically.

The parameter

$$\sigma = \frac{\pi^2}{60}\frac{k^4}{c^2\hbar^3} = 5.67\times 10^{-8}\ {\rm W m^{-2} K^{-4}},$$

is called the Stefan-Boltzmann constant.

We can use the Stefan-Boltzmann law to estimate the temperature of the Earth from first principles. The Sun is a ball of glowing gas of radius $R_\odot \simeq 7\times 10^5$ km and surface temperature $T_\odot \simeq 5770°$ K. Its luminosity is

$$L_\odot = 4\pi R_\odot^2\,\sigma T_\odot^4,$$

according to the Stefan-Boltzmann law. The Earth is a globe of radius $R_\oplus \sim 6000$ km located an average distance $\tau_\oplus \simeq 1.5\times 10^8$ km from the Sun. The Earth intercepts an amount of energy

$$P_\oplus = L_\odot \frac{\pi R_\oplus^2/\tau_\oplus^2}{4\pi}$$

per second from the Sun's radiative output: i.e., the power output of the Sun reduced by the ratio of the solid angle subtended by the Earth at the Sun to the total solid angle 4π. The Earth absorbs this energy, and then re-radiates it at longer wavelengths. The luminosity of the Earth is

$$L_\oplus = 4\pi R_\oplus^2\,\sigma T_\oplus^4,$$

according to the Stefan-Boltzmann law, where $T_\oplus$ is the average temperature of the Earth's surface. Here, we are ignoring any surface temperature variations between polar and equatorial regions, or between day and night. In steady-state, the luminosity of the Earth must balance the radiative power input from the Sun, so equating $L_\oplus$ and $P_\oplus$ we arrive at

$$T_\oplus = \left(\frac{R_\odot}{2\tau_\oplus}\right)^{1/2} T_\odot.$$

Remarkably, the ratio of the Earth's surface temperature to that of the Sun depends only on the Earth-Sun distance and the solar radius. The above expression yields $T_\oplus \sim 279°$ K or 6 °C (or 43 °F). This is slightly on the cold side, by a few degrees, because of the greenhouse action of the Earth's atmosphere, which was neglected in our calculation. Nevertheless, it is quite encouraging that such a crude calculation comes so close to the correct answer.

CONDUCTION ELECTRONS IN A METAL

The conduction electrons in a metal are non-localized (i.e., they are not tied to any particular atoms). In conventional metals, each atom contributes a single such electron. To a first approximation, it is possible to neglect the mutual interaction of the conduction electrons, since this interaction is largely shielded out by the stationary atoms. The conduction electrons can, therefore, be treated as an ideal gas. However, the concentration of such electrons in a metal far exceeds the concentration of particles in a conventional gas. It is, therefore, not surprising that conduction electrons cannot normally be analyzed using classical statistics: in fact, they are subject to Fermi-Dirac statistics (since electrons are fermions).

Recall, that the mean number of particles occupying state *s* (energy $\in_s$) is given by

$$\bar{n}_s = \frac{1}{e^{\beta(c_s-\mu)}+1},$$

according to the Fermi-Dirac distribution. Here,

$$\mu \equiv -kT\,\alpha$$

is termed the Fermi energy of the system. This energy is determined by the condition that

$$\sum_{\tau} \bar{n}_{\tau} = \sum_{\tau} \frac{1}{e^{\beta(c_{\tau}-\mu)+1}} = N,$$

where N is the total number of particles contained in the volume V. It is clear, from the above equation, that the Fermi energy μ is generally a function of the temperature T.

Let us investigate the behaviour of the Fermi function

$$F(\in) = \frac{1}{e^{b(c-\mu)}+1}$$

as $\in$ varies. Here, the energy is measured from its lowest possible value $\in = 0$. If the Fermi energy μ is such that $\beta\,\mu << 1$
then $\beta(\in - \mu) >> 1$, and F reduces to the Maxwell-Boltzmann distribution. However, for the case of conduction electrons in a metal we are interested in the opposite limit, where

$$\beta\mu \equiv \frac{\mu}{kT} >> 1.$$

In this limit, if $\in << \mu$ then $\beta(\in - \mu) << 1$, so that $F(\in) = 1$. On the other hand, if $\in >> \mu$ then $\beta(\in - \mu) >> 1$, so that $F(\in) = \exp[-\beta(\in - \mu)]$ falls off exponentially with increasing $\in$, just like a classical Boltzmann distribution. Note that $F = 1/2$ when $\in = \mu$. The transition region in which F goes from a value close to unity to a value close to zero corresponds to an energy interval of order kT, centred on $\in = \mu$.

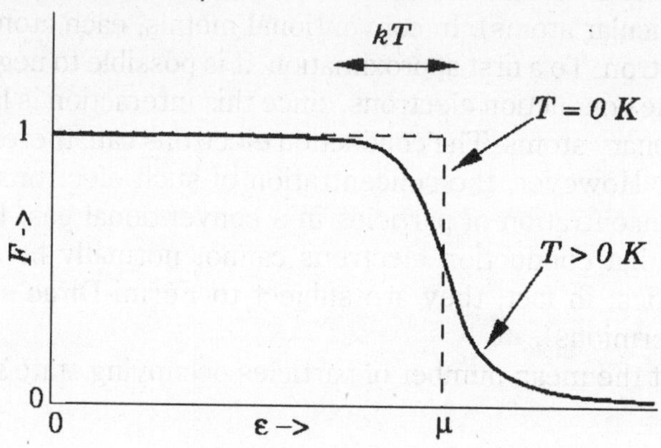

Fig. The Fermi Function.

In the limit as $T \to 0$, the transition region becomes infinitesimally narrow. In this case, $F = 1$ for $\in < \mu$ and $F = 0$ for $\in > \mu$. This is an obvious result, since when $T = 0$ the conduction electrons attain their lowest energy, or ground-state, configuration. Since the Pauli exclusion principle requires that there be no more than one electron per single-particle quantum state, the lowest energy configuration is obtained by piling electrons into the lowest available unoccupied states until all of the electrons are used up. Thus, the last electron added to the pile has quite a considerable energy, $\in = \mu$, since all of the lower energy states are already occupied. Clearly, the exclusion principle implies that a Fermi-Dirac gas possesses a large mean energy, even at absolute zero.

Let us calculate the Fermi energy $\mu = \mu_0$ of a Fermi-Dirac gas at $T = 0$. The energy of each particle is related to its momentum $p = \hbar k$ via

$$\in = \frac{p^2}{2m} = \frac{\hbar^2}{2m},$$

where k is the de Broglie wave-vector. At $T = 0$ all quantum states whose energy is less than the Fermi energy μ_0 are filled. The Fermi energy corresponds to a Fermi momentum $P_F = \hbar\, k_F$ which is such that

$$\mu_0 = \frac{p_F^2}{2m} = \frac{\hbar^2 k_F^2}{2m}.$$

Thus, at $T = 0$ all quantum states with $k < k_F$ are filled, and all those with $k > k_F$ are empty.

Now, we know, by analogy with previous equation that there are $(2\pi)^{-3} V$ allowable translational states per unit volume of k-space. The volume of the sphere of radius k_F in k-space is $(4/3)\pi k_F^2$ .It follows that the Fermi sphere of radius k_F contains $(4/3)\ \pi k_F^2\ (2\pi)^{-3}V$ translational states.

The number of quantum states inside the sphere is twice this, because electrons possess two possible spin states for every possible translational state. Since the total number of occupied states (i.e., the total number of quantum states inside the Fermi sphere) must equal the total number of particles in the gas, it follows that

$$2\frac{V}{(2\pi)^3}\left(\frac{4}{3}\pi k_F^3\right) = N.$$

The above expression can be rearranged to give

$$k_F = \left(3\pi^2 \frac{N}{V}\right)^{1/3}.$$

Hence,

which implies that the de Broglie wavelength λ_F corresponding to the Fermi energy is of order the mean separation between particles $(V/N)^{1/3}$. All quantum states with de Broglie wavelengths $l \equiv 2\pi/k > \lambda_F$ are occupied at $T = 0$, whereas all those with $\lambda < \lambda_F$ are empty.

According to given equation the Fermi energy at $T = 0$ takes the form

$$\mu_0 = \frac{\hbar^2}{2m}\left(3\pi^2 \frac{N}{V}\right)^{1/3}.$$

It is easily demonstrated that $\mu_0 >> kT$ for conventional metals at room temperature.

The majority of the conduction electrons in a metal occupy a band of completely filled states with energies far below the Fermi energy. In many cases, such electrons have very little effect on the macroscopic properties of the metal. Consider, for example, the contribution of the conduction electrons to the specific heat of the metal. The heat capacity C_V at constant volume of these electrons can be calculated from a knowledge of their mean energy $\bar{E}(T)$ as a function of T: i.e.,

$$C_V = \left(\frac{\partial \bar{E}}{\partial T}\right)_V.$$

If the electrons obeyed classical Maxwell-Boltzmann statistics, so that $F \propto \exp(-\beta\epsilon)$ for all electrons, then the equipartition theorem would give

$$\bar{E} = \frac{3}{2}NkT,$$

$$C_V = \frac{3}{2}Nk.$$

However, the actual situation, in which F has the form is very different. A small change in T does not affect the mean energies of the majority of the electrons, with $\in << \mu$, since these electrons lie in states which are completely filled, and remain so when the temperature is changed. It follows that these electrons contribute nothing whatsoever to the heat capacity.

On the other hand, the relatively small number of electrons N_{eff} in the energy range of order kT, centred on the Fermi energy, in which $F$is significantly different from 0 and 1, do contribute to the specific heat. In the tail end of this region $F \propto \exp(-\beta\in)$, so the distribution reverts to a Maxwell-Boltzmann distribution. Hence, from previous equation we expect each electron in this region to contribute roughly an amount $(3/2)k$ to the heat capacity. Hence, the heat capacity can be written

$$C_V \simeq \frac{3}{2}N_{eff}k.$$

However, since only a fraction kT/μ of the total conduction electrons lie in the tail region of the Fermi-Dirac distribution, we expect

$$N_{\text{eff}} \simeq \frac{kT}{\mu}N.$$

It follows that

$$C_V \simeq \frac{3}{2}N\,k\frac{kT}{\mu}.$$

Since $kT << \mu$ in conventional metals, the molar specific heat of the conduction electrons is clearly very much less than the classical value $(3/2)R$. This accounts for the fact that the molar specific heat capacities of metals at room temperature are about the same as those of insulators. Before the advent of quantum mechanics, the classical theory predicted incorrectly that the presence of conduction electrons should raise the heat capacities of metals by 50 per cent [i.e., $(3/2)R$] compared to those of insulators.

Note that the specific heat is not temperature independent. In fact, using the superscript e to denote the electronic specific heat, the molar specific heat can be written

$$C_V^{(e)} = \gamma T,$$

where γ is a (positive) constant of proportionality. At room temperature $C_V^{(e)}$ is completely masked by the much larger specific heat $C_V^{(L)}$ due to lattice vibrations. However, at very low temperatures $C_V^{(L)} = AT^3$, where A is a (positive) constant of proportionality. Clearly, at low temperatures $C_V^{(L)} = AT^3$ approaches zero far more rapidly that the electronic specific heat, as T is reduced. Hence, it should be possible to measure the electronic contribution to the molar specific heat at low temperatures.

The total molar specific heat of a metal at low temperatures takes the form

$$c_V = C_V^{(e)} + c_V^{(L)} = \gamma T + AT^3.$$

Hence,

$$\frac{c_V}{T} = \gamma + AT^2.$$

If follows that a plot of c_V/T versus T^2 should yield a straight line whose intercept on the vertical axis gives the coefficient γ. The fact that a good straight line is obtained verifies that the temperature dependence of the heat capacity predicted by previous equation is indeed correct.

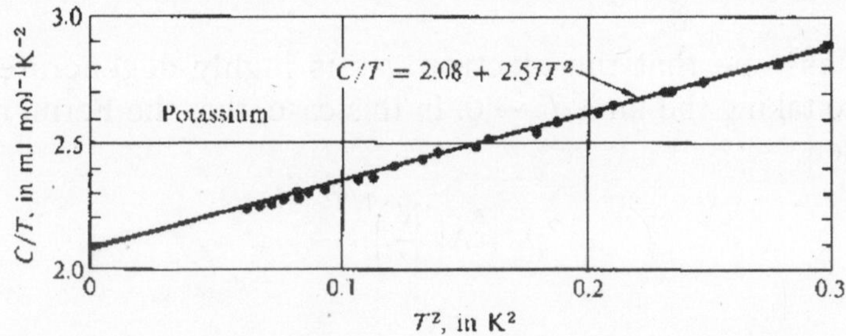

Fig. The Low Temperature Heat Capacity of Potassium, Plotted as C_V/T versus T^2. From C. Kittel, and H. Kroemer, Themal Physics.

White-Dwarf Stars

A main-sequence hydrogen-burning star, such as the Sun, is maintained in equilibrium via the balance of the gravitational attraction tending to make it collapse, and the thermal pressure tending to make it expand. Of course, the thermal energy of the star is generated by nuclear reactions occurring deep inside its core. Eventually, however, the star will run out of burnable fuel, and, therefore, start to collapse, as it radiates away its remaining thermal energy. What is the ultimate fate of such a star?

A burnt-out star is basically a gas of electrons and ions. As the star collapses, its density increases, so the mean separation between its constituent particles decreases. Eventually, the mean separation becomes of order the de Broglie wavelength of the electrons, and the electron gas becomes degenerate. Note, that the de Broglie wavelength of the ions is much smaller than that of the electrons, so the ion gas remains non-degenerate.

Now, even at zero temperature, a degenerate electron gas exerts a substantial pressure, because the Pauli exclusion principle prevents the mean electron separation from becoming significantly smaller than the typical de Broglie wavelength. Thus, it is possible for a burnt-out star to maintain itself against complete collapse under gravity via the degeneracy pressure of its constituent electrons. Such stars are termed white-dwarfs. Let us investigate the physics of white-dwarfs in more detail.

The total energy of a white-dwarf star can be written

$$E = K + U,$$

where K is the total kinetic energy of the degenerate electrons (the kinetic energy of the ion is negligible) and U is the gravitational potential energy. Let us assume, for the sake of simplicity, that the density of the star is uniform. In this case, the gravitational potential energy takes the form

$$U = -\frac{3}{5}\frac{G M^2}{R},$$

where G is the gravitational constant, M is the stellar mass, and R is the stellar radius.

Let us assume that the electron gas is highly degenerate, which is equivalent to taking the limit $T \to 0$. In this case, that the Fermi momentum can be written

$$p_F = \Lambda\left(\frac{N}{V}\right)^{1/3},$$

where

$$\Lambda = \left(3p^2\right)^{1/3} \hbar.$$

Here,

$$V = \frac{4\pi}{3} R^3$$

is the stellar volume, and N is the total number of electrons contained in the star. Furthermore, the number of electron states contained in an annular radius of P-space lying between radii P and $P + dp$ is

$$dN = \frac{3V}{\Lambda^3} p^2 \, dp.$$

Hence, the total kinetic energy of the electron gas can be written

$$K = \frac{3V}{\Lambda^3}\int_0^{p_F} \frac{p^2}{2m} p^2 = \frac{3}{5}\frac{V}{\Lambda^3}\frac{p_F^5}{2m},$$

where m is the electron mass. It follows that

$$K = \frac{3}{5} N \frac{V^2}{2m}\left(\frac{N}{V}\right)^{2/3}.$$

The interior of a white-dwarf star is composed of atoms like C^{12} and C^{16} which contain equal numbers of protons, neutrons, and electrons. Thus,

$$M = 2\,N\,m_p,$$

where m_p is the proton mass.

Equations can be combined to give

$$E = \frac{A}{R^2} - \frac{B}{R},$$

where

$$A = \frac{3}{20}\left(\frac{9\pi}{8}\right)^{2/3}\frac{\hbar^2}{m}\left(\frac{M}{m_p}\right)^{5/3},$$

$$B = \frac{3}{5}G\,M^2.$$

The equilibrium radius of the star R_* is that which minimizes the total energy E. In fact, it is easily demonstrated that

$$R_* = \frac{2A}{B},$$

which yields

$$R_* = \frac{(9\pi)^{2/3}\hbar^2}{8}\frac{\hbar^2}{m}\frac{1}{G\,m_p^{5/3}M^{2/3}}.$$

The above formula can also be written

$$\frac{R_*}{R_\odot} = 0.010\left(\frac{M_\odot}{M}\right)^{1/3},$$

where $R_\odot = 7 \times 10^5$ kmis the solar radius, and $M_\odot = 2 \times 10^{30}$ kg is the solar mass. It follows that the radius of a typical solar mass white-dwarf is about 7000 km: i.e., about the same as the radius of the Earth. The first white-dwarf to be discovered (in 1862) was the companion of Sirius. Nowadays, thousands of white-dwarfs have been observed, all with properties similar to those described above.

THE CHANDRASEKHAR LIMIT

One curious feature of white-dwarf stars is that their radius decreases as their mass increases. It follows, from Eq. that the mean energy of the degenerate electrons inside the star increases strongly as the stellar mass increases: in fact, $K \propto M^{4/3}$. Hence, if M becomes sufficiently large the electrons become relativistic, and the above analysis needs to be modified. Strictly speaking, the non-relativistic analysis is only valid in the low mass limit $M << M_\odot$. Let us, for the sake of simplicity, consider the ultra-relativistic limit in which $p >> m\,c$.

The total electron energy (including the rest mass energy) can be written

$$K = \frac{3V}{\Lambda^3}\int_0^{PF} (p^2c^2 + m^2c^4)^{1/2}\, p^2\, dp,$$

by analogy with previous equation. Thus,

$$K \simeq \frac{3V\,c}{\Lambda^3}\int_0^{PF}\left(p^3 + \frac{m^2c^2}{2}p + \cdots\right) dp,$$

giving

$$E \simeq \frac{3\,V\,c}{4\,\Lambda^3} + \left[p_F^4 + m^2\,c^2 p_F^2 + \cdots\right].$$

It follows, from the above, that the total energy of an ultra-relativistic white-dwarf star can be written in the form

$$E \simeq \frac{A - B}{R} + C\,R,$$

where

$$A = \frac{3}{8}\left(\frac{9\pi}{8}\right)^{1/3} \hbar\, c\left(\frac{M}{m_p}\right)^{4/3},$$

$$B = \frac{3}{5} G\, M^2,$$

$$C = \frac{3}{4}\frac{1}{(9\pi)^{1/3}}\frac{m^2c^3}{\hbar}\left(\frac{M}{m_p}\right)^{2/3}.$$

As before, the equilibrium radius R_* is that which minimizes the total energy E. However, in the ultra-relativistic case, a non-zero value of R_* only exists for $A - B > 0$. When $A - B < 0$ the energy decreases monotonically with decreasing stellar radius: in other words, the degeneracy pressure of the electrons is incapable of halting the collapse of the star under gravity. The criterion which must be satisfied for a relativistic white-dwarf star to be maintained against gravity is that

$$\frac{A}{B} > 1.$$

This criterion can be re-written

$$M < M_C,$$

where

$$M_C = \frac{15}{64}(5\pi)^{1/2}\frac{(\hbar\, c/G)^{1/2}}{m_p^2} = 1.72\, M_\odot$$

is known as the Chandrasekhar limit, after A. Chandrasekhar who first derived it in 1931. A more realistic calculation, which does not assume constant density, yields

$$M_C = 1.4\, M_\odot.$$

Thus, if the stellar mass exceeds the Chandrasekhar limit then the star in question cannot become a white-dwarf when its nuclear fuel is exhausted, but, instead, must continue to collapse. What is the ultimate fate of such a star?

6

Nuclear and Particle Physics

NUCLEAR PHYSICS

The relevant length scale for measuring nuclear size is the femtometer 1 fm = 10^{-15} meters. Physicists usually call this length one *fermi*. Nuclei vary from about one to a few fermis in radius. Recall that the Bohr radius of Hydrogen is of order 10^{-10} meters, so the nucleus is far smaller than the atom. Nuclear size was first measured by Rutherford, by noting how close α-particles came to the nucleus before the scattering ceased to be pure Coulomb repulsion (at which point they were actually hitting the nuclear surface). Despite its small size, the nucleus has about 99.97 per cent of the mass of the atom (a bit less for hydrogen).

Nuclear physics is a vast branch of physics. It deals with the properties of an atomic nucleus, such as mass, charge, size, nucleons (neutrons and protons) and their binding energies, nuclear structure, bound and excited energy levels, spin states, interaction behaviour with gamma rays, with other particles, and with other nuclei, variation of such behaviour with energy and angle of interaction, the nuclear stability, radioactivity of excited and unexcited nucleus, decay half-lives, decay particles and their energies, nuclear transmutations, and so on. The aim of the subject is to understand the properties of the nuclei, both quantitatively and qualitatively. A variety of highly sophisticated nuclear physics experiments aid such an understanding.

A detailed understanding needs quantum mechanical interpretations and advanced mathematics. A thorough understanding will facilitate exploitation of nuclear properties for the benefit of mankind.

We present below some basic notions of nuclear physics, that help us appreciate the principles and mechanisms of nuclear-energy production. It has been amply understood that substantial nuclear energy is released in breaking a heavy nucleus into lighter ones (known as fission) and in joining lighter nuclei together to form a heavier one (known as fusion). Which heavy nucleus to break? How to break it? Which light nuclei to fuse? How to fuse them? What are the arrangements and precautions needed? How to extract the energy released and make it available for human use? Is nuclear energy production economical, practicable and safe? Are there different options available? Such and many more questions arise.

The purpose of this report is to describe the scientific and technical developments in this country since 1940 directed towards the military use of energy from atomic nuclei. Although not written as a "popular" account of the subject, this report is intended to be intelligible to scientists and engineers generally and to other college graduates with a good grounding in physics and chemistry. The equivalence of mass and energy is chosen as the guiding principle in the presentation of the background material of the "Introduc-tion."

CONSERVATION OF MASS AND OF ENERGY

There are two principles that have been cornerstones of the structure of modern science. The first—that matter can be neither created nor destroyed but only altered in form—was enunciated in the eighteenth century and is familiar to every student of chemistry; it has led to the principle known as the law of conservation of mass.

The second—that energy can be neither created nor destroyed but only altered in form—emerged in the nineteenth century and has ever since been the plague of inventors of perpetual-motion machines; it is known as the law of' conservation of energy.

These two principles have constantly guided and disciplined the development and application of science. For all practical purposes they were unaltered and separate until some five years ago.

For most practical purposes they still are so, but it is now known that they are, in fact, two phases of a single principle for we have discovered that energy may sometimes be converted into matter and matter into energy. Specifically, such a conversion is observed in the phenomenon of nuclear fission of uranium, a process in which atomic nuclei split into fragments with the release of an enormous amount of energy. The military use of this energy has been the object of the research and production projects described in this report.

EQUIVALENCE OF MASS AND ENERGY

One conclusion that appeared rather early in the development of the theory of relativity was that the inertial mass of a moving body increased as its speed increased. This implied an equivalence between an increase in energy of motion of a body, that is, its kinetic energy, and an increase in its mass.

To most practical physicists and engineers this appeared a mathematical fiction of no practical importance. Even Einstein could hardly have foreseen the present applications, but as early as 1905 he did clearly state that mass and energy were equivalent and suggested that proof of this equivalence might be found by the study of radioactive substances. He concluded that the amount of energy, E, equivalent to a mass, m, was given by the equation,

$$E = mc^2$$

where c is the velocity of light. If this is stated in actual numbers, its startling character is apparent. It shows that one kilogram (2.2 pounds) of matter, if converted entirely into energy, would give 25 billion kilowatt hours of energy. This is equal to the energy that would be generated by the total electric power industry in the United States (as of 1939) running for approximately two months. Compare this fantastic figure with the 8.5 kilowatt hours of heat energy which may be produced by burning an equal amount of coal.

The extreme size of this conversion figure was interesting in several respects. In the first place, it explained why the equivalence of mass and energy was never observed in ordinary chemical combustion.

We now believe that the heat given off in such a combustion has mass associated with it, but this mass is so small that it cannot be detected by the most sensitive balances available. (It is of the order of a few billionths of a gram per mole.) In the second place, it was made clear that no appreciable quantities of matter were being converted into energy in any familiar terrestrial processes, since no such large sources of energy were known. Further, the possibility of initiating or controlling such a conversion in any practical way seemed very remote. Finally, the very size of the conversion factor opened a magnificent field of speculation to philosophers, physicists, engineers, and comic-strip artists. For twenty-five years such speculation was unsupported by direct experimental evidence, but beginning about 1930 such evidence began to appear in rapidly increasing quantity.

Before discussing such evidence and the practical partial conversion of matter into energy that is our main theme, we shall review the foundations of atomic and nuclear physics. General familiarity with the atomic nature of matter and with the existence of electrons is assumed. Our treatment will be little more than an outline which may be elaborated by reference to books such as Pollard and Davidson's *Applied Nuclear Physics*and Stranathan's *The "Particles" of Modern Physics.*

RADIOACTIVITY AND ATOMIC STRUCTURE

First discovered by H. Becquerel in 1896 and subsequently studied by Pierre and Marie Curie, E. Rutherford, and many others, the phenomena of radioactivity have played leading roles in the discovery of the general laws of atomic structure and in the verification of the equivalence of mass and energy.

WHAT IS THE NUCLEUS MADE OF?

The simplest nucleus, that of hydrogen, is a single proton: an elementary particle of mass about 940 MeV, carrying positive charge exactly opposite to the electron's charge, having a spin of one half and being a fermion (so no two protons can be in the same quantum state).

The next simplest nucleus, called the *deuteron*, is a bound state of a proton and a neutron. The neutron, like the proton, is a spin one-half fermion, but it has no electric charge, and is slightly heavier (by 1.3 MeV) than the proton. The binding energy of the deuteron (analogous to the 13.6 eV for the Hydrogen atom) is 2.2MeV.

A photon of this energy could "ionize" the deuteron into a separated proton and neutron. However, it is not necessary to actually do this experiment to establish how tightly the deuteron is bound. One need only weigh the deuteron accurately. It has a mass of 1875.61MeV. The proton has a mass of 938.27MeV, the neutron 939.56MeV, so together (but some distance apart!) they have a mass of 1877.93MeV, 2.2MeV *more* than the deuteron. Thus, when a proton and a neutron come together to form a deuteron, they must unload 2.2MeV of energy, which they do by emitting a photon (called a γ-ray at these energies). Both protons and neutrons, being fermions, obey the exclusion principle, two protons with spin up cannot be in the same state, although two with opposite spin directions could, and a proton and a neutron can occupy the same spot at the same time!

Protons and neutrons are referred to as *nucleons*. The total number of nucleons in a nucleus is usually denoted by A, where $A = Z + N$, Z protons and N neutrons. The chemical properties of an atom are determined by the number of electrons, the same as the number of protons Z. This is called the *atomic number*. Nuclei can have the same atomic number, but different numbers of neutrons. These nuclei are called *isotopes*, the Greek for "same place", since they are in the same place in the periodic table.

These nucleons attract each other with a short range but very strong force, called the *nuclear force*. The situation here is different from that for electrons in the atom, where the strong central force tends to dominate. In the nucleus the nucleons are attracted mainly by their immediate neighbours. Nevertheless, it is a useful beginning to think of this attractive force as being a potential well, as seen by an individual nucleon, and think in terms of the nuclei as filling the lowest available quantum states in this well, just as we did for electrons in the atom.

We find, for example, that the Helium nucleus, $2p+2n$, (a.k.a. the α-particle) is tightly bound—the four nucleons can all occupy the lowest state in the well. However, some larger nuclei, like C, O, Fe are actually a little more tightly bound even than He (about 8.5 MeV per nucleon as opposed to about 7.5 for He) because each nucleon is attracted to its close partners, and there are more close partners in these larger nuclei. It has also been argued that some of these higher nuclei strongly resemble bound states of á-particles.

The total binding energy (usually expressed per nucleon) of any nucleus is easy to find—just as for the deuteron above, the mass of the nucleus is found accurately, and subtracted from the sum of the masses of the separate nucleons.

COULOMB REPULSION AND α-DECAY

As one goes to really large nuclei, the binding per nucleon actually decreases. This is a consequence of the electrostatic repulsion between the protons. The essential point is that the Coulomb repulsion is *long range*, but the nuclear attraction is short range. Considering a proton at the surface of a nucleus, it is bound to the nucleus by the attractive forces of its nearest neighbours, say five or six of them, and this number is little affected as one goes up the periodic table from, say, Fe to U. On the other hand, the proton on the surface is being pushed away by the electrostatic repulsion from the other protons, and this increases substantially as the number of *other* protons goes from 25 in iron to 91 in uranium, even though they may be a little further away on average. In fact, this is why there is a limit on the size of atoms.

A large nucleus does not, however, tend to spit out protons—there's a much more economical way to unload charge. It ejects an α-particle. The binding energy of the nucleons in an α is almost exactly the same as it is in a U nucleus, so little nuclear binding has to be traded for lowering the electrostatic potential energy.

Thus the most common form of Uranium, U^{238} (A=238, Z=92) goes to Thorium (A=234, Z=90) by α-decay. The surprising thing is, it takes a time comparable to the age of the universe to do so! If it's energetically desirable, why does it take so long?

The answer is—there's a barrier in the way. It is a good model to imagine the U to have the a particle inside it, confined to a deep square well potential representing the attractive nuclear force. However, the α-particle wavefunction will penetrate some distance into the wall of this finite square well potential, where the α will be away from the nuclear force, but not away from the long-range electrostatic repulsion of the other protons in the nucleus. The picture is, then, that outside the well, the ground is not flat it's like a well at the top of a hill. The implication of this is that if the energy level of the a inside the well is higher than the flat land surrounding the hill, the α-wavefunction after dying away underground as it tunnels out from the well will re-emerge at some distance, oscillating. In other words, there is some finite probability of finding the α shooting away from the nucleus some distance away.

The actual probability of this happening depends on how much the wavefunction decayed as it tunneled underground away from the well, which depended on how far its energy was below the ground. For U^{238}, this decay on tunneling away from the well is so rapid that the á takes on average billions of years to escape. However, many different nuclei undergo α decay, the quickest in millionths of a second. The energies of the emitted α's can be measured, and confirm this explanation of quantum mechanical tunneling, first put forward in 1928. The quickest decays correspond to the most energetic α's, just as the tunneling picture would suggest.

β-DECAY

These heavier nuclei have more neutrons than protons, because the Coulomb repulsion makes it harder to bind protons. However, a price is paid—thinking of a shell-type model, the extra neutrons go into higher energy shells. Thus for nuclei of a given size, there is an optimum ratio of neutrons to protons, close to one for small nuclei where the Coulomb effects are small enough to be neglected, close to 1.5 for the largest nuclei.

Note that if a large nucleus decays by emitting an α, the nucleus left behind has an even greater ratio of neutrons to protons, since $\alpha = 2p+2n$. This implies that it might be able to go to a lower energy state if it could trade a neutron for a proton. In fact, this can sometimes be done. The neutron itself is not a stable particle. A neutron alone in space will only last about ten minutes before ejecting an electron, leaving a proton behind. This is energetically possible, because the mass of the neutron exceeds the sum of the masses of the proton and electron.

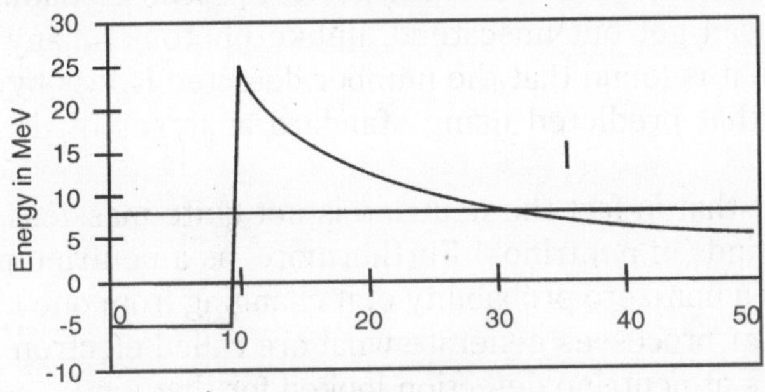

Fig. Distance in Fermis

The process is called β-decay. It can also occur in a nucleus, but now the different binding energies of the initial neutron and the final proton must be factored in to decide if it will go. Other things being equal, the more neutrons there are present in the nucleus, the more likely it is to β-decay. Actually, β-decay was a big puzzle when it was first carefully observed, because it was discovered that identical nuclei β-decaying to the same final states ejected electrons over a range of energies, up to a maximum corresponding to the energy difference between the initial and final nuclei. This led to the suggestion that maybe, after all, energy wasn't conserved in every process in nature.

But physicists are a conservative lot when it comes to energy conservation, and Pauli suggested that maybe $n \rightarrow p+e$ wasn't the whole story some other particle was being created at the same time, carrying away the balance of the energy, but *undetected*. This meant it had no charge, no mass, and, unlike the photon, no electric interaction.

Fermi called it the *neutrino*, the little neutral one. Another reason the neutrino seemed necessary was that the initial neutron had spin one-half, as did the final proton *and* the emitted electron. It was difficult to see how angular momentum could be conserved if no other particle were emitted, because even if the final proton and electron had some orbital angular momentum, that comes in whole-number amounts, so how could two half-integer spins and an integer angular momentum add to be equal to the initial half-integer spin?

It turned out they couldn't, the neutrino really was there all along, it had spin one-half itself, but it was years before it was detected.

Detecting neutrinos is still a difficult business, since they are not sensitive to the nuclear (strong) force, nor to electromagnetic fields. The force that causes neutron (and other particle) decay is called the weak force.

The only way to detect a neutrino is if it collides with a particle and interacts weakly essentially reversing one of these decays. The problem is that a neutrino moving downwards is extremely unlikely to interact with a nucleus, it will almost certainly just pass through the earth.

Therefore, to detect neutrinos takes large detectors and a strong flux of neutrinos. Much effort has been expended detecting neutrinos from the sun, since they are our only window into the nuclear processes taking place deep in the sun—they can get out unscathed, unlike photons or any other particle emitted. In fact, it is found that the number detected is less by a factor of two or three from that predicted using standard analyses of the solar nuclear processes.

It turns out that in fact the neutrino is not quite massless, and there are actually three kinds of neutrinos! Furthermore, as a neutrino passes through matter, there is a non-zero probability of it changing from one kind to another. The solar nuclear processes generate what are called electron neutrinos, and the initial efforts at neutrino detection looked for that kind.

More recent detectors count the other two kinds (mu neutrinos and tau neutrinos) as well, and the total number of neutrinos detected agrees well with the prediction of the total number generated in the sun, so it is now believed that some of the original electron neutrinos change types traveling from their point of origin to the sun's surface.

Anyway, back to the radioactive decay of U^{238}. It emits an α to become Th^{234}, then β decays twice in succession to become U^{234}. This is now relatively proton-rich, and in fact α decays five times in succession to become a form of lead,

$$U^{234} \to Th^{230} \to Ra^{226} \to Rn^{222} \to Po^{218} \to Pb^{214}$$

Pb^{214} undergoes two β decays, an α decay and two more β decays before it reaches stability at Pb^{206}.

The sequence above is through metallic elements with one exception—Rn, radon, is a gas because, atomically speaking, it has completely filled electron shells, and so is chemically unreactive, and gaseous at ordinary temperatures. It emits its α particle in a period of days, which gives it time to diffuse out of the rock into somebody's basement.

γ-DECAY

The three forms of radiation emitted by decaying nuclei, α, β and γ, were named before the particles being emitted were fully identified. It was only realized later that the β-rays were electrons. The γ rays are just high-energy photons, of order 100 keV to a few MeV.

Emission of γ rays is similar (but at much higher energies) to emission of photons by excited states of atoms. The nucleus can be excited by having just emitted an α or β, or by colliding with another nucleus, or being bombarded by neutrons, say. All these events can lead to a nucleus in which the charge distribution is oscillating, and electromagnetic radiation ensues.

THE DROPLET MODEL

In some ways, the binding of nucleons in a nucleus is reminiscent of that of molecules in a drop of water—the force is attractive between near neighbours. Notice how different this is from the electrons in an atom, held in place by a large central attracting object. This "droplet" model of the nucleus is useful for understanding nuclear fission. If the droplet is somehow caused to vibrate vigourously, it might break into two, or more, smaller nuclei.

For the U^{235} nucleus (only 0.7 per cent in naturally occurring uranium), if a slow neutron gets too close, the attractive nuclear force pulls it towards the nucleus so strongly that in the resulting collision, the nucleus breaks into two smaller ones, and two or three neutrons are ejected (since the smaller nuclei formed have lower proportions of neutrons). There is also substantial energy release—about 200 MeV. The much more common U^{238} nucleus also attracts free neutrons, but not so strongly, and it survives neutron capture without fission. A neutron has to be moving very slowly to be captured by U^{235}. On the other hand, when U^{235} fissions the emitted neutrons are moving very fast. To get a *chain reaction* in a nuclear reactor, the main problem is to slow down the emitted neutrons before they leave the reactor.

If they can be slowed down successfully, neutrons emitted by one fission will initiate further fissions, giving continuous—perhaps increasing—energy output. However, the neutrons bounce off the heavy uranium nuclei with almost no energy loss—to get a particle to lose energy in a collision, it must collide with something approximately its own size.

Water is often used, the neutrons lose energy bouncing off the protons in the hydrogen atoms. Unfortunately, the neutrons often stick to the protons, forming deuterons, so to keep the reaction going the uranium must be enriched—the U^{235} content boosted to one to four per cent. Of course, care must be taken to ensure that the chain reaction isn't *too* successful moderators, which absorb neutrons, are put into the reactor if the rate of fission increases beyond the planned level. Needless to say, moderators are not used in military applications of this concept.

THE BEGINNINGS OF NUCLEAR PHYSICS

The First World War lasted from 1914 to 1918. Geiger and Marsden were both at the Western front, on opposite sides. Rutherford had a large water tank installed on the ground floor of the building in Manchester, to carry out research on defence against submarine attack.

Nevertheless, occasional research on alpha scattering continued. Scattering from heavy nuclei was fully accounted for by the electrostatic repulsion, so Rutherford concentrated on light nuclei, including hydrogen and nitrogen. In 1919, Rutherford established that an alpha impinging on a nitrogen nucleus can cause a *hydrogen* atom to appear! Newspaper headlines blared that Rutherford had "split the atom".

Shortly after that experiment, Rutherford moved back to Cambridge to succeed J. J. Thomson as head of the Cavendish laboratory, working with one of his former students, Chadwick, who had spent the war years interned in Germany. They discovered many unusual effects with alpha scattering from light nuclei.

In 1921, Chadwick and co-author Bieler wrote: "The present experiments do not seem to throw any light on the nature of the law of variation of the forces at the seat of an electric charge, but merely show that the forces are of great intensity ... It is our task to find some field of force which will reproduce these effects."

In fact, Rutherford was beginning to focus his attention on the actual construction of the nucleus and the alpha particle. He coined the word "proton" to describe the hydrogen nucleus, it first appeared in print in 1920. At first, he thought the alpha must be made up of four of these protons somehow bound together by having two electrons in the middle-this would get the mass and charge right, but of course nobody could construct a plausible electrostatic configuration.

Then he had the idea that maybe there was a special very tightly bound state of a proton and an electron, much smaller than an atom. By 1924, he and Chadwick were discussing how to detect this neutron. It wasn't going to be easy-it probably wouldn't leave much of a track in a cloud chamber. In fact, Chadwick did discover the neutron, but not until 1932, and it wasn't much like their imagined proton-electron bound state. But it did usher in the modern era in nuclear physics.

REACTIVITY IN PHYSICS

Reactivity is the name of the quantity which determines the rate of change of reactor power. For reactor power to be steady the reactivity must be zero.

Maintaining zero reactivity requires moving"control rods" to maintain a balance among the rates of leakage and absorption of neutrons in nuclear fuel and in other reactor materials.

Because the absorption and leakage of neutrons have been extensively studied, these balancing dependencies are relatively well understood. Positive reactivity causes power to rise exponentially at a rate proportional to the reactivity. Negative reactivity causes power to decrease.

To change power in a planned manner, reactivity is adjusted by moving the control rods, either manually or by means of automatic controls. Partially removing a control rod is expected to increase the reactivity, causing the power to rise to a new level. Inserting a properly designed control rod farther into the reactor should, but doesn't always, lower the power level.

Example shows the effect of a relatively large initial reactivity, ignoring several important limiting effects, leading to a rapid rise to a"blowup." The unacceptable behaviour which would result from a continuation of this exponential response is modified in real reactors by two important effects.

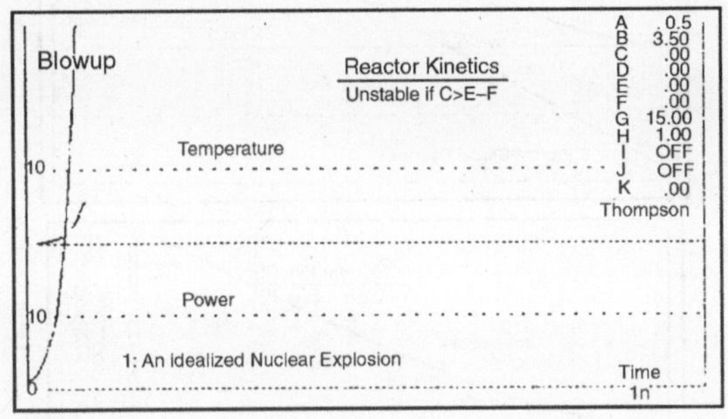

DELAYED NEUTRONS

The first nuclear effect is that of"delayed neutrons." Of all neutrons formed in fission a small fraction (.007) are delayed by significant time periods after absorption of the fission neutrons.

The delayed neutrons, not being available immediately to support a transient rise in power, act as a damping factor on transients, providing the strong beneficial effect of limiting sudden changes in power. Example below show the classical"damping effect" of delayed neutrons on the rise of power, ignoring the temperature coefficient of reactivity and mechanical effects. Example 3 shows the reaction to"prompt critical" reactivity, in which by definition the initial reactivity is numerically equal to the fraction of delayed neutrons (G=F=15).

The initial rate of power rise continues indefinitely on a straight line. In Example 2the addition of reactivity greater than prompt critical (G/F = 15/14) causes an exponential rise of power faster than in Example 3. Example 4 shows a"sub-critical" reaction, in which a value of reactivity less than the fraction of delayed neutrons,

$$(G/F = 15/16)$$

leads to the leveling-off of power at a new value,

$$P/P0=F/(G-F) = 16.$$

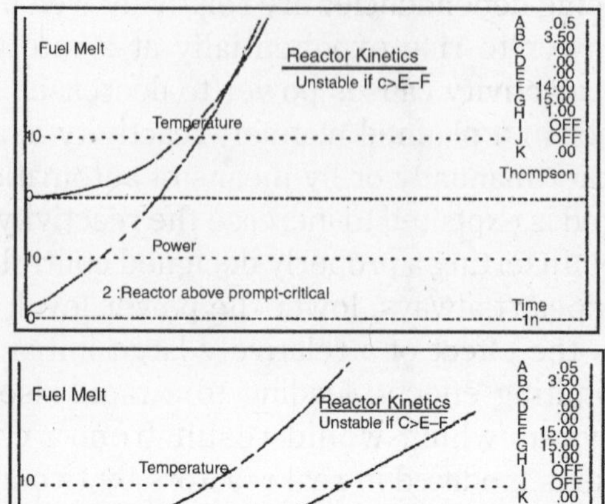

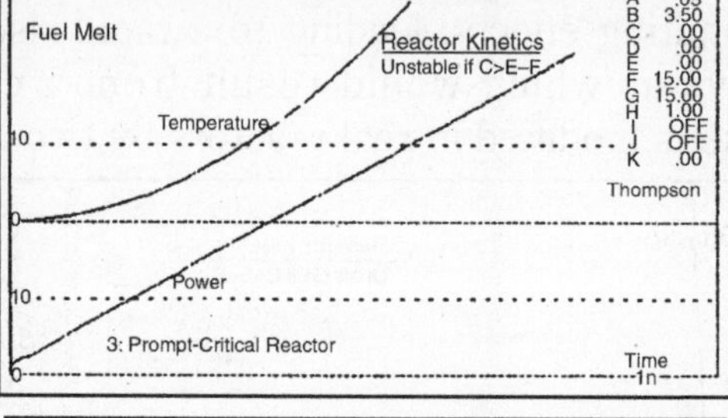

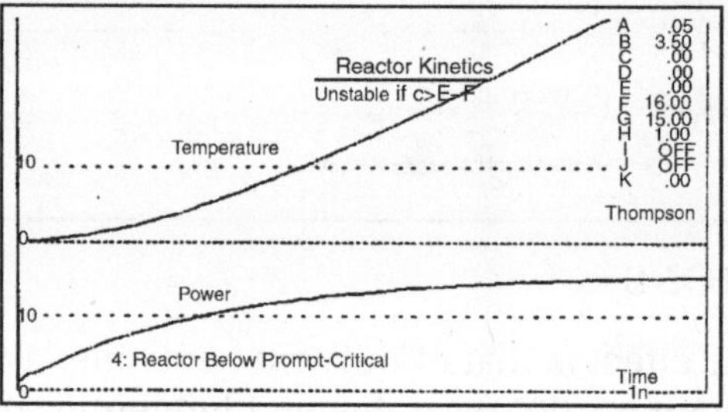

TEMPERATURE COEFFICIENT OF REACTIVITY

The rise in temperature which accompanies a rapid rise in power generally changes the reactivity. The"temperature coefficient of reactivity" measures the effect of changing temperature on reactivity. Because power stability requires a rise in temperature to decrease the reactivity, a"negative temperature coefficient" is necessary for stability. Examples show the effects on power stability of including delayed neutron damping and temperature coefficients of reactivity, but neglecting all other effects.

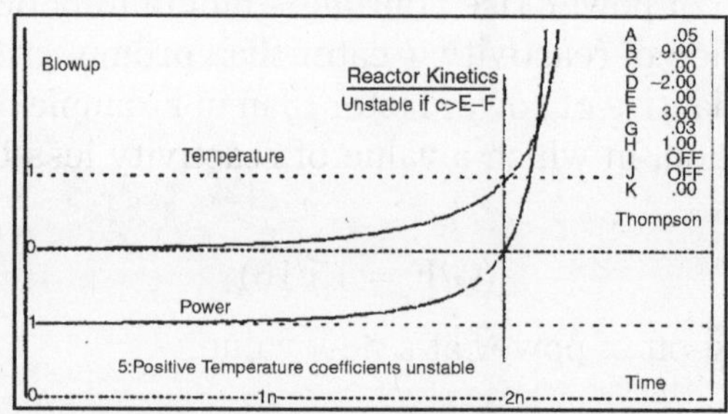

Example shows the destabilizing effect on reactor power of a positive temperature coefficient (D= –2.0), leading to"blowup."

Example shows an oscillation of power at nuclear frequency, with a negative temperature coefficient (D=2.0) and with delayed neutron damping (F=0.20).

Because of neutron damping the oscillations decrease in magnitude with time. The effects of the nuclear factors, reactivity, delayed neutrons and temperature coefficients, are described by a non-linear second-order differential equation.

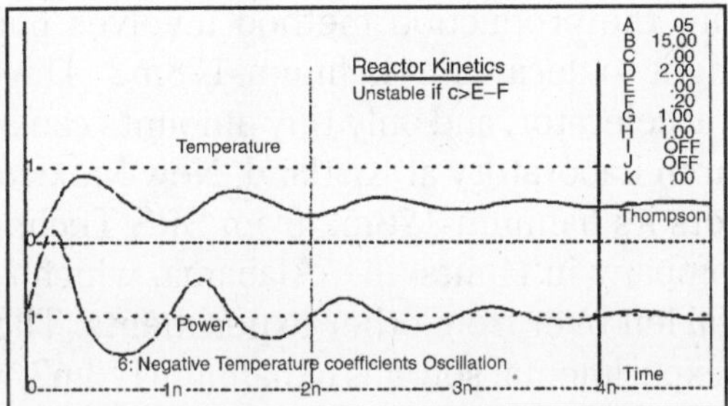

This relatively simple equation predicts stable reactor power if the temperature coefficient of reactivity is negative.

Neutron damping in reactors is generally relatively large, providing a strong tendency towards stability for a simple nuclear system.

The real situation is more complicated, and more difficult than can be described in terms only of nuclear factors.

NUCLEAR ISOMERISM

The explosive works by stimulating the release of energy from the nuclei of certain elements but does not involve nuclear fission or fusion.

The energy, emitted as gamma radiation, is thousands of times greater than that from conventional chemical explosives. The technology has already been included in the Department of Defence's Militarily Critical Technologies List, which says: "Such extraordinary energy density has the potential to revolutionise all aspects of warfare."

Scientists have known for many years that the nuclei of some elements, such as hafnium, can exist in a high-energy state, or nuclear isomer, that slowly decays to a low-energy state by emitting gamma rays. For example, hafnium178m2, the excited, isomeric form of hafnium-178, has a half-life of 31 years.

The possibility that this process could be explosive was discovered when Carl Collins and colleagues at the University of Texas at Dallas demonstrated that they could artificially trigger the decay of the hafnium isomer by bombarding it with low-energy X-rays. The experiment released 60 times as much energy as was put in, and in theory a much greater energy release could be achieved.

Before hafnium can be used as an explosive, energy has to be "pumped" into its nuclei. Just as the electrons in atoms can be excited when the atom absorbs a photon, hafnium nuclei can become excited by absorbing high-energy photons. The nuclei later return to their lowest energy states by emitting a gamma-ray photon. Nuclear isomers were originally seen as a means of storing energy, but the possibility that the decay could be accelerated fired the interest of the Department of Defence, which is also investigating several other candidate materials such as thorium and niobium.

For the moment, the production method involves bombarding tantalum with protons, causing it to decay into hafnium-178m2. This requires a nuclear reactor or a particle accelerator, and only tiny amounts can be made. Currently, the Air Force Research Laboratory at Kirtland, New Mexico, which is studying the phenomenon, gets its hafnium-178m2 from SRS Technologies, a research and development company in Huntsville, Alabama, which refines the hafnium from nuclear material left over from other experiments. The company is under contract to produce experimental sources of hafnium178m2, but only in amounts less than one ten-thousandth of a gram.

But in future there may be cheaper ways to create the hafnium isomer - by bombarding ordinary hafnium with high-energy photons, for example. Hill Roberts, chief scientist at SRS, believes that technology to produce gram quantities will exist within five years. The price is likely to be high- similar to enriched uranium, which costs thousands of dollars per kilogram- but unlike uranium it can be used in any quantity, as it does not require a critical mass to maintain the nuclear reaction. The hafnium explosive could be extremely powerful. One gram of fully charged hafnium isomer could store more energy than 50 kilograms of TNT. Miniature missiles could be made with warheads that are far more powerful than existing conventional weapons, giving massively enhanced firepower to the armed forces using them.

The effect of a nuclear-isomer explosion would be to release high-energy gamma rays capable of killing any living thing in the immediate area. It would cause little fallout compared to a fission explosion, but any undetonated isomer would be dispersed as small radioactive particles, making it a somewhat "dirty" bomb. This material could cause long-term health problems for anybody who breathed it in.

There would also be political fallout. In the 1950s, the US backed away from developing nuclear mini-weapons such as the "Davy Crockett" nuclear bazooka that delivered an explosive punch of 18 tonnes of TNT. These weapons blurred the divide between the explosive power of nuclear and conventional weapons, and the government feared that military commanders would be more likely to use nuclear weapons that had a similar effect on the battlefield to conventional weapons. By ensuring that the explosive power of a nuclear weapon was always far greater, it hoped that they could only be used in exceptional circumstance when a dramatic escalation of force was deemed necessary. Then in 1994, the US confirmed this policy with the Spratt-Furse law, which prevents US military from developing mini-nukes of less than five kilotons. But the development of a new weapon that spans the gap between the explosive power of nuclear and conventional weapons would remove this restraint, giving commanders a way of increasing the amount of force they can use in a series of small steps. Nuclear-isomer weapons could be a major advantage to armies possessing them, leading to the possibility of an arms race.

A Concern for National Defence

DOE was interested in the hafnium claims because, if verified, they presented new national security issues as well as potential scientific and energy applications.

Nuclear isomers with a long lifetime and a high energy release offer the potential to be a stand-alone energy source. However, if scientists could accelerate an isomer's decay so that gamma rays are emitted in an instantaneous burst, they could also use these isomers to develop propellants and explosives.

"Many applications might be possible if we could duplicate the gamma-ray burst reported in the original research," says Livermore physicist John Becker, who led the tri-laboratory effort. "For example, we might consider developing a gamma-ray laser or a nuclear battery that could power a spacecraft. But we did a textbook nuclear physics experiment, and our results did not match the original claim."

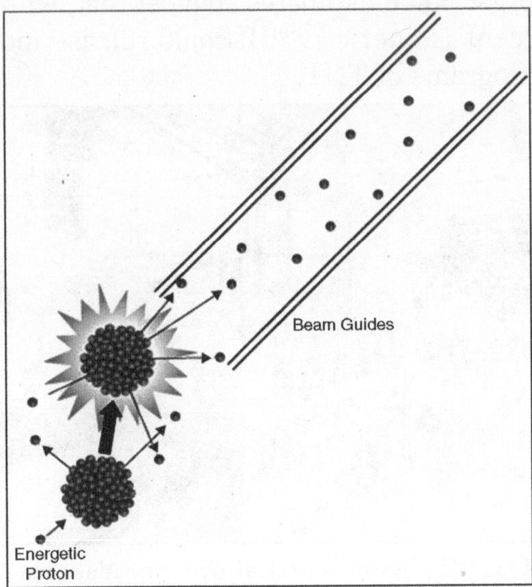

When a high-energy proton bombards a heavy atomic nucleus, the nucleus becomes excited and expels 20 to 30 neutrons—a process known as the spallation reaction.

A Nucleus with Energy to Burn?

A nuclear isomer is an atom whose nucleus is in a higher energy state than its ground state. This excited state is very long-lived compared with the usual lifetimes of excited nuclear states. The long lifetime results because transition to the ground state would require a large change either in the spatial structure of the atom's nucleus (for a shape isomer) or in the angular momentum (a measure of the spin of the nucleus) between the isomer and the nuclear ground state (for a spin isomer). Both types of isomers release energy, usually as electromagnetic radiation, when the nuclei transition from a high energy state to a lower one. Nuclear isomers have a wide range of lifetimes, radiating away the extra energy anywhere from about a picosecond (one-trillionth of a second) to years. A common nuclear isomer is tantalum-180m (^{180m}Ta), whose half-life is about 1015 years (a quadrillion years). In its ground state, tantalum-180 (^{180}Ta) is very unstable and decays to other nuclei in about 8 hours. Tantalum-180 is never found in natural samples, which are billions of years old, but its isomeric state, ^{180m}Ta, is found in natural samples. This metastable excited state has an excitation energy 75 kiloelectronvolts (keV) higher than ^{180}Ta, and its decay is inhibited because the angular momentum of the isomer's nucleus is so different from that of the ground state's nucleus.

Hafnium-178m, the isomer studied by Becker's team, has a long half-life of 31 years and a high excitation energy of 2.4 megaelectronvolts (MeV). As a result, 1 kilogram of pure ^{178m}Hf contains approximately 1 million megajoules (1012 joules) of energy. Some estimates suggest that, with accelerated decay, 1 gram of 100-per cent isomeric ^{178m}Hf could release more energy than the detonation of 200 kilograms of TNT.

Fig. The Advanced Photon Source at Argonne National Laboratory.

Reality Check

To verify the results of the earlier experiments, the tri-laboratory collaboration used Argonne's Advanced Photon Source (APS) as the x-ray trigger for the energy release. This light source is 100,000 times more intense than the x-ray machine used in the 1990s. Thus, if results from those experiments were valid, Becker's team would easily see gamma-ray emissions characteristic of the accelerated decay of ^{178m}Hf.

Scientists at the Los Alamos Neutron Science Center fabricated the isomer samples for the Argonne experiments. Using the center's proton linear accelerator, they bombarded the atomic nucleus of tantalum with 800-MeV photons to induce spallation—a nuclear reaction that causes neutrons to be knocked out of the atom's nucleus, or spalled. The team then separated this material to produce hafnium samples containing about 4 parts ^{178m}Hf to 10,000 parts ^{178}Hf.

With Argonne's APS, the tri-laboratory team generated a broadband x-ray beam with a wide range of wavelengths, called a white beam, to stimulate the isomer samples. "The precise x-ray energy required to induce the hafnium decay had not been cited in the earlier results," says Becker. "We chose to bombard the sample with a white beam of x rays so that our experiments would include all the energies at which enhanced decay had been reported."

If the previous results were valid, the tri-laboratory experiments would result in the prompt decay of the isomer samples—that is, decay time would decrease from 31 years to less than seconds. Instead, the collaboration's results were consistent with expected prediction of nuclear physics. In fact, the team's cross-section limits, which represent the probability of an interaction event between the x-rays and ^{178m}Hf, were more than 100,000 times lower in the relevant energy regions than those previously reported. Says Becker, "We conducted a classic experiment, but we saw no evidence of triggered decay."

These findings can allay DOE's concern about potential applications of the purported isomer energy source. X-ray induced decay of the hafnium isomer does not present a new concern for national security. It also is not a viable alternative as a stand-alone energy source. Nuclear physicists at Livermore and throughout the world continue to investigate such unusual nuclear processes.

INTERNAL CONVERSION

Internal conversion is a radioactive decay process wherein an excited nucleus interacts electromagnetically with one of its electrons. This causes the electron to be emitted (ejected) from the atom. Thus, in an internal conversion process, a high-energy electron is emitted from the radioactive atom, not from the nucleus. For this reason, the high-speed electrons resulting from internal conversion are not beta particles, since, the latter come from beta decay, where they are newly created in the nuclear decay process. Since, no beta decay takes place during internal conversion, the element atomic number does not change, and thus (as is the case with gamma decay) no transmutation of one element to another takes place. However, since, an electron is lost, an otherwise neutral atom becomes ionized. Also, no neutrino is emitted during internal conversion.

Internally converted electrons do not have the energetically spread spectrum characteristic of beta particles. The spread spectrum of beta particles results from the decay of a neutron into a proton, a beta particle (electron), and an electron antineutrino..

Varying amounts of decay energy are carried off by the antineutrino during beta decay, resulting in the spectrum of beta electrons' energy. Internally converted electrons, however, carry a fixed fraction of the characteristic decay energy, hence, they have a discrete energy. The energy spectrum of a beta particle thus plots as a broad hump, extending from essentially zero (a bound electron that does not even have enough energy to escape the atom) to a maximum decay energy value. By contrast, the energy spectrum of internally converted electrons plots as a single sharp peak.

INTERNAL CONVERSION

The excess energy of radioactive nuclei in excited states is usually relieved through gamma emission. However, if the wave function of an orbital electron is such that it can exist close to or in the nucleus, the excess energy can be transferred directly to the orbital electron. Hence, as an alternative to gamma emission, the excited nucleus may return to a lower state or to the ground state by ejecting an orbital electron. This is known as internal conversion and results in the emission of a mono-energetic electron. Following the internal conversion, outer orbital electrons fill the deeper atomic energy levels, leading to characteristic X-ray emission or, alternatively, to the emission of Auger electrons.

The internal conversion process can be described by:

$$ {}_{Z}^{A}P^{*} \rightarrow \left[{}_{Z}^{A}P \right]^{*} + e^{-} $$

The ratio of internal conversion to gamma emission photons is known as the internal conversion coefficient α where,

$$ \alpha = \frac{N_e}{N_\gamma} $$

where N_e is the number of de-excitations via electron emission and N_γ the number of de-excitations via gamma emission. The kinetic energy E_e of the ejected electron is almost equal to the excitation energy E^* less the binding energy of the orbital electron E_B , *i.e.*

$$ E_e = E^{*} - E_B $$

The internal conversion coefficent alpha depends on the atomic number and the excitation energy, *i.e.*

$$\alpha \propto \frac{Z^3}{E^*}$$

Internal conversion is more important in heavy nuclei in particular in the decay of low-lying excited states (small E^*). Gamma emission predominated in light nuclei. Example:

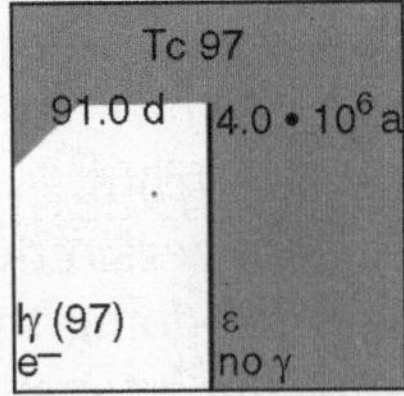

Fig. Tc97m in the Karlsruhe Nuclide Chart.

Consider the decay of the isomeric state Tc97m (half-life 91 days). The box is mainly white indicating isomeric transition as the main decay mode and the small red triangle indicates electron capture to the daughter Mo97. The nuclide data shows $I\gamma$ (97) and e^-. The symbol e^- indicates that the converison coefficient is > 1 *i.e.* the state de-excites mainly by electron conversion rather than by gamma emission.

In addition, the fact that the energy of the isomeric transition is in backets *i.e.* (97) implies that gamma emission occurs in less that 1percent of the de-excitations. Hence, the electron conversion occurs is over 99percent over the de-excitations. The actual conversion coefficient is not given in the chart, but can be obtained from the literature e.g nuclear Data sheets or from Nucleonica.

Internal Conversion

The nucleus can be de-excited by γ-ray emission, but also by emitting a closely bound electron from the atom. Usually a *K*-electron is emitted, but an electron in the *L*-shell (or a higher shell) can also be emitted.

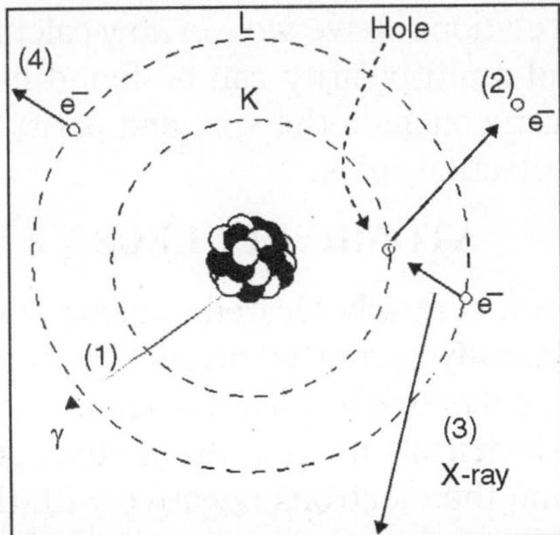

Fig. A schematic picture of different ways to de-excite an atomic nucleus.

The electron hole that appears will soon be filled by another electron. This can result in the emission of an X-ray photon, or the emission of an Auger electron. The internal conversion coefficient is defined as,

$$\alpha_{ik} = \frac{I_e}{I_\gamma}$$

where I_e is the intensity of the conversion electrons and I_γ is the intensity of the gamma radiation. I_γ can easily be determined by calculating the area of the corresponding peak in the energy spectrum.
To determine the intensity of the conversion electrons we will not measure the electrons directly in this case (for doing this we would need an electron detector). Instead, we will measure the intensity of the X-ray radiation that is sent out when the electron holes (created by sending out the conversion electrons) are filled by electrons from higher lying shells. This intensity must however be corrected to account for the Auger effect, since, this process competes with X-ray emission.
The ratio between the X-ray intensity I_X and the intensity of internal conversion I_{ik} is given as,

$$\eta = \frac{I_X}{I_{ik}}$$

Diagrams for η can be found in appendix A. Another problem is that we must compensate for the efficiency of the detector. This efficiency is energy dependent, so an efficiency curve can be plotted by measuring the intensities of some γ-ray peaks for a number of calibration sources. Using the efficiency curve, the ratio between the detector efficiencies for the X-rays and γ-rays of interest can be measured as:

$$\in = \frac{\in_\gamma}{\in_X}$$

By combining the relations above we can now calculate α_{ik}. Diagrams that relate α_{ik} to energy and multi-polarity can be found in the laboratory. From this multi-polarity we can conclude the spin and parity for the 662 keV state of ^{137}Ba by using the selection rules.

ATOMIC NUCLEUS

An atom consists of a centrally located nucleus surrounded by electrons revolving in certain physically permitted orbits. The nucleus itself is made up ofneutrons and protons, collectively called nucleons.

The neutrons are electrically neutral, the protons positive (with 1.6×10^{-19} coulomb of charge) and the electrons negative with the same magnitude of charge. The nuclear dimension is in the range of 10^{-13} 10^{-12} cm, while the atomic dimension is about 10^{-8} cm.

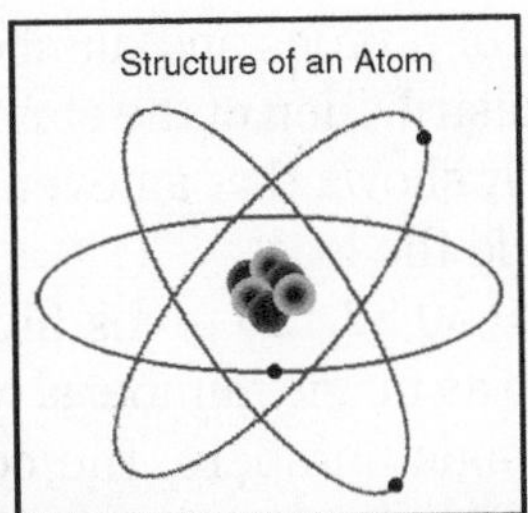

That is, the diametre of an atom is over 10,000 times the diametre of its nucleus. 1 cm is nearly the width of our little finger. The atomic diametre is about one hundred millionth the width of our little finger. The nuclear size is 1 billionth the width of our little finger.

This nuclear range is called a short-range, wherein the nucleons attract each other with a force, much larger than the electrostatic repulsion between the positively charged protons. This means, two neutrons attract each other though they are neutral; two protons attract each other though they are both positively charged; a proton and a neutron attract each other though they are not of opposite charges. In other words the nucleons bind themselves together when they happen to come closer than 10^{-12} cm. Within this range, the closer they are the stronger their bond is.

The electron orbit need not be circular, but may be oval in shape. Generally, there are as many electrons around the nucleus as there are protons in the nucleus, thus making the atom as a whole, neutral.

Please note that the atomic neutrality is with respect to a point sufficiently away from the atom, since the influences of electrons and protons cancel out. The respective charge would be felt, very close to the electron or the nucleus. If one or more electrons are removed from the orbit, the atom has a net positive charge, and is said to be positively ionized. Similarly, if electrons are added to the orbits, the atom becomes negatively ionized.

The number of protons (Z) is called the atomic number and the total number (A) of nucleons in a nucleus is called the atomic (or nuclear) mass number. Occasionally the number of neutrons (A-Z) is represented as N.

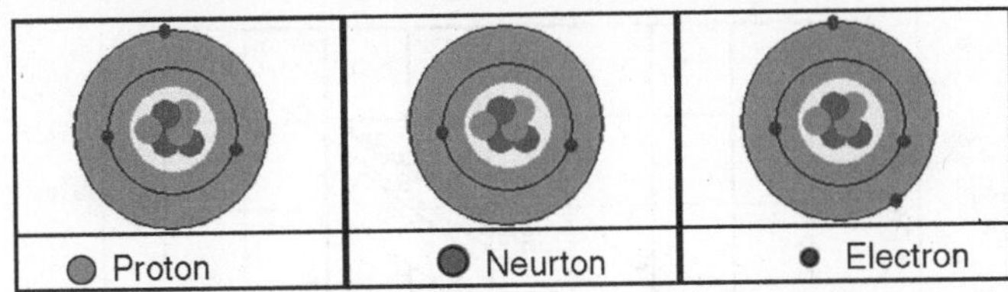

There are about 266 stable nuclides, and 65 radioactive nuclides in nature. A radioactive nucleus is one in an unstable configuration.

It approaches stable configuration(s), by emitting radiation in the form of gamma energy, or particles like alpha, beta, neutron, etc., or by splitting into fragments. The internal energy state of a nucleus is responsible for the stability of a nucleus, and is affected by the number of protons and number of neutrons in the nucleus. The following table gives the distribution of the stable nuclei with respect to even/odd combinations of Z and N. It shows that an even-even combination of Z and N gives high stability, and odd-odd the least.

Further, Z or N = 2,8,20,50,82,126 gives high stability to a nucleus, and these numbers are referred to as magic numbers. If Z and N of a nucleus happen to be individually equal to magic numbers, the configuration is called doubly magic leading to exceptional stability.

The magic numbers are not mythical, though they appear to be so. A nucleus is understood to be formed in a shell-structure, with each shell having a capacity for filling with neutrons or protons. Each shell gets closed when it is filled to full capacity. Quantum mechanical arguments show that the capacities of different shells correspond to the (so called) magic numbers. If all the shells of a nucleus are closed, it becomes very stable.

As said, the stability of a nucleus is influenced by the numbers of protons and neutrons in it. Hydrogen $^1H_1$ has one proton and no neutron. This is the only nucleus without a neutron. An isotope of hydrogen, *viz.* heavy hydrogen $^2H_1$, some times called deuterium $^2D_1$, has a proton and a neutron. Light stable nuclei, except $^1H_1$, have generally equal number of protons and neutrons. As the nucleus becomes heavier, the N/Z ratio, or the neutron-to-proton ratio of stable nuclides increases.

As said earlier, the nuclear short-range attractive force is the same between any type of nucleons. As the number of protons increases, the electric repulsive force becomes more and more significant. Neutrons have no electric repulsion but only attraction. Hence adequate presence of neutrons will help overcome the average repulsive force per nucleon in the nucleus.

Thus stability of a high-Z nucleus demands that the number of neutrons well exceeds the number of protons. In other words, more neutrons than protons are needed to support the binding of the positively charged protons within the nucleus, as the proton number gets larger. One must not jump to the conclusion that the more the neutrons, the more stable the nucleus is.

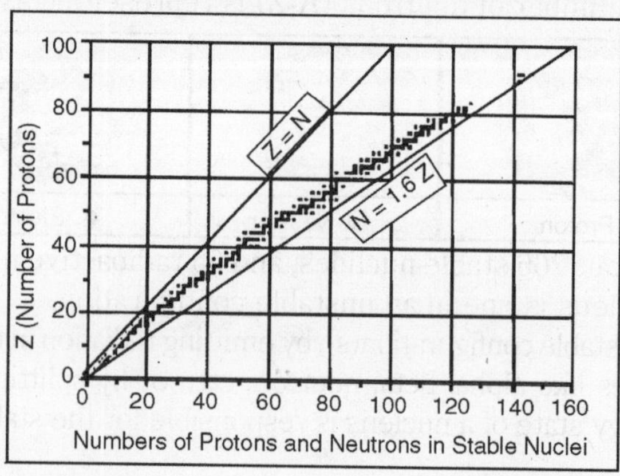

The picture below shows the numbers of neutrons and protons in stable nuclei and illustrates how the N/Z ratio deviates from unity. The black squares indicating stable nuclides form an approximate line called the Stability Line. Sometimes it is called the beta stability line.

Excessive or inadequate number of neutrons, as compared to the stability-line, leads to instability of the nucleus, leading to its radioactivity. Some of the very stable (magic) nuclei (*e.g.* lead, with 82 protons and 126 neutrons), are shown as open squares.

The masses of sub-atomic particles are often expressed in atomic mass unit (amu. This shows that a neutron is slightly heavier than a proton, and an electron is lighter by a factor of about 1840.

According to Einstein, energy and mass are related. Energy = mass} c^2, where c is the speed of light = 2.998} 10^8 m/s. The unit of energy often chosen in nuclear physics is an electron-volt (eV). 1 eV is the energy associated with an electron moving across a potential difference of 1 Volt. 1 million eV is denoted as MeV. 1 MeV = 1.602×10^{-13} Joules. 1 amu = 931.5 MeV.

LOW-ENERGY NUCLEAR PHYSICS

The field of low-energy nuclear physics, which concentrates mainly on structure of and low-energy reaction on nuclei, has become one of the smaller parts of nuclear physics (apart from in the UK). Notable results have included better understanding of the nuclear medium, high-spin physics, superdeformation and halo nuclei. Current experimental interest is in those nuclei near the "driplines" which are of astrophysical importance, as well as of other interest.

MEDIUM-ENERGY NUCLEAR PHYSICS

Medium energy nuclear physics is interested in the response of a nucleus to probes at such energies that we can no longer consider nucleons to be elementary particles. Most modern experiments are done by electron scattering, and concentrate on the role of QCD in nuclei, the structure of mesons in nuclei and other complicated questions.

HIGH-ENERGY NUCLEAR PHYSICS

This is not a very well-defined field, since particle physicists are also working here. It is mainly concerned with ultra-relativistic scattering of nuclei from each other, addressing questions about the quark-gluon plasma. It should be nuclear physics, since we consider "dirty" systems of many particles, which are what nuclear physicists are good at.

MESONS, LEPTONS AND NEUTRINOS

In 1934 Yukawa introduces a new particle, the pion (π), which can be used to describe nuclear binding. He estimates it's mass at 200 electron masses. In

1937 such a particle is first seen in cosmic rays. It is later realised that it interacts too weakly to be the pion and is actually a lepton (electron-like particle) called the μ. The is found (in cosmic rays) and is the progenitor of the μ's that were seen before:

$$\pi^+ \rightarrow \mu^+ + \nu_\mu$$

The next year artificial pions are produced in an accelerator, and in 1950 the neutral pion is found,

$$\pi^0 \rightarrow \gamma\gamma .$$

This is an example of the conservation of electric charge. Already in 1938 Stuckelberg had found that there are other conserved quantities: the number of baryons (n and p and...) is also conserved!

After a serious break in the work during the latter part of WWII, activity resumed again. The theory of electrons and positrons interacting through the electromagnetic field (photons) was tackled seriously, and with important contributions of (amongst others) Tomonaga, Schwinger and Feynman was developed into a highly accurate tool to describe hyperfine structure.

Experimental activity also resumed. Cosmic rays still provided an important source of extremely energetic particles, and in 1947 a "strange" particle (K^+ was discovered through its very peculiar decay pattern. Balloon experiments led to additional discoveries: So-called V particles were found, which were neutral particles, identified as the Λ^0 and K^0. It was realised that a new conserved quantity had been found. It was called strangeness. The technological development around WWII led to an explosion in the use of accelerators, and more and more particles were found. A few of the important ones are the antiproton, which was first seen in 1955, and the Δ, a very peculiar excited state of the nucleon, that comes in four charge states $\Delta^{++}, \Delta^+, \Delta^0, \Delta^-$.

Theory was develop-ping rapidly as well. A few highlights: In 1954 Yang and Mills develop the concept of gauged Yang-Mills fields. It looked like a mathematical game at the time, but it proved to be the key tool in developing what is now called "the standard model". In 1956 Yang and Lee make the revolutionary suggestion that parity is not necessarily conserved in the weak interactions. In the same year "madam" CS Wu and Alder show experimentally that this is true: God is weakly left-handed! In 1957 Schwinger, Bludman and Glashow suggest that all weak interactions (radioactive decay) are mediated by the charged bosons $W^\pm$. In 1961 Gell-Mann and Ne'eman introduce the "eightfold way": a mathematical taxonomy to organize the particle zoo.

THE SUB-STRUCTURE OF THE NUCLEON (QCD)

In 1964 Gell-mann and Zweig introduce the idea of quarks: particles with spin 1/2 and fractional charges. They are called up, down and strange and have charges 2/3,-1/3,-1/3 times the electron charge. Since it was found (in 1962) that electrons and muons are each accompanied by their own neutrino, it is proposed to organize the quarks in multiplets as well:

$$e^{v} e(u,d)\mu v\mu(s,c)$$

This requires a fourth quark, which is called charm.

In 1965 Greenberg, Han and Nambu explain why we can't see quarks: quarks carry colour charge, and all observe particles must have colour charge 0. Mesons have a quark and an antiquark, and baryons must be build from three quarks through its peculiar symmetry. The first evidence of quarks is found (1969) in an experiment at SLAC, where small pips inside the proton are seen. This gives additional impetus to develop a theory that incorporates some of the ideas already found: this is called QCD. It is shown that even though quarks and gluons (the building blocks of the theory) exist, they cannot be created as free particles. At very high energies (very short distances) it is found that they behave more and more like real free particles. This explains the SLAC experiment, and is called asymptotic freedom.

The J/ψ meson is discovered in 1974, and proves to be the $c\bar{c}$ bound state. Other mesons are discovered (D0, $\bar{\mu}$ c) and agree with QCD. In 1976 a third lepton, a heavy electron, is discovered (τ). This was unexpected! A matching quark (b for bottom or beauty) is found in 1977. Where is its partner, the top? It will only be found in 1995, and has a mass of 175 GeV/c^2 (similar to a lead nucleus...)! Together with the conclusion that there are no further light neutrinos (and one might hope no quarks and charged leptons) this closes a chapter in particle physics.

THE $W^{\pm}$ AND Z BOSONS

On the other side a electro-weak interaction is developed by Weinberg and Salam. A few years later 't Hooft shows that it is a well-posed theory. This predicts the existence of three extremely heavy bosons that mediate the weak force: the Z^0 and the $W^{\pm}$. These have been found in 1983. There is one more particle predicted by these theories: the Higgs particle. Must be very heavy!

GUTS, Supersymmetry, Supergravity

This is not the end of the story. The standard model is surprisingly inelegant, and contains way to many parameters for theorists to be happy. There is a dark mass problem in astrophysics-most of the mass in the universe is not seen! This all leads to the idea of an underlying theory. Many different ideas have been developed, but experiment will have the last word! It might already be getting some signals: researchers at DESY see a new signal in a region of particle that are 200 GeV heavy-it might be noise, but it could well be significant! There are several ideas floating around: one is the grand-unified theory, where we try to comine all the disparate forces in nature in one big theoretical frame. Not unrelated is the idea of supersymmetries: For every "boson" we have a "fermion". There are some indications that such theories may actually be able to make useful predictions.

Extraterrestrial Particle Physics

One of the problems is that it is difficult to see how e can actually build a microscope that can look a a small enough scale, *i.e.*, how we can build an accelerator that will be able to accelarte particles to high enough energies? The answer is simple-and has been more or less the same through the years: Look at the cosmos. Processes on an astrophysical scale can have amazing energies.

Balloon Experiments

One of the most used techniques is to use balloons to send up some instrumentation. Once the atmosphere is no longer the perturbing factor it normally is, one can then try to detect interesting physics. A problem is the relatively limited payload that can be carried by a balloon.

Ground Based Systems

These days people concentrate on those rare, extremely high energy processes (of about 10^{29} eV), where the effect of the atmosphere actually help detection. The trick is to look at showers of (lower-energy) particles created when such a high-energy particle travels through the earth's atmosphere.

Dark Matter

One of the interesting cosmological questions is whether we live in an open or closed universe. From various measurements we seem to get conflicting indications about the mass density of (parts of) the universe. It seems that the ration of luminous to non-luminous matter is rather small. Where is all that "dark mass": Mini-jupiters, small planetoids, dust, or new particles....

(Solar) Neutrinos

The neutrino is a very interesting particle. Even though we believe that we understand the nuclear physics of the sun, the number of neutrinos emitted from the sun seems to anomalously small. Unfortunately this is very hard to measure, and one needs quite a few different experiments to disentangle the physics behind these processes. Such experiments are coming on line in the next few years. These can also look at neutrinos coming from other astrophysical sources, such as supernovas, and enhance our understanding of those processes. Current indications from Kamiokande are that neutrinos do have mass, but oscillation problems still need to be resolved.

PARTICLE ACCELERATORS

Beams of high-energy particles are useful for both fundamental and applied research in the sciences. For the most basic enquiries into the dynamics and structure of matter, space, and time, physicists seek the simplest kinds of interactions at the highest possible energies.

These typically entail particle energies of many GeV, and the interactions of the simplest kinds of particles: leptons (*e.g.* electrons and positrons) and quarks for the matter, or photons and gluons for the field quanta. Since isolated quarks are experimentally unavailable due to colour confinement, the simplest available experiments involve the interactions of, first, leptons with each other, and second, of leptons with nucleons, which are composed of quarks and gluons.

To study the collisions of quarks with each other, scientists resort to collisions of nucleons, which at high energy may be usefully considered as essentially 2-body interactions of the quarks and gluons of which they are composed.

Thus elementary particle physicists tend to use machines creating beams of electrons, positrons, protons, and anti-protons, interacting with each other or with the simplest nuclei (eg, hydrogen or deuterium) at the highest possible energies, generally hundreds of GeV or more.

At a higher level of complexity, nuclear physicists and cosmologists may use beams of bare atomic nuclei, stripped of electrons, to investigate the structure, interactions, and properties of the nuclei themselves, and of condensed matter at extremely high temperatures and densities, such as might have occurred in the first moments of the Big Bang. These investigations often involve collisions of heavy nuclei--of atoms like iron or gold--at energies of several GeV per nucleon. At lower energies, beams of accelerated nuclei are also used in medicine, as for the treatment of cancer. Besides being of fundamental interest, high energy electrons may be coaxed into emitting extremely bright and coherent beams of high energy photons--ultraviolet and *X* ray--via synchrotron radiation, which photons have numerous uses in the study of atomic structure, chemistry, condensed matter physics, biology, and technology.

Examples include the ESRF in Europe, which has recently been used to extract detailed 3-dimensional images of insects trapped in amber. Thus there is a great demand for electron accelerators of moderate (GeV) energy and high intensity.

LOW-ENERGY MACHINES

Everyday examples of particle accelerators are those found in television sets and X-ray generators. Low-energy accelerators such as cathode ray tubes and X-ray generators use a single pair of electrodes with a DC voltage of a few thousand volts between them. In an X-ray generator, the target itself is one of the electrodes. A low-energy particle accelerator called an ion implanter is used in the manufacture of integrated circuits.

High-energy Machines

In early particle accelerators a Cockcroft-Walton voltage multiplier was responsible for voltage multiplying. This piece of the accelerator helped in the development of the atomic bomb. Built in 1937 by Philips of Eindhoven it currently resides in the National Science Museum in London, England.

DC accelerator types capable of accelerating particles to speeds sufficient to cause nuclear reactions are Cockcroft-Walton generators or voltage multipliers, which convert AC to high voltage DC, or Van de Graaff generators that use static electricity carried by belts.

The largest and most powerful particle accelerators, such as the RHIC, the Large Hadron Collider (LHC) and the Tevatron, are used for experimental particle physics. Particle accelerators can also produce proton beams, which can produce "proton-heavy" medical or research isotopes as opposed to the "neutron-heavy" ones made in fission reactors. An example of this type of machine is LANSCE at Los Alamos.

ELEMENTARY PARTICLES

The first elementary particle, the electron, was discovered by Thompson. But elementary particle physics as a separate area of physics was really started in the 1940's and 1950's, some time after the discovery of proton, neutron, and positron, that are all elementary particles (at least they were regarded as such until about 1970). There were two important reasons for the blossoming of elementary physics, one experimental and one theoretical.

The experimental impetus was given by the observation that very high energy particles, first observed in cosmic radiation, are able to initiate reactions, similar to nuclear interactions, when collide with the nuclei of O, N, H... of the atmosphere. It was observed, using photographic emulsions flown at high altitude, that when particles of the cosmic radiation (mostly protons) collide with nuclei they produce a large number of new particles, at very narrow forward angles, sometimes a far larger number than the number of nucleons in the target nucleus.

The large cross section for these collisons and the shear number of particles produced were a proof that the interaction between the nucleus and the incoming cosmic ray particle is very strong. Thus, it was immediately thought that most of the particles produced were not nucleons, but rather the quanta of the strong force.

The existence of these quanta, called p-mesons, was predicted on purely theoretical grounds by Hideki Yukawa in the 1930's. It was soon proved that indeed most of the particles produced of these high energy collisions are p-mesons. Their approximate mass was also predicted successfully based on the structure of Yukawa interactions,

$$V_{Yukawa} = - g^2 e^{- m\, r\, c/\, hbar/}\, r = - g^2 e^{- r/\, R}$$

where m is the mass of the particle communicating the interaction. Note that electromagnetic interactions are similar if we take $g^2 \to k\, e^2$ and m=0 (the mass of the photon is 0!). By estimating the range of nuclear interactions, which is equal to R = hbar/(m c) = hbar c/ (m c^2) one is able to come up with a number,

$$mc^2 = \text{hbar } c/\, R = 197.3 \text{ MeV f}/\, 1.2 \text{ f} = 164 \text{ MeV}.$$

The true mass of p-mesons is 138 MeV, remarkably close to this crude estimate. The investigation of interactions of nucleons and p-mesons lead to the development of elementary particle physics. The other ingredient in the emergence of elementary particle physics was the success of quantum field theory that became and still is the only successful framework of describing the interactions (creation, annihilation, and scattering) of elementary particles. Quantum field theory was started by Dirac around 1930 from the marriage of relativity and quantum mechanics.

Though the theoretical framework was understood quanum field theory was useless for most practical reasons because calculations, beyond a trivial first approximation cannot be made any sense, they were plagued by divergences (infinities) that could not be eliminated. It took almost two decades until Tomonaga, Schwinger, and Feynman could device a self-consistent method of eliminating these divergences. This method is called renormalization.

The basic idea behind renormalization can be understood easiest in is the framework of Quantum electrodynamics, the quantum field theory describing the interaction of photons with charged particles, such as electrons. Quantum field theory implies that the vacuum is not really empty, but filled with "virtual" particles that are keep being produced and annihilated. Conservation of energy momentum is satisfied because the mass-energy of these virtual particles does not equal to the mass-energies observed for physical particles traveling in the vacuum.

Virtual particles live for a very short time, before annihilating, thus the uncertainty relation between time and energy requires that their mass-energy is undetermined to a large degree. The presence of these virtual particles on real electrons is profound and twofold. The mass-energy of the real electron is changed substantially by interactions with the bath of virtual particles around it. Note that even classically the self energy of a pointlike electron is infinite. It is also infinite, due to interactions with virtual particles, in quantum field theory.

These interactions will give an infinite contribution to the mass energy of electrons. It turns out, however that if the mass-energy of the electron (the so-called bare mass-energy) without the interaction with virtual particles is chosen to be infinitely large and negative then the sum of contributions can be made finite and one can end up with a finite mass in a consistent series of approximations of arbitrary precision. Similarly, the other parameter describing electrons, their charge, can also be treated in a similar manner. Note that the presence of the negatively charged electron polarizes virtual particles, the total charge of which must be equal to zero.

Thus, the electron sits in a polarized dielectric medium. Then, observed from practically infinite distance (that is what experimental observations practically do) the charge of the electron seems considerable smaller due to screening. In fact if, starting form infinity, one would be able to go closer and closer to the electron, one would observe a larger and larger negative charge. One can show that if one wishes to keep the observed charge (the one observed from infinite distance) finite then the charge one would observe at zero distance from the electron (the so-called bare charge), when all the layers of screening are peeled away, would have to be infinitely large negative. Conversely, chosing the "bare charge" to be infinite one can build up a scheme of systematic calculations that can in principle performed to arbitrary precision. Calculations performed following the method of Tomonaga, Schwinger, Feynman and Dyson had a tremendous success in explaining experiments of ultimate precision, such as the anomalous magnetic moment of electrons and the Lamb shift.

CLASSIFICATION OF ELEMENTARY PARTICLES

Two types of statistics are used to describe elementary particles, and the particles are classified on the basis of which statistics they obey. Fermi-Dirac statistics apply to those particles restricted by the Pauli exclusion principle; particles obeying the Fermi-Dirac statistics are known as fermions. Leptons and quarks are fermions. Two fermions are not allowed to occupy the same quantum state.

Bose-Einstein statistics apply to all particles not covered by the exclusion principle, and such particles are known as bosons. The number of bosons in a given quantum state is not restricted. In general, fermions compose nuclear and atomic structure, while bosons act to transmit forces between fermions; the photon, gluon, and the *W* and *Z* particles are bosons. Basic categories of particles have also been distinguished according to other particle behaviour. The strongly interacting particles were classified as either mesons or baryons; it is now known that mesons consist of quark-antiquark pairs and that baryons consist of quark triplets.

The meson class members are more massive than the leptons but generally less massive than the proton and neutron, although some mesons are heavier than these particles. The lightest members of the baryon class are the proton and neutron, and the heavier members are known as hyperons.

In the meson and baryon classes are included a number of particles that cannot be detected directly because their lifetimes are so short that they leave no tracks in a cloud chamber or bubble chamber. These particles are known as resonances, or resonance states, because of an analogy between their manner of creation and the resonance of an electrical circuit.

The fundamental particles may be classified into groups in several ways. First, all particles are classified into fermions, which obey Fermi-Dirac statistics and bosons, which obey Bose-Einstein statistics. Fermions have half-integer spin, while bosons have integer spin.

All the fundamental fermions have spin 1/2. Electrons and nucleons are fermions with spin 1/2. The fundamental bosons have mostly spin 1. This includes the photon. The pion has spin 0, while the graviton has spin 2. There are also three particles, the W^+, W^- and Z_0 bosons, which are spin 1. They are the carriers of the weak interactions.

We can also classify the particles according to their interactions.

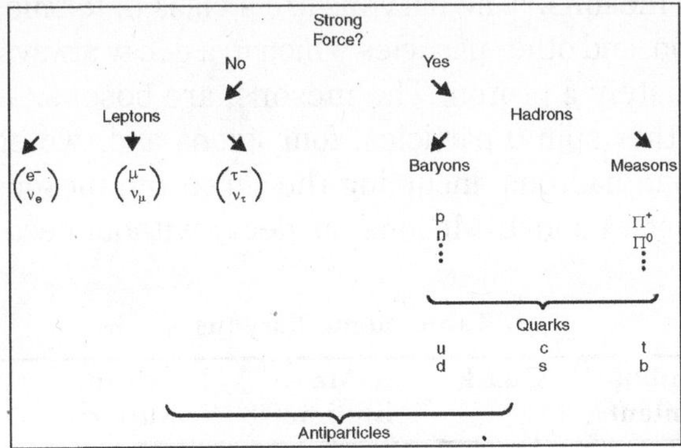

The electron and the neutrino are members of a family of leptons. Originally leptons meant "light particles", as opposed to baryons, or heavy particles, which referred initially to the proton and neutron.

The pion, or pi-meson, and another particle called the muon or mu-meson, were called mesons, or medium-weight particles, because their masses, a few hundred times heavier than the electron but six times lighter than a proton, were in the middle. But that distinction turned out not to be very useful. We now recognise the muon to be almost the same as an electron, and the leptons now consist of three "generations" of pairs of particles,

$$\begin{pmatrix} e^- \\ \nu_e \end{pmatrix}, \begin{pmatrix} \mu^- \\ \nu_\mu \end{pmatrix}, \begin{pmatrix} \tau^- \\ \nu_\tau \end{pmatrix}$$

with the heaviest of these, the tau lepton τ^-, being almost twice as massive as the proton

The leptons are distinguished from other particles called hadrons in that leptons do not participate in strong interactions.

The bottom lepton in each of the three "doublets" shown above is not only neutral, but also has a very small mass. Neutrinos had been considered massless for many years, but more recent experiments have shown their mass to be non-zero.

Table. Leptons

	Particle			**Associated Neutrino**	
Name	**Charge(e)**	**Mass(MeV)**	**Name**	**Charge(e)**	**Mass(MeV)**
Electron (e^-)	–1	0.511	Electron neutrino (ν_e)	0	< 0.000003
Muon (μ^-)	–1	105.6	Muon	0	< 0.19
		neutrino (ν_μ)			
Tau (τ^-)	-1	1777	Tau	0	< 18.2
		neutrino (ν_τ)			

Hadrons are strongly interacting particles. They are divided into baryons and mesons. The baryons are a class of fermions, including the proton and neutron, and other particles which in a decay always produce another baryon, and ultimately a proton. The mesons, are bosons. In addition to the pion, there are other spin 0 particles, four kaons and two eta mesons, and a number of spin one hadrons, including the three rho mesons, which like the pion come in charges 1 and 0. Mesons can decay without necessarily producing other hadrons.

Table. Some Baryons

Particle	**Symbol Content**	**Quark**	**Mass lifetime(s)**	**Mean MeV/c^2**	**Decays to**
Proton	p	uud	938.3	Stable	Unobserved
Neutron	n	ddu	939.6	885.7±0.8	$p + e^- + \nu_e$
Delta	Δ^{++}	uuu	1232	6×10^{-24}	$\Pi^+ + p$
Delta	Δ^+	uud	1232	6×10^{-24}	$\Pi^+ + n$
				or	$\Pi^0 + p$

Delta	Δ^0	udd	1232	6×10^{-24}	$\Pi^0 + n$
				or	$\Pi^- + p$
Delta	Δ^-	ddd	1232	6×10^{-24}	$\Pi^- + n$
Lambda	Λ^0	uds	1115.7	2.60×10^{-10}	$\Pi^- + p$
				or	$\Pi^0 + n$
Sigma	Σ^+	uus	1189.4	0.8×10^{-10}	$\Pi^0 + p$
				or	$\Pi^+ + n$
Sigma	Σ^0	uds	1192.5	6×10^{-20}	$\Lambda^0 + \gamma$
Sigma	Σ^-	dds	1197.4	1.5×10^{-10}	$\Pi^- + n$
Xi	Ξ^0	uss	1315	2.9×10^{-10}	$\Lambda^0 + \Pi^0$
Xi	Ξ^-	dss	1321	1.6×10^{-10}	$\Lambda^0 + \Pi^-$
Omega	Ω^-	sss	1672	0.82×10^{-10}	$\Lambda^0 + K^-$
				or	$\Xi^0 + \Pi^-$

Each elementary particle is associated with an antiparticle with the same mass and opposite charge. Some particles, such as the photon, are identical to their antiparticle. Such particles must be neutral, but not all neutral particles are identical to their antiparticle. Particle-antiparticle pairs can annihilate each other if they are in appropriate quantum states, releasing an amount of energy equal to twice the rest energy of the particle. They can also be produced in various processes, if enough energy is available.

The minimum amount of energy needed is twice the rest energy of the particle, if momentum conservation allows the particle-antiparticle pair to be produced at rest. Most often the antiparticle is denoted by the same symbol as the particle, but with a line over the symbol. For example, the antiparticle of the proton p, is denoted by p.

Table. Some Mesons

Particle	**Symbol**	**Anti -particle**	**Quark Content**	**Mass MeV/c²**	**Mean lifetime(s)**	**Principal decays**	
Charged	Π^+	Π''	ud	139.6	2.60×10^{-8}	$\mu^+ + \nu_\mu$	Pion
Neutral	Π^0	Self	uu – dd	135.0	0.84×10^{-16}	2γ	Pion
Charged	K^+	K^-	$u\bar{s}$	493.7	1.24×10^{-8}	$\mu^+ + \nu_\mu$	Kaon
		or					
$\Pi^+ + \Pi_0$							
Neutral	K^0	K^0	ds	497.7			Kaon
Eta	H	Self	uu + dd – 2ss	547.8	5×10^{-19}		
Eta Prime	η'	Self	uu + dd + ss	957.6	3×10^{-21}		

Protons and neutrons are made of still smaller particles called quarks. At this time it appears that the two basic constituents of matter are the leptons and the quarks. There are believed to be six types of each.

Each quark type is called a flavour, there are six quark flavours. Each type of lepton and quark also has a corresponding antiparticle, a particle that has the same mass but opposite electrical charge and magnetic moment. An isolated quark has never been found, quarks appear to almost always be found in pairs or triplets with other quarks and antiquarks. The resulting particles are the hadrons, more than 200 of which have been identified. Baryons are made up of 3 quarks, and mesons are made up of a quark and an anti-quark. Baryons are fermions and mesons are bosons. Two theoretically predicted five-quark particles, called pentaquarks, have been produced in the laboratory. Four- and six-quark particles are also predicted but have not been found.

The six quarks have been named up, down, charm, strange, top, and bottom. The top quark, which has a mass greater than an entire atom of gold, is about 35 times more massive than the next biggest quark and may be the heaviest particle nature has ever created. The quarks found in ordinary matter are the up and down quarks, from which protons and neutrons are made. A proton consists of two up quarks and a down quark, and a neutron consists of two down quarks and an up quark. The pentaquark consists of two up quarks, two down quarks, and the strange antiquark. Quarks have fractional charges of one third or two thirds of the basic charge of the electron or proton. Particles made from quarks always have integer charge.

Table. Quarks

Name	Symbol	Charge(e)	Spin	Mass MeV/c^2	Strangeness	Baryon number	Lepton number
up	u	+2/3	1/2	1.7-3.3	0	1/3	0
down	d	-1/3	1/2	4.1-5.8	0	1/3	0
strange	s	-1/3	1/2	101	-1	1/3	0
charm	c	+2/3	1/2	1270	0	1/3	0
bottom	b	-1/3	1/2	4190-4670	0	1/3	0
top	t	+2/3	1/2	172000	0	1/3	0

In the current theory, known as the Standard Model there are 12 fundamental matter particle types and their corresponding antiparticles. In addition, there are gluons, photons, and *W* and *Z* bosons, the force carrier particles that are responsible for strong, electromagnetic, and weak interactions respectively. These force carriers are also fundamental particles.

	Ferminos			Bosons	
Quarks	*u* up	*c* charm	*t* top	γ photon	Force Carriers
Quarks	*d* down	*s* strange	*b* bottom	Z Z boson	Force Carriers
Leptons	ν_e electron neutrino	ν_μ muon neutrino	ν_τ tau neutrino	W W boson	Force Carriers
Leptons	e electron	μ muon	τ tau	g gluon	Force Carriers
	I	II	III		

Three Families of Matters

All we know is that quarks and leptons are smaller than 10^{-19} meters in radius. As far as we can tell, they have no internal structure or even any size. It is possible that future evidence will, once again, show our understanding to be incomplete and demonstrate that there is substructure within the particles that we now view as fundamental.

THE DISCOVERY OF ELEMENTARY PARTICLES

The first subatomic particle to be discovered was the electron, identified in 1897 by J. J. Thomson. After the nucleus of the atom was discovered in 1911 by Ernest Rutherford, the nucleus of ordinary hydrogen was recognised to be a single proton. In 1932 the neutron was discovered.

An atom was seen to consist of a central nucleus containing protons and neutrons, surrounded by orbiting electrons. However, other elementary particles not found in ordinary atoms immediately began to appear.

In 1928 the relativistic quantum theory of P. A. M. Dirac predicted the existence of a positively charged electron, or positron, which is the antiparticle of the electron. It was first detected in 1932. Difficulties in explaining beta decay led to the prediction of the neutrino in 1930, and by 1934 the existence of the neutrino was firmly established in theory, although it was not actually detected until 1956. Another particle was also added to the list, the photon, which had been first suggested by Einstein in 1905 as part of his quantum theory of the photoelectric effect.

The next particles discovered were related to attempts to explain the strong interactions, or strong nuclear force, binding nucleons together in an atomic nucleus. In 1935 Hideki Yukawa suggested that a meson, a charged particle with a mass intermediate between those of the electron and the proton, might be exchanged between nucleons.

The meson emitted by one nucleon would be absorbed by another nucleon. This would produce a strong force between the nucleons, analogous to the force produced by the exchange of photons between charged particles interacting through the electromagnetic force. It is now known that the strong force is mediated by the gluon.

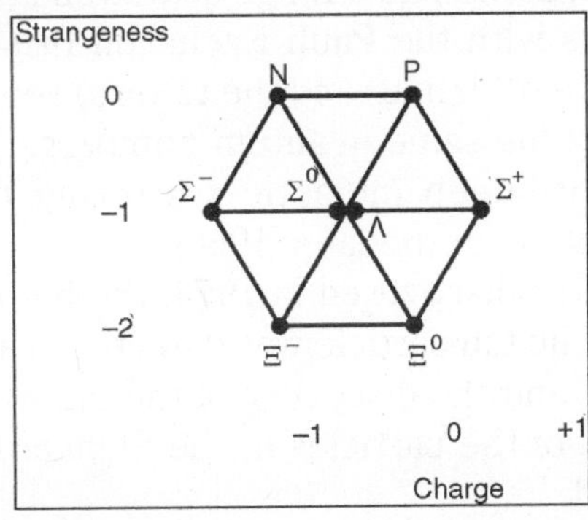

The following year a particle of approximately the required mass, about 200 times that of the electron, was discovered and named the mu-meson, or muon. However, its behaviour did not conform to that of the theoretical particle. In 1947 the particle predicted by Yukawa was finally discovered and named the pi-meson, or pion. Both the muon and the pion were first observed in cosmic rays. Further studies of cosmic rays turned up more particles. By the 1950s these elementary particles were also being observed in the laboratory as a result of particle collisions in particle accelerators.

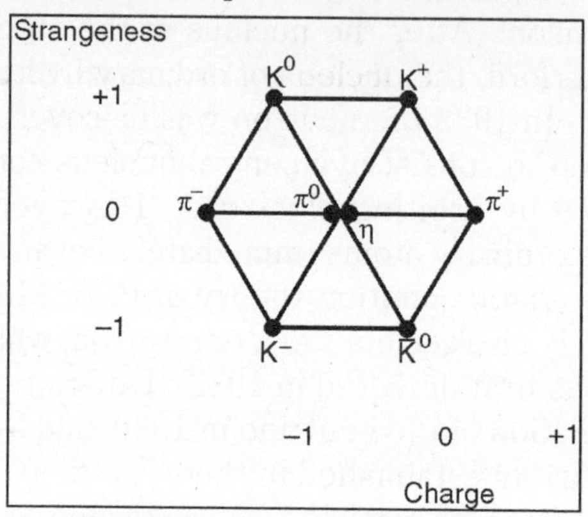

Fig. Example patterns for some baryons and mesons (The Eightfold Way)

By the early 1960s over 30 "fundamental particles" had been found. A rigourous way of classifying them was needed. Were there any symmetries or patterns? Murray Gell-Mann believed that a framework for such patterns could be found in the mathematical structure of groups. A symmetry group called SU(3) offered patterns he was looking for. In 1961, after grouping the known particles, he predicted the existence of the ç particle which was needed to complete a pattern. The ç particle was discovered a few months later.

After finding the patterns an explanation was needed. In 1964 Gell-Mann published a short article showing that the patterns could be produced if the known particles were viewed as combinations of 3 fundamental subunits with fractional charge, the up, down, and strange quarks and their antiquarks. There were however problems with the Pauli Exclusion Principle. The quarks are spin 1/2 fermions and the Δ^{++}(uuu) and the Ω^{-}(sss) seemed to contain at least two quarks with exactly the same quantum numbers. The quark theory was not really accepted until deep inelastic scattering experiments revealed structure inside the protons in the later 1960s.

The charm quark was discovered in 1974, the bottom quark in 1977, and the top quark in 1995. The tau particle was detected in a series of experiments between 1974 and 1977 and the discovery of the tau neutrino was announced in 2000. It was the last of the particles in the Standard Model of elementary particles to be detected.

One of the current frontiers in the study of elementary particles concerns the interface between that discipline and cosmology. The known quarks and leptons, for instance, are typically grouped in three families, where each family contains two quarks and two leptons. Investigators have wondered whether additional families of elementary particles might be found. Recent work in cosmology pertaining to the evolution of the universe has suggested that there could be no more families than four, and the cosmological theory has been substantiated by experimental work at the Stanford Linear Accelerator and at CERN, which indicates that there are no families of elementary particles other than the three that are known today. For example, detailed studies of Z_0 decays at CERN revealed that there can be no more than three different kinds of neutrino. If there was a fourth, or a fifth, further decay routes would be open to the Z_0 which would affect its measured lifetime.

PARTICLE INTERACTIONS, RESONANCES, AND QUARKS

After the discovery of mesons and hyperons experimentalists found a large number of excited states of these particles as well. The typical example of these resonances is the Δ particle that has four charge states (isotopic spin, I=3/2) Δ^{++}, Δ^{+}, Δ^{0}, and Δ^{-}· This resonance was found by Fermi in an experiment in which he bombarded protons (hydrogen) by a newly created p-meson been at the Bevatron accelerator in Berkeley. The reaction he observed was,

$$\pi^- + \pi \rightarrow \Delta^0 \rightarrow \pi^- + p$$

Now the intermediate state, Δ^0 is of extremely short lifetime. In fact, t ~ 10^{-23} s. Direct observation of such a particle is impossible. The uncertainty relation between time and energy, however, implies that the energy of the Δ is uncertain to about 100 MeV. The Δ^- p cross section increases when the total energy-momentum squared $E^2 = (E_p + E_\pi)^2 - c^2 (p_p + p_\pi)^2 = m_\Delta{}^2 c^4$ This is so because there is a new channel throgh which the scattering can happen.

Due to the uncertainty of m_Δ, $\delta\, m_\Delta \sim$ hbar/ $(c^2\, \tau)$, the maximum in the p^- p cross section is not sharp, but smeared over a region of about 100Mev in the primary energy of π^-· Thus, if we plot the cross section as a function of the energy of the pion (performing the experiment at a series of different energies) then we will see a large bump at $E^2 = m_\Delta{}^2 c^4$. The width of the bump provides the lifetime of the Δ particle. This particle is a baryon if you require baryon number conservation. It also turns out that the spin of this particle is s =3/2. The mass of Δ is m_Δ=1232 MeV/ c^2

As time went by many more particle resonances were discovered. These resonances can be regarded as new particles on their own right. They just happen to have large enough masses so that conservation laws allow them to decay via strong interactions into strongly interacting particles.

That is the reason why their lifetime are so short. As an example the mass of Δ happens to satisfy $m_\Delta > m_\pi + m_p$ so Δ is allowed to decay into π + p. After literally hundeds of resonances with exactly the same quantum numbers as some other particles were discovered the situation became untenable.

This looked more or more as the a system of states with the excitations of the same system, like an atom or a nucleus. The existence of such excitations is a sure sign of composite nature. Clearly atoms and nuclei are composite.

This gave the idea to Nambu and Gell-Mann to postulate the existence of constituents for strongly interacting particles: quarks. The most economic choice (due to Gell-Mann) turned out to be very strange: All normal and strange strongly interacting particles in every charge mode can be constructed out of three quarks, and their antiparticles (antiquarks). The three quarks are u (up), d (down), and s (strange) quarks. Their charge are not integer times e.

quark	*charge*	*spin*	*Baryon number*	*Strangeness*	*Isotopic spin*	I_3
u	2/3 e	1/2	1/3	0	1/2	1/2
d	–1/3 e	1/2	1/3	0	1/2	–1/2
s	– 1/3 e	1/2	1/3	–1	0	0

If one uses these quantum numbers then all states known at that time could be built from the combination of these quarks. Examples: $\pi^- = du^c$ (where I denote antiparticles by superscript c. Ususal notation is a bar above the u, or d, or s), p = uud, n = udd, Δ^{++} = uuu, $K^- = su^c$, Λ^0 = sdu, etc. Note that one gets the right spins and isotopic spins as well.

Baryons, made of three quarks, must be fermions. In the ground state (zero orbital momentum) they must have spins of s = 1/2 or s = 3/2. Indeed the nucleons have s=1/2 and the D baryons have s = 3/2. Since both of these states are made of three I=1/2 quarks (u and d) their isospin must be 1/2 or 3/2 as well. Of course, we know that I(N) = 1/2 and I(Δ) = 3/2, because they have 2 and four charge states respectively.

Anyway, the existence of Δ^{++} demands I = 3/2, because I_3 is an additive quantum number and π^{++} = uuu. Each u has I_3 = 1/2 so the total I_3=3/2. But in that case I cannot be smaller than I=3/2. Similarly, Σ^+ = uus, so, since I(s) = 0, it must have I = 1. Indeed it has three charge states.

Finally, the starngeness of Ξ S(Ξ) = -2 so it must contain 2 strange quarks. Concequently, Ξ^- = ssd, Ξ^0 = ssu, they have isospin I(Ξ) =1/2, because they contain one I = 1/2 quark (u or d) and two I = 0 quarks (s). The above list conatins all ground states of baryons constructed from the three quarks u, d, and s, except sss. This state was predicted by Gell-Mann and subsequently discovered and was called Ω^-. After the identification of the ground state baryons the previously discovered resonances of higher mass and frequently higher spin were also identified as excited states of these meson and baryon states, with non-zero orbital angular momenta between quarks. Subsequently, heavier and heavier states were investigated, mostly in $e^+ - e^-$ colliders.

On theoretical grounds, the existence of another quark, the charmed quark was predicted. c- anti c bound states were found as tremendously high peaks in the $e^+ - e^-$ cross sections near total energy (in the CMS) 3 GeV. These peaks were very narrow O(100 keV), which cannot be explained any other way but by the interpretation that such a new state, so called charmonium was produced (Richter and Ting got the Nobel prize for the discovery).

The fact is that the decay is from strongly interacting particles to strongly interacting particles should be much faster (*i.e.* have larger width). QCD, the theory of strong interactions. While the u d and s quarks are light, M(c) ~ 1.5 GeV/c^2 Subsequently, and unexpectedly, two more quarks, b (bottom) and t (top) (other two Nobel prizes) were discovered in similar typ of experiments.

Their masses are large M(b) = 5 GeV/c^2 and M(t) = 170 GeV/c^2 (as heavy as a A=180 nucleus). Furthermore, a third lepton family was also discovered (also Nobel prize winner), with the t (positive and negative) lepton and corresponding neutrino. There is fairly strong (astrophysical, and accelerator physics) evidence that there are no more quarks or leptons. The known leptons and quarks can be grouped into three families.

First family: d, u, e, ν_e, Second family: s, c, m, ν_μ and the third family: b, t, t, ν_τ. These are listed in order of increasing mass (except possible neutrinos, whose mass is consistent with zero). All ordinary matter, including radioactive decay products are made of the members of the first family. As we see it, the world would not be substantially different if the second and third families would not exist.

It is still a mistery why the first family is duplicated twice. The additive quantum numbers distinguishing the six quarks, charge, strangness, char, and the bottom and top quantum numbers are called flavours. They are all conserved in strong and electromagnetic interactions, but with the exception of charge they all can be violated in weak interactions.

THE STANDARD MODEL

There is a substantial problem with the quark assignments of baryons. Take for example the Ω^- hyperon. Its assignment is sss and its spin is s(Ω^-) = 3/2 This spin state is completely symmetric for the exchange of the s quarks (all the three spins are parallel). Consequently the wave function of the Ω^- hyperon is a completely symmetric combination of the wavefunctions of the three s quarks. Similarly the Δ baryons have isospin I = 3/2 and spin I = 3/2, both of these states are completely symmetric, i.e the three quark wavefunctions are completely symmetric.

In fact, assuming that the three quark wavefunctions are completely symmetric one can prove that the baryon ground states formed from threee u and d quarks are exactly the observed isospin I=3/2 and spin I=3/2 state (D) and an I=1/2 and spin I=1/2 state (the nucleons). This is very troublesome because it requires the symmetrizations of wavefuncions in fermions, while according to the Pauli exclusion principle the wavefunctions should be completely antisymmetric.

Now the symmetry properties of wavefunctions and their relation to the bosonic and fermionic nature of particles is a very basic law of physics and should not be violated. This gave a lot of headache to theorists, who finally came up with the right solution. The only way to solve the situation is that the the three s quarks that make up Ω^-, or the three u quarks that make up Δ^{++} differ in an additional, hitherto unknown, quantum number, so they can be antisymmetrized in that quantum number.

Gell-Mann called this quantum number colour (he also named quarks). Thus there are three colours: say red, blue, and yellow and then the wavefunctions of the three quarks inside the baryon can be completely antisymmetrized in colour, requiring that the three quarks are completely symmetrized in the remainging quantum numbers, such as spin and isospin. One can show using group theory, that such an antisymmetric combination of colour states is "colourless" does not have any colour quantum numbers. In a similar way mesons are constructed by equal combination red-antired, blue-antiblu, and yellow-antiyellow quarks, that is also a colourless state. Furthermore, one can see that the baryon and meson states of this kinds are the only colourless states one can construct from three quarks and a quark antiquark pair respectively.

Since free quarks have never been seen in nature the challange is to construct a theory of quarks that allows the existence of colourless physical states constructed from quarks only. Such a theory is quantum chromodynamics (QCD). Quantum chromodynamics is built based on the analogy of Quantum electrodynamics, the theory of interaction of photons with charged particles. The strength of electromagnetic interactions is given by e. In field theory, so a similar "strong coupling constant," g, is defined.

Since g is much larger than e, strong interactions are stronger than electromagnetic interactions. The role of photons is to generate interactions among charged particles. There should be particles call gluons, that create interactions among particles that have the colour quantum number. The difference is that while there is only one kind of a conserved charge in electromagnetism, in QCD there are three colours to play a similar role.

Accordingly, there are much more kinds of gluons (eight) that communicate interactions among objects having colour quantum numbers. While in Quantum electrodynamics an photon can turn into an electron positron pair, gluons can turn into quark-antiquark pairs.

So there is one gluon that can turn into a blue-antired quark pair, etc. The great difference is that while photons themselves are not charged (the total charge of the electron positron pair is zero), gluons themselves carry colour. Since photons interact with charged objects only, tnot being charged, they do not directly interact with each other. At the same time gluons, carrying colour charges do have direct interactions with each other. The fact that gluons interact directly with each other has far reaching consequences in strong interactions. Remember, that the vacuum screening of charges gets stronger and stronger as one goes farther and farther away from the electron.

The consequence is that the effective charge is decreasing with distance. Alternatively, using the uncertainty relation that says that large distance corresponds to low energy and short distance corresponds to high energy. The one can show that the effective charge has an energy dependence, *i.e.* at low energies a lower effective charge can be used in calculations (of cross sections, decay rates, etc.), but that increases at high energies the effective charge is larger. Now in QCD, due to the self interaction of gluons the screening situation is entirely different.

The gluons that pop out from the vacuum generate antisceening, because the like colour charges and anticharges repel each other. The end result is that at large distances (or at low momentum) the effective colour charge (g) increases, while at short distances (high momentum) the effective colour charge, g decreases. One can also show that in the limit of of infinite momentum (or energy) the effective charge vanishes. This property is called asymptotic freedom.

If the effective charge is very small at high energies then the basic calculational method of field theory of expansion in a power series of charge works very well. Indeed very highenergy experiments confirm this they agree completely with the predictions of QCD, though the precision is far less than the experimental agreement with predictions of QED. At low energies the effective charge is large and the calculations break down.

The only results at low energies are from crude numerical symulations that confirm the property of confinement: all states that carry a colour quantum number (single quark, single gluon, diquark) are non- existent in the free form, quarks and gluons are permanently confined inside mesons and baryons. The picture of strongly interactive particles is then the following: If we bombard a proton with very high energy probes then it behaves as a collection of three free quarks. This can be clearly seen in experiements. The very high energy probes (think of the uncertainty relation again) probe the proton at very short distances. So if the quarks are close they interact very little.

Then when we try to separate them their interaction is getting stronger. When they are at a larger distance, the attractive energy between themm increases proportionally to the distance. To separate them to infinite distance we must supply infinite energy. Thus, they are permanently confined inside mesons and baryons.

If quarks are never seen in nature in the free form the question can be raised whether they exist at all or they are some kind of a mathematical concept only, a mnemonic that tells you how to construct elementary particles. There is extensive evidence for the existence of quarks. One is the bombardment of protons by very high energy electrons. Electrons do not participate in strong interactions, only in electromagnetic and weak interactions.

Negelecting weak interactions, the scattering is purely electromagnetic. At very high energy the proton looks like a collection of free quarks. Scattering on free quarks can be easily calculated (something like Rutherford scattering) and the calculated cross sections agree precisely with the experimentally observed ones. More direct and specttacular evidence comes from high energy $e^+ - e^-$colliders. When these particles collide in th eCMS (which is the laboratory) they create a virtual photon at rest in the laboratory, but having an enormous mass. The photon (if it is not polarized) loses the directionality of the original e^+ and e^- comlpetely.

Then it is spectacular to see that the large number of strongly interacting particles produced and observed in the final state come out in two narrowly focused jets in oppositie direction.

Without quarks one would expect a more or less isotropic distribution for these produced particles. Now how do quarks explain the existence of jets? At very high energy quarks interact with the photon (they are charged!) but their strong interactions are "weak" so the only thing that happens is that the virtual photon can decay into quark-antiquark pairs, which, due to energy-momentum conservation fly out in opposite directions with very high momentum. As long as the quarks are not very far from each other they continue to fly and have still weak interactions.

When the distance between them is increasing the potential energy also increases and try to stop them. Instead of stopping them virtual quark -antiquark pairs and gluons are produced from the vacuum and join the initial quarks in a way that they become colourless object that are now free to go out to infinite distance.

Since this happens after the original quarks were produced their directionality is conserved and two jets are produced. More precise calculations in QCD confirm this scenario. There is no other known theory that could produce jets as QCD does.

PARTICLES AND ANTIPARTICLES

PARTICLE

In the physical sciences, a particle is a small localised object to which can be ascribed several physical or chemical properties such asvolume or mass. The word is rather general in meaning, and is refined as needed by various scientific fields. Something that is composed of particles may be referred to as being particulate.

Fig. Arc welders need to protect themselves from welding sparks, which are heated metal particles that fly off the welding surface.

However, the term particulate is most frequently used to refer to pollutants in the Earth's atmosphere, which are a suspension of unconnected particles, rather than a connected particle aggregation.

Whether objects can be considered particles depends on the scale of the context; if an object's own size is small or negligible, or ifgeometrical properties and structure are irrelevant, then it can often be considered a particle. For example, grains of sand on a beachcan be considered particles because the size of one grain of sand (~1 mm) is negligible compared to the beach, and the features of individual grains of sand are usually irrelevant to the problem at hand. However, grains of sand would not be considered particles if compared to buckyballs (~1 nm).

PARTICLES AND ANTIPARTICLES

The material world is complicated, yet it is also very simple and symmetric. Analogous to two kinds of charge are two kinds of particles—particle and antiparticle. One of the beta decay processes is the emission of positrons to reduce the nuclear positive charge. The positron is actually an antiparticle of electron, but we did not mention the antiparticle concept at that time.

The Antiparticle Concept

Paul Dirac rearranged Einstein's equation of relative mass for a moving particle in 1928, and obtained an equation showing that the kinetic energy of certain type of particles became more negative as they move faster. He called such particles antiparticles.

Antiparticles and their counter parts have the same rest mass, but opposite charge, and opposite magnetic moment if they possess these properties.

Nothing is observed when energy states are empty or fully occupied.

An electron occupying a positive energy state and an empty state in an otherwise occupied energy state are observable. This is a pair of electron and antielectron.

An electron annihilates an antielectron and the excess energy is released as a pair of photons. The positive energy states become empty and the negative energy state is full after the annihilation.

The Discovery of Antielectrons

While studying cosmic-ray tracks in his cloud chamber in 1932, Carl Anderson saw tracks similar to those of electrons but these particles bent in opposite direction in the presence of a magnetic field. He discovered antielectrons prescribed by Dirac, and called them positron.

Annihilation and Pair Production

At the end of the positron tracts, Anderson noticed two thin tracts that can be attributed to gamma rays, generated when the positrons were annihilated when colliding with electrons.

A year later, the creation of an electron-positron pair using high-energy photon was confirmed, and this process is called pair production.

Today, positrons and electrons are created routinely using high-energy photons. These particles are created for particle accelerator. Electrons and positrons move in opposite directions in circular accelerators accelerated using electric and magnetic fields. When high-energy electrons and positrons collide, particles with rest mass heavier than electrons are produced.

Simhony argued that electrons and positrons form a elecron-positron lattice, (EPOLA), which is unobservable. Absorption of 1.02 MeV liberates a pair of electron and positron. Thus, energy does not convert to particles. It is absorbed, and as a result, particles are liberated.

ANTIPARTICLE

Corresponding to most kinds of particles, there is an associated antiparticle with the same mass and opposite charge (including electric charge). For example, the antiparticle of the electron is the positively charged electron, or positron, which is produced naturally in certain types of radioactive decay.

The laws of nature are very nearly symmetrical with respect to particles and antiparticles. For example, an antiproton and a positroncan form an antihydrogen atom, which is believed to have the same properties as a hydrogen atom. This leads to the question of why the formation of matter after the Big Bang resulted in a universe consisting almost entirely of matter, rather than being a half-and-half mixture of matter and antimatter. The discovery of Charge Parity violation helped to shed light on this problem by showing that this symmetry, originally thought to be perfect, was only approximate.

Particle-antiparticle pairs can annihilate each other, producing photons; since, the charges of the particle and antiparticle are opposite, total charge is conserved. For example, the positrons produced in natural radioactive decay quickly annihilate themselves with electrons, producing pairs of gamma rays, a process exploited in positron emission tomography.

Antiparticles are produced naturally in beta decay, and in the interaction of cosmic rays in the Earth's atmosphere. Because charge is conserved, it is not possible to create an antiparticle without either destroying a particle of the same charge (as in beta decay) or creating a particle of the opposite charge. The latter is seen in many processes in which both a particle and its antiparticle are created simultaneously, as in particle accelerators. This is the inverse of the particle-antiparticle annihilation process.

Although particles and their antiparticles have opposite charges, electrically neutral particles need not be identical to their antiparticles.

The neutron, for example, is made out of quarks, the antineutron from antiquarks, and they are distinguishable from one another because neutrons and antineutrons annihilate each other upon contact. However, other neutral particles are their own antiparticles, such as photons, the hypothetical gravitons, and some WIMPs.

7

Energy and its Transformations

MASS AND POTENTIAL ENERGY

Suppose now at the far end of the tube we have a hydrogen atom at rest. This atom is essentially a proton having an electron bound to it by electrostatic attraction. It is known that a flash of light with total energy 13.6eV is just enough to tear the electron away, so in the end the proton and electron are at rest far away from each other.

The energy of the light was used up dragging the proton and electron apart—that is, it went into potential energy. (It should be mentioned that the electron also loses kinetic energy in this process, 13.6 ev is the net energy required to break up the atom.)

Now, the light is absorbed by this process, so from our argument above the right hand end of the tube must become heavier. That is to say, a proton at rest plus a (distant) electron at rest weigh more than a hydrogen atom by E/c^2, with E equal to 13.6eV.

Thus, Einstein's box forces us to conclude that increased *potential* energy in a system also entails the appropriate increase in mass. It is interesting to consider the hydrogen atom dissociation in reverse—if a slow moving electron encounters an isolated proton, they may combine to form a hydrogen atom, emitting 13.6eV of electromagnetic radiation energy as they do so. Clearly, then, the hydrogen atom remaining has that much less energy than the initial proton + electron.

The actual mass difference for hydrogen atoms is about one part in 10^8. This is typical of the energy radiated away in a violent chemical reaction—in fact, since most atoms are an order of magnitude or more heavier than hydrogen, a part in 10^9 or 10^{10} is more usual. However, things are very different in nuclear physics, where the forces are stronger so the binding is tighter.

We shall discuss this later, but briefly mention an example: a hydrogen nucleus can combine with a lithium nucleus to give two helium nuclei, and the mass shed is 1/500 of the original.

This reaction has been observed, and all the masses involved are measurable. The actual energy emitted is 17 MeV. This is the type of reaction that occurs in hydrogen bombs. Notice that the energy released is at least a million times more than the most violent chemical reaction. As a final example, let's make a ballpark estimate of the change in mass of a million tons of TNT on exploding.

The TNT molecule is about a hundred times heavier than the hydrogen atom, and gives off a few eV on burning. So the change in weight is of order $10^{-10} \times 10^{6}$ tons, about a hundred grams.

In a hydrogen bomb, this same mass to energy conversion would take about fifty kilograms of fuel.

FOOTNOTE: EINSTEIN'S BOX IS A FAKE

Although Einstein's box argument is easy to understand, and gives the correct result, it is based on a physical fiction—the rigid box. If we had a rigid box, or even a rigid stick, all our clock synchronization problems would be over—we could start clocks at the two ends of the stick simultaneously by nudging the stick from one end, and, since it's a rigid stick, the other end would move instantaneously.

Actually there are no such materials. All materials are held together by electromagnetic forces, and pushing one end causes a wave of compression to travel down the stick. The electrical forces between atoms adjust at the speed of light, but the overall wave travels far more slowly because each atom in the chain must accelerate for a while before it moves sufficiently to affect the next one measurably.

So the light pulse will reach the other end of the box before it has begun to move! Nevertheless, the wobbling elastic box *does* have a net recoil momentum, which it does lose when the light hits the far end. So the basic point is still valid.

French gives a legitimate derivation, replacing the box by its two (disconnected) ends, and finding the centre of mass of this complete system, which of course remains at rest throughout the process.

ENERGY AND MOMENTUM IN LORENTZ TRANSFORMATIONS

TOTAL ENERGY OF A PARTICLE DEPEND ON SPEED

We have a formula for the total energy E = K.E. + rest energy,

$$E = mc^2 = \frac{m_0 c^2}{\sqrt{1 - v^2 / c^2}}$$

so we can see how total energy varies with speed.

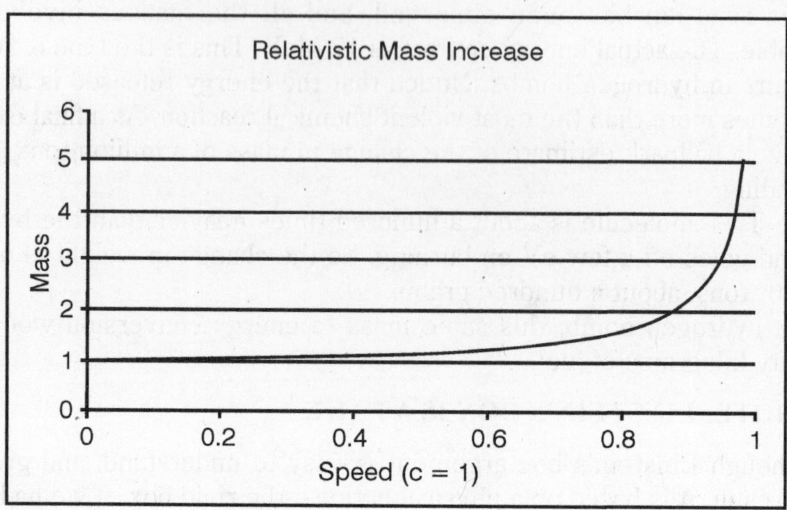

The momentum varies with speed as

$$p = mv = \frac{m_0 v}{\sqrt{1 - v^2/c^2}}$$

TOTAL ENERGY OF A PARTICLE DEPEND ON MOMENTUM

It turns out to be useful to have a formula for E in terms of p. Now

$$E^2 = m^2c^4 = \frac{m_0^2 c^4}{1 - v^2/c^2}$$

so

$$m^2c^4 (1 - v^2/c^2) = m_0^2 c^4$$

$$m^2c^4 - m^2v^2c^2 = m_0^2 c^4$$

$$m^2c^4 = E^2 = m_0^2 c^4 + m^2c^2v^2$$

hence using $p = mv$ we find

$$E = \sqrt{m_0^2 c^4 + c^2 p^2}\ .$$

If p is very small, this gives

$$E \approx m_0 c^2 + \frac{p^2}{2m_0},$$

the usual classical formula.

If p is very large, so $c^2p^2 >> m_0{}^2c^4$, the approximate formula is $E = cp$.

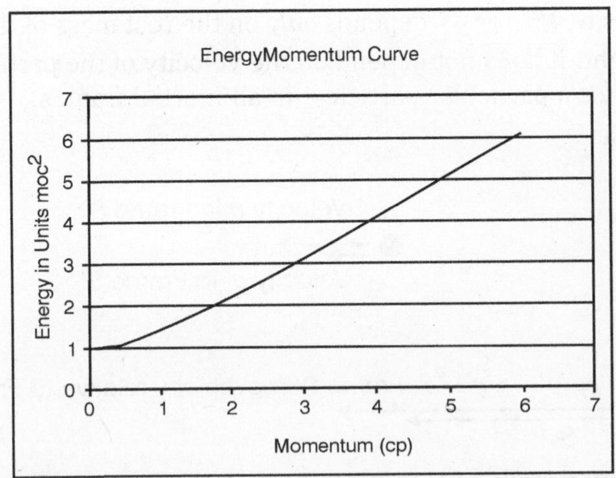

THE HIGH KINETIC ENERGY LIMIT: REST MASS BECOMES UNIMPORTANT!

Notice that this high energy limit is just the energy-momentum relationship Maxwell found to be true for light, for all p. This could only be true for *all* p if $m_0{}^2c^4 = 0$, that is, $m_0 = 0$.

Light is in fact composed of "photons"—particles having zero "rest mass".

The "rest mass" of a photon is meaningless, since they're never at rest—the energy of a photon

$$E^2 = mc^2 = \frac{m_0 c^2}{\sqrt{1 - v^2/c^2}}$$

is of the form 0/0, since $m_0 = 0$ and $v = c$, so "m" can still be nonzero. That is to say, the mass of a photon is really all K.E. mass.

For very fast electrons, such as those produced in high energy accelerators, the additional K.E. mass can be thousands of times the rest mass. For these particles, we can neglect the rest mass and take $E = cp$.

TRANSFORMING ENERGY AND MOMENTUM TO A NEW FRAME

We have shown

$$\vec{p} = m\vec{v} = \frac{m_0 \vec{v}}{\sqrt{1 - v^2/c^2}}$$

$$E^2 = mc^2 = \sqrt{m_0^2 c^4 + c^2 - \vec{p}^2}\,.$$

Notice we can write this last equation in the form

$$E^2 - c^2\vec{p}^2 = m_0^2 c^4\,.$$

That is to say, $E^2 - c^2\vec{p}^2$ depends *only* on the rest mass of the particle and the speed of light. It does not depend on the velocity of the particle, so it must be the same—for a particular particle—in all inertial frames.

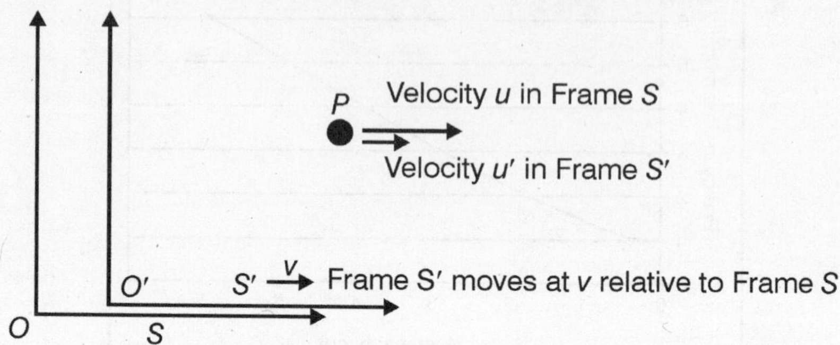

This is reminiscent of the invariance of $\vec{x}^2 - c^2t^2$, the interval squared between two events, under the Lorentz transformations. One might guess from this that the laws governing the transformation from E, p in one Lorentz frame to E', p' in another are similar to those for x,t. We can actually derive the laws for E, p to check this out.

As usual, we consider all velocities to be parallel to the x-axis.

We take the frame S' to be moving in the x-direction at speed v relative to S. Consider a particle of mass m_0 (rest mass) moving at u' in the x' direction in frame S', and hence at u along x in S, where

$$u = \frac{u' + v}{1 + vu'/c^2}$$

The energy and momentum in S' are

$$E' = \frac{m_0c^2}{\sqrt{1 - u'^2/c^2}},\ p' = \frac{m_0u'}{\sqrt{1 - u'^2/c^2}}.$$

and in S:

$$E = \frac{m_0c^2}{\sqrt{1 - u^2/c^2}},\ p = \frac{m_0u}{\sqrt{1 - u^2/c^2}}$$

Thus

$$E = \frac{m_0c^2}{\sqrt{1 - \left(\dfrac{u' + v}{1 + vu'/c^2}\right)^2 / c^2}}$$

giving

$$E = \frac{m_0 c^2 (1 + vu'/c^2)}{\sqrt{(1 - v^2/c^2)(1 - u'^2/c^2)}}$$

from which it is easy to show that

$$E = \frac{1}{\sqrt{1 - v^2/c^2}} (E' + vp').$$

Similarly, we can show that

$$p = \frac{p' + vE'/c^2}{\sqrt{1 - v^2/c^2}}$$

These are the Lorentz transformations for energy and momentum of a particle—it is easy to check that

$$E^2 - c^2 p^2 = E'^2 - c^2 p'^2 = m_0^2 c^4 .$$

PHOTON ENERGIES IN DIFFERENT FRAMES

For a zero rest mass particle, such as a photon, $E = cp, E^2 - c^2 p^2 = 0$ in all frames.

Thus

$$E = \frac{E' + vp'}{\sqrt{1 - v^2/c^2}} = \frac{E' + vE'/c}{\sqrt{1 - v^2/c^2}} = E' \sqrt{\frac{1 + v/c}{1 - v/c}}$$

Since $E = cp, E' = cp'$ we also have

$$p = p' \sqrt{\frac{1 + v/c}{1 - v/c}} .$$

Notice that the ratios of photon *energies* in the two frames coincides with the ratio of photon *frequencies* found in the Doppler shift. The photon energy is proportional to the frequency, so these two must of course transform in identical fashion. But it's interesting to see it come about this way. Needless to say, relativity gives us no clue on what the constant of proportionality (Planck's constant) is: it must be measured experimentally. But the *same* constant plays a role in all quantum phenomena, not just those concerned with photons.

KINETIC MOLECULAR THEORY

The Kinetic Molecular Theory of Gases begins with five postulates that describe the behaviour of molecules in a gas. These postulates are based upon some simple, basic scientific notions, but they also involve some simplying assumptions. In reading a postulate, do two things. First, try to understand and appreciate the basic physical idea embodied in the postulate; this idea will ultimately be important in understanding the macroscopic properties of the gas in terms of the behaviour the microscopic molecules making up the gas. Second, identify possible weakness or flaws in the postulates. Inaccurate predictions by a theory derive from flawed postulates used in the derivation of the theory.

The ideal gas law was developed from macroscopic observations, with no knowledge of the behaviour of a gas at the molecular level. We now attempt to interpret the ideal gas law in terms of a reasonable molecular model (simple physical picture) of gas behaviour. What are the individual molecules doing, and why does their behaviour manifest itself in the ideal gas law? We now enter the domain of theory — a mental interpretation of experimental results. The model of gas behaviour currently accepted by scientists is the Kinetic Molecular model. The theory of gas behaviour built on this model is called the Kinetic Molecular Theory (KMT). This theory is one of the oldest and most resoundingly successful in science.

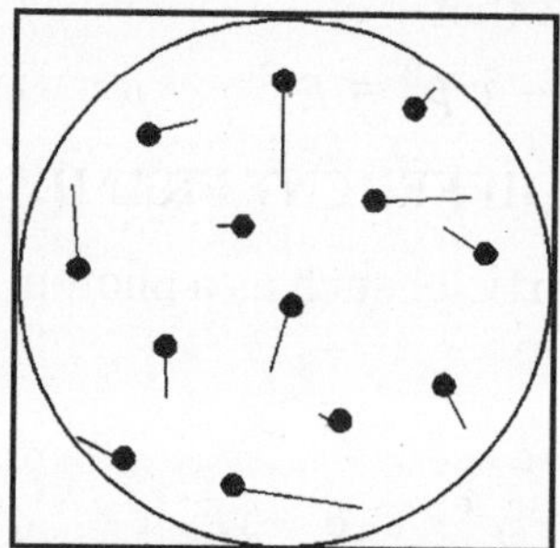

Fig. Dynamic Model of the Gas Phase

First, we assume that gases consist of very small molecules. Since a gas occupies the entire volume of its container, and since gases readily diffuse, the molecules must move about in space. Consequently they have kinetic energy — energy of motion. Since gases are compressible, the molecules must be far apart. Focussing on an individual molecule, it seems reasonable to assume that it is a small particle that moves about at random, occasionally colliding with a wall of the container and with other gas molecules. The molecule sometimes gains and sometimes loses energy in these collisions, so that it frequently changes speed. Sometimes it moves rapidly, sometimes slowly. Thus our picture of the gas is dynamic rather than static. An attempt to portray this simple physical model of the gas phase. These statements about gas molecules are consistent with the properties of gases listed. None of them have been proven, but they seem reasonable and constitute the postulates of the Kinetic Molecular Theory:

- Gases consist of molecules that behave like tiny hard spheres;
- The molecules have no volume — they may be treated as points;
- There are no forces of attraction between molecules;
- Collisions between molecules are elastic (kinetic energy is conserved);
- The molecules are in ceaseless random motion, frequently colliding

with the container walls and with each other. A collection of gas molecules is highly disordered.

- The average kinetic energy of the molecules is proportional to Kelvin temperature.

We now pursue the consequences of these postulates. Postulate 6, the origin of which is not obvious. Our aim is to develop an expression for pressure. At the molecular level, pressure must result from collisions of gas molecules with the container walls. A molecule hitting the wall exerts a force on it. The collection of all such forces on a unit area of wall during a given time interval constitutes the pressure. We should be able to calculate P from the force exerted by each collision, multiplied by the number of collisions per unit area of wall:

P = Force/Area = Force/collision × Collisions/area

From Newton's second law, force is the rate of change of momentum. Thus

F = momentum change/time-collision × collisions/area

We move the time factor over to the second term to obtain equation.

F = momentum change/collision × collisions/time-area

To obtain the momentum change per collision, envision a molecule moving directly towards a wall with velocity v, as shown in Figure.

Since momentum is conserved in any collision, the change in momentum of the molecule is momentum change = momentum after collision–momentum before= m(-v)–m(v) = -2mv The change of momentum of the wall must therefore be 2mv to give a total change of zero. We now have the first term in equation:

momentum change/collision = 2mv

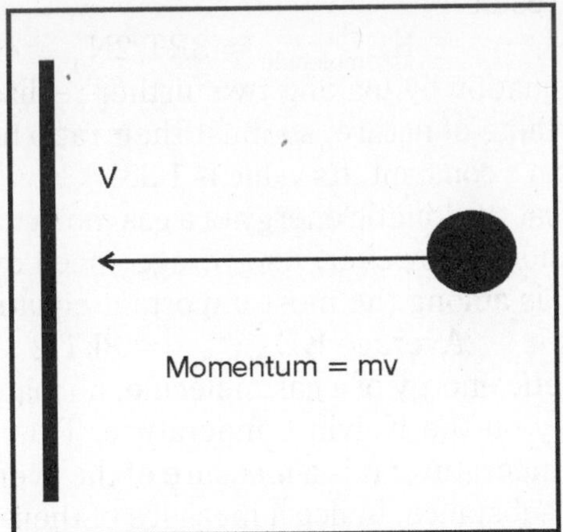

Fig. Collision of Gas Molecule with Container Wall

We take an intuitive approach to the second term. The number of collisions per unit time per unit area of wall should depend on 1) the number of molecules per unit volume in the container, N/V (the more there are, the more collisions there should be with the walls); and 2) the speed v at which a molecule moves (the faster the movement, the more collisions per unit time that should occur). Equation is based on these ideas:

$$\text{collisions/time-area} = (N/V)(v)$$

If we assume for simplicity that the container is cubical, there are 6 walls over which these collisions must be spread. The collisions with a particular wall are then

$$\text{collisions/time-area} = Nv/6V$$

(Despite our restrictive assumptions about the shape of the container, equation turns out to be valid for a container of any shape!) We now multiply the expressions in equations to obtain the pressure:

$$P = (2mv)(Nv/6V) = Nmv^2/3V$$

Rearrangment gives equation.

$$PV = Nmv^2/3$$

This is as far as our molecular model takes us. We are now at the interface between theory and experiment. Experiment (the ideal gas law) relates pressure and volume to temperature:

$$PV = nRT$$

Theory, equation, relates pressure and volume to the mass and speed of a gas molecule. To bring experiment and theory into correspondence, we must equate the right sides of the preceding two equations. In order for KMT to successfully explain the ideal gas law, it is necessary that

$$nRT = Nmv^2/3$$

Equation can be simplified by replacing n with N/N_o (N_o is Avogadro's Number); $mv^2/2$ by KE (the kinetic energy of an average gas molecule); and solving for KE. The result is

$$KE_{molecule} = 3RT/2N_o$$

We refine this equation by making two further realizations. First, since R and N_o are both constants of nature, so must their ratio be. It is symbolized k, and is called Boltzmann's constant. Its value is 1.381×10^{-23} J/K-particle. Second, we have recognized that the kinetic energy of a gas molecule changes frequently as it undergoes collisions. However, its average speed over time is constant. The refined equation is among the most important equations in science:

$$\text{Average } KE_{molecule} = 3kT/2$$

The average kinetic energy of a gas molecule, no matter what its chemical identity, depends only on the Kelvin temperature. This gives a deep insight into the concept of temperature: it is a measure of the average kinetic energies of the molecules of a substance, hence a measure of their average speed. As T is increased, molecules move faster; as it is lowered, they move slower. The temperature at which molecular motion ceases is absolute zero. Temperatures lower than absolute zero are impossible because a molecule may not have negative kinetic energy. Equation is simple and profound. Is there an experiment that can be done to test its validity? There is in fact a simple experiment that verifies equation in a rearranged form. If we replace the average kinetic energy of a molecule with the expression $m(v^2)_{avg}/2$, where $(v^2)_{avg}$ is the average of the square of the speed, and solve the resulting expression for $(v^2)_{avg}$, we obtain equation.

$$(v^2)_{avg} = 3kT/m$$

The square root of $(v^2)_{avg}$ is called the root mean square speed, and is for our purposes approximately equal to the average molecular speed. Making this equality gives equation.

$$v_{avg} = (3kT/m)^{1/2}$$

Multiplying both numerator and denominator of the argument on the right side of this equation by Avogadro's number N_o gives the useful variant in equation.

$$v_{avg} = (3RT/MM)^{1/2}$$

The implication of equations is that a heavy gas molecule moves more slowly than a light one, in a quantifiable way. The ratio of the speeds of the light (L) and heavy (H) molecules should be the square root of the inverse ratio of their molar masses:

$$(v_H/v_L)_{avg} = (MM_L/MM_H)^{1/2}$$

In 1846, Thomas Graham measured the rates of diffusion of various gases. Diffusion is the process by which gases disperse in space via random molecular motion.

The results of Graham's experiments are summarized in equation rate diffusion of gas A/rate of diffusion of gas B =(density of gas B/density of gas A)$^{1/2}$ But we have previously seen that gas density is proportional to MM. If we make the logical assumption that a gas diffuses at a rate that is directly proportional to the average speed of its molecules, then Graham's Law of Diffusion is in exact agreement with equation.

Example: Calculate the ratio of the speeds of H_2 and CO_2 molecules at 25°C.

Solution: As long as temperature is the same for both molecules, its actual value is unimportant.

$$v_{H2}/v_{CO2} = (MM_{CO2}/MM_{H2})^{1/2} = (44.0/2.02)^{1/2} = 4.67$$

Example. Calculate the average speed of a molecule of nitrogen gas at room temperature, 25°C.

Solution. $v_{N2} = (3RT/MM)^{1/2} = (3(8.314\ J/K\text{-mole})(298\ K)/(0.02801\ kg/mole))^{1/2} = 5.15\times10^2$ m/s

This is 1150 miles per hour, or somewhere between the speeds of a commercial jet liner and a fighter jet.

Please note that in applying equation in Example, values of R and MM with appropriate units have been used, so that speed comes out with appropriate units. Use of R in units of, for example, atm-L, produces a speed with absurd units. The curve is a plot of the fraction of molecules having speed s, f(s) (we use s for speed, v for velocity), versus the possible values of speed between zero and infinity.

The curve has several noticeable features:

- The curve has the expected shape — low on both ends, indicating that few molecules have extremes of speed; and peaking in the middle, indicating that most molecules have speeds somewhere between the extremes.
- The curve is unsymmetrical because there is a definite lower limit (zero) but no definite upper limit to speed.
- The total area under the curve is the total number of gas molecules.
- The height of the curve at a particular speed is proportional to the fraction of molecules having speed s_1.
- The area under the curve between two speeds s_1 and s_2 (shaded in the figure) is the fraction of molecules haveing speeds in the range s_1 to s_2.
- The mathematical expression for f(s) obtained by Maxwell and Boltzmann is equation:

$$f(s) = (\text{constant})(s^2) \exp(-ms^2/2kT)$$

The s^2 term increases with increasing s and the exponential term decreases with increasing s. Thus f(s) first increases, but then peaks and decreases with increasing s as the exponential term takes over. The speed at the maximum of the curve, s_{mp}, is the most probable speed, because the greatest fraction of molecules have it.

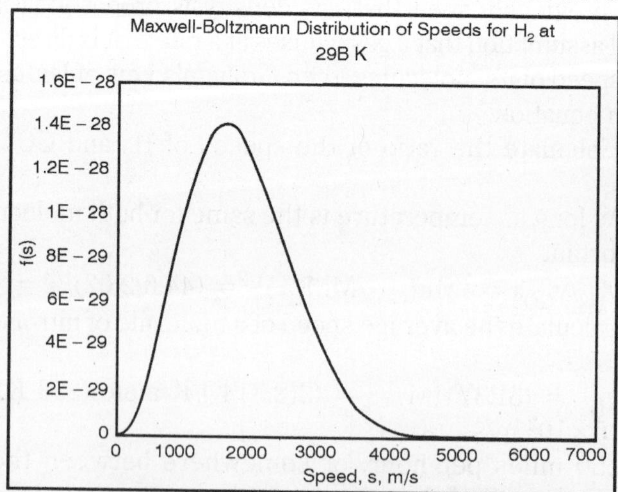

Fig. Maxwell-Boltzmann Distribution

The Maxwell-Boltzmann Distribution Law. We have said that in a gas the molecules move randomly, and that the speed of a molecule changes frequently. In the mid 1800's, James Clerk Maxwell in England, and Ludwig Boltzmann in Austria, were concerned with using statistical methods to describe, precisely and mathematically, the distribution of molecular speeds. The result of their efforts is called the Maxwell- Boltzmann Distribution Law. They showed that at a given temperature, the molecular speed distribution follows a curve like that in Figure.

- As T increases, the curve flattens and the maximum moves to higher speed, because the speeds of all molecules tend to increase with increasing T. The area under the curve stays constant, however, as long as the number of gas molecules is unchanged.

We now do two calculations using equation. First, we calculate the most probable speed, s_{mp}. This is a good approximation to the average speed, but the average will be somewhat larger because the curve is biased to higher speeds. s_{mp} is the speed at which f(s) is maximum. At this maximum, the slope of the plot — the derivative of f(s) — is zero:

$$df(s)/ds = d/ds\,[(s^2)(\exp(-ms^2/2kT))] = 0$$
$$= ds^2/ds\,\exp(-ms^2/2kT) + s^2\,d[\exp(-ms^2/2kT)]/ds = 0$$
$$= 2s\,\exp(-ms^2/2kT) - s^2(ms/kT)\exp(-ms^2/2kT) = 0$$

Dividing by the exponential, we obtain

$$2s - ms^3/kT = 0$$

This is readily solved for s_{mp}:

$$s_{mp} = (2kT/m)^{1/2}$$

As expected, this is a bit smaller than the root-mean-square speed in equation.

Next we calculate the average KE of a gas molecule by a method of averaging; we multiply each possible kinetic energy by the number of molecules with that kinetic energy, and divide by the total number of molecules. The required mathematical procedure is integration.

$$KE_{molecule} = \text{Integral from 0 to Infinity off}(s)(ms^2/2)ds = 3kT/2$$

Even if you are not yet able to carry out the integration, you should note that the result is the same as equation, which came from KMT. This is gratifying. It reinforces the validities of both KMT and the M-B Distribution Law.

Finally, we calculate the KE per mole of gas from the average $KE_{molecule}$:

$$KE_{mole} = N_o \times KE_{molecule} = 3RT/2$$

The kinetic energy of a mole of gas depends only on temperature. For this reason, kinetic energy of molecules is called thermal energy. Summary of the Interpretation of the Ideal Gas Law in Terms of Molecular Behaviour. The relationship between the macroscopic and microscopic views of gas behaviour can be summarized in several statements.

- Gases are compressible because they consist of small molecules that are far apart.
- Molecules of a gas are in constant motion, allowing gases to diffuse and to flow.

- Pressure results from collisions of gas molecules with the container walls.
- Pressure increases as volume decreases (Boyle's Law). A decrease in V increases the number of molecules per unit volume, leading to more molecular collisions with a unit area of the wall per unit time.
- An increase of temperature causes an increase in pressure (the Law of Gay-Lussac). KE of the molecules increases with T. Faster molecules cause an increase in pressure for two reasons:
  - There are more collisions with the wall per unit time;
  - Each collision exerts a greater force, since the molecule is moving faster.
- An increase in the amount (moles) of gas causes an increase in pressure at constant T (Avogadro's hypothesis). More gas means larger N/V. Higher N/V means more collisions per unit time and higher pressure.
- The ratio of diffusion rates of two gases is proportional to the square root of the inverse ratio of their molar masses (Graham's Law). This is a consequence of the average kinetic energy of a gas molecule being dependent only on T. At a particular T, a heavy molecule moves slower than a light one, but its larger mass offsets its smaller speed, giving the same kinetic energy. An important consequence of this is that a sample of gas with heavy molecules will exert the same pressure as a sample of gas with the same number of light molecules, as long as their temperatures and volumes are the same. Both samples satisfy the ideal gas law, independent of what their molecules are like. Although heavy molecules move slowly and strike the wall less often, they strike it with bigger mass, and cause the same pressure.

KINETICS AND EQUILIBRIUM

The connection between the equilibrium constant for a reaction and the rate constants of the elementary steps by which it occurs is an important one. We begin by stating the relationship. We will then demonstrate, for a specific case, that it is true.

The equilibrium constant for a net reaction is the ratio of the product of the forward rate constants for all steps in the mechanism to the product of the reverse rate constants for all steps in the mechanism:

$$K_{eq} = k_1 k_2 k_3 \ldots / k_{-1} k_{-2} k_{-3} \ldots$$

We now show for a specific net reaction and a specific (proposed) mechanism that this statement is true. Consider again the reaction of NO_2 and F_2 to produce NO_2F, reaction:

$$2NO_2 + F_2 \Leftrightarrow 2NO_2F$$

We have written the reaction with the double arrow because we are now interested in the forward and reverse rates at equilibrium. At equilibrium, the overall forward rate and overall reverse rate must be equal:

$$rate_f = rate_r$$

This must be true regardless of the detailed pathway by which the reaction occurs. All available experimental evidence indicates that the mechanism is the two-step process:

$$NO_2 + F_2 \Leftrightarrow NO_2F + F\ [k_1, k_{-1}]$$

$$NO_2 + F \rightarrow NO_2F\ [k_2, k_{-2}]$$

When the overall reaction is in equilibrium, so must each elementary step be. Consequently, we have used double arrows in the elementary steps as well. The equality of the forward and reverse rates of the individual steps is expressed mathematically as follows.

$$k_1[NO_2]_e[F_2]_e = k_{-1}[NO_2F]_e[F]_e$$

$$k_2[NO_2]_e[F]_e = k_{-2}[NO_2F]_e$$

Eliminating the concentration of the intermediate, F.

$$[NO_2F]_e^{\,2}/[NO_2]_e[F_2]_e = k_1k_2/k_{-2}k_{-2} = K_{eq}$$

The statement made at the outset is thus seen to be true for this specific case: the equilibrium constant is the ratio of the product of forward rate constants to the product of reverse rate constants. For the general overall reaction occurring by an n-step mechanism,

$$K_{eq} = k_1k_2 \ldots k_n/k_{-1}k_{-2} \ldots k_{-n}$$

Objection is frequently made to the method of derivation that we have just used, because the mechanism for is not known with certainty; it is and will remain a hypothesis, however well based in experiment. If it is found that the two step mechanism is indeed NOT correct, then the relationship is invalid. This is certainly true. However, even if the mechanism above is incorrect, Reaction must occur by some mechanism. At equilibrium each step in the mechanism must be balanced in rate, enabling us to eliminate concentrations of any and all intermediates just as we did above, to arrive at a modified version. The validity of the general relationship is independent of the details of mechanism. Thus the intuitively-expected connection between k and K does indeed exist. In fact, there are many examples in which the equilibrium constant for a reaction has been obtained from kinetics studies. This process can never be reversed however; it is not possible to obtain k by performing equilibrium studies. Before leaving this matter, we make one more very important point.

There is a tendency to believe that reactions with large equilibrium constants are fast, whereas those with small equilibrium constants are slow. However, neither of these beliefs is true. A large value for K_{eq} means that the product of forward rate constants is much larger than the product of reverse rate constants. It does not follow, however, that the forward rate constants are LARGE. It means only that they are larger than the reverse rate constants.

$k_1k_2...k_n >> k_{-1}k_{-2}...k_{-n}$ does not mean that $k_1,k_2,...,k_n$ are large in the absolute sense.

A large K_{eq} can result from the ratio of a slow forward rate to an even slower reverse rate.

POTENTIAL ENERGY

Maxwell's analysis solves the problem of finding the statistical velocity distribution of molecules of an ideal gas in a box at a definite temperature T: the relative probability of a molecule having velocity $\vec{v}$ is proportional to $e^{-mv^2/2T} = e^{-E/kT}$. The *position* distribution is taken to be uniform: the molecules are assumed to be equally likely to be anywhere in the box.

But how is this distribution affected if in fact there is some kind of potential pulling the molecules to one end of the box? In fact, we've already solved this problem, in the discussion earlier on the isothermal atmosphere.

Consider a really big box, kilometers high, so air will be significantly denser towards the bottom. Assume the temperature is uniform throughout. We found under these conditions that with Boyles Law expressed in the form

$$\rho = CP$$

the atmospheric density varied with height as

$$P = P_0e^{-Cgh}\text{, or equivalenty } \rho = \rho_0e^{-Cgh}.$$

Now we know that Boyle's Law is just the fixed temperature version of the Gas Law $PV = nRT$, and the density

$$\rho = \text{mass/volume} = Nm/V$$

with N the total number of molecules and m the molecular mass,

$$CP = \rho = Nm/V$$

Rearranging,

$$PV = Nm/C = nN_Am/C,$$

for n moles of gas, each mole containing Avogadro's number N_A molecules.

Putting this together with the Gas Law,

$$PV = nN_Am/C = nRT,$$

so

$$C = N_Am/RT = mkT$$

where Boltzmann's constant $k = R/N_A$ as discussed previously.

The dependence of gas density on height can therefore be written

$$\rho = \rho_0e^{-Cgh} = \rho_0e^{-mgh/kT}.$$

The important point here is that mgh is the potential energy of the molecule, and the distribution we have found is exactly parallel to Maxwell's velocity distribution, the potential energy now playing the role that kinetic energy played in that case. We're now ready to put together Maxwell's velocity distribution with this height distribution, to find out how the molecules are distributed in the atmosphere, *both* in velocity space *and* in ordinary space. In other words, in a *six*-dimensional space!

Our result is:

$$f(x, y, z, v_x, v_y, v_z) = f(h, v) \propto e^{-mv^2/2kT}$$
$$= e^{-(1/2)mv2+\text{mgh})/kT} = e^{-E/kT}.$$

That is, the probability of a molecule having total energy E is proportional to $e^{-E/kT}$. This is the Boltzmann, or Maxwell-Boltzmann, distribution. It turns out to be correct for any type of potential energy, including that arising from forces between the molecules themselves.

DEGREES OF FREEDOM AND EQUIPARTITION OF ENERGY

By a "degree of freedom" we mean a way in which a molecule is free to move, and thus have energy—in this case, just the x, y, and z directions. Boltzmann reformulated Maxwell's analysis in terms of degrees of freedom, stating that there was an average energy $\frac{1}{2}kT$ in each degree of freedom, to give total average kinetic energy $3.\frac{1}{2}kT$, so the specific heat per molecule is presumable $1.5k$, and given that $k = R/N_A$, the specific heat per mole comes out at $1.5R$. In fact, this is experimentally confirmed for monatomic gases. However, it is found that diatomic gases can have specific heats of $2.5R$ and even $3.5R$. This is not difficult to understand—these molecules have more degrees of freedom. A dumbbell molecule can rotate about two directions perpendicular to its axis. A diatomic molecule could also vibrate. Such a simple harmonic oscillator motion has both kinetic and potential energy, and it turns out to have total energy kT in thermal equilibrium.

Thus, reasonable explanations for the specific heats of various gases can be concocted by assuming a contribution $\frac{1}{2}k$ from each degree of freedom. But there are problems. Why shouldn't the dumbbell rotate about its axis? Why do monatomic atoms not rotate at all? Even more ominously, the specific heat of hydrogen, $2.5R$ at room temperature, drops to $1.5R$ at lower temperatures. These problems were not resolved until the advent of quantum mechanics.

BROWNIAN MOTION

One of the most convincing demonstrations that gases really *are* made up of fast moving molecules is Brownian motion, the observed constant jiggling around of tiny particles, such as fragments of ash in smoke. This motion was first noticed by a Scottish botanist, who initially assumed he was looking at living creatures, but then found the same motion in what he knew to be particles of inorganic material. Einstein showed how to use Brownian motion to estimate the size of atoms.

TRANSFORMERS

A magnetic field is a construct having to do with the observable fact that objects in the neighbourhood of the bar may be affected by the presence of the bar. Figure is a drawing of a bar magnet together with a schematic representation of its magnetic field. Its effect on an object varies depending on where the object is with respect to the bar. This fact is represented by the statement that the strength of the magnetic field varies from place to place around the bar. The lines in Fig. called lines of force, are a schematic representation of the spatial distribution of the strength of the field. The closer the lines, the stronger the field. Since this method of representing a magnetic field facilitates the explanation of many phenomena associated with magnetism, it will be used throughout the chapter.

When a steady current is passed through a coil of wire that is wrapped around an iron core, a magnetic field is set up in and around the core. This magnetic field, represented in Fig, is identical to the field of a permanent bar magnet; the same method of generating magnetism has already been mentioned in connection with meter movements and relays. The magnetic field generated by a steady current is fixed and unchanging even though the electrons ultimately responsible for its maintenance are themselves moving through the coil. However, as soon as the amount of current flowing through the coil is changed in any way, the field changes accordingly.

A transformer is a device that transfers electrical energy from one circuit to another through inductively coupled conductors — the transformer's coils or "windings". Except for air-core transformers, the conductors are commonly wound around a single iron-rich core, or around separate but magnetically-coupled cores. A varying current in the first or "primary" winding creates a varying magnetic field in the core (or cores) of the transformer. This varying magnetic field induces a varying electromotive force (EMF) or "voltage" in the "secondary" winding. This effect is called mutual induction.

If a load is connected to the secondary, an electric current will flow in the secondary winding and electrical energy will flow from the primary circuit through the transformer to the load. In an ideal transformer, the induced voltage in the secondary winding (V_S) is in proportion to the primary voltage (V_P), and is given by the ratio of the number of turns in the secondary to the number of turns in the prima

$$\frac{V_S}{V_P} = \frac{N_S}{N_P}$$

By appropriate selection of the ratio of turns, a transformer thus allows an alternating current (AC) voltage to be "stepped up" by making N_S greater than N_P, or "stepped down" by making N_S less than N_P.

Transformers come in a range of sizes from a thumbnail-sized coupling transformer hidden inside a stage microphone to huge units weighing hundreds of tons used to interconnect portions of national power grids.

All operate with the same basic principles, although the range of designs is wide. While new technologies have eliminated the need for transformers in some electronic circuits, transformers are still found in nearly all electronic devices designed for household ("mains") voltage. Transformers are essential for high voltage power transmission, which makes long distance transmission economically practical.

First Steps: Experiments with Induction Coils

What would become the "transformer principle" was revealed in 1831 by Michael Faraday in his demonstration of electromagnetic induction, but without recognition of its future role in manipulating EMF. The first "induction coils" to see wide use were invented by Rev. Nicholas Callan of Maynooth College, Ireland in 1836, one of the first researchers to realize that the more turns the secondary winding has in relation to the primary winding, the larger the increase in EMF. Induction coils evolved from scientists' and inventors' efforts to get higher voltages from batteries.

Rather than alternating current (AC), their action relied upon a vibrating "make-and-break" mechanism that regularly interrupted the flow of direct current (DC) from the batteries. Between the 1830s and the 1870s, efforts to build better induction coils, mostly by trial and error, slowly revealed the basic principles of transformers. Efficient, practical designs did not appear until the 1880s, but within a decade the "transformer" would be instrumental in the "War of Currents", and in seeing AC distribution systems triumph over their DC counterparts, a position in which they have remained dominant ever since.

In 1876, Russian engineer Pavel Yablochkov invented a lighting system based on a set of induction coils where the primary windings were connected to a source of alternating current and the secondary windings could be connected to several "electric candles" (arc lamps) of his own design. The coils used in the system behaved as primitive transformers. The patent claimed the system could "provide separate supply to several lighting fixtures with different luminous intensities from a single source of electric power".

In 1878, the engineers of the Ganz Company in Hungary assigned part of its extensive engineering works to the manufacture of electric lighting apparatus for Austria-Hungary, and by 1883 made over fifty installations. It offered an entire system consisting of both arc and incandescent lamps, generators, and other accessories. Lucien Gaulard and John Dixon Gibbs first exhibited a device with an open iron core called a "secondary generator" in London in 1882, then sold the idea to the Westinghouse company in the United States. They also exhibited the invention in Turin, Italy in 1884, where it was adopted for an electric lighting system. Induction coils with open magnetic circuits are inefficient for transfer of power to loads. Various methods of adjusting the cores or bypassing magnetic flux around part of a coil were developed, since until about 1880 the paradigm for AC power transmission from a high voltage supply to a low voltage load was a series circuit.

In practice, several coils with a ratio near 1:1 were connected with their primaries in series to allow use of a high voltage for transmission while presenting a low voltage to the lamps. The inherent flaw in this method was that turning off a single lamp affected all the others on the circuit, and many adjustable coil designs were introduced in an effort to accommodate this problematic characteristic of the series circuit.

Between 1884 and 1885, Hungarian engineers Zipernowsky, Bláthy and Déri from the Ganz company in Budapest created the efficient "ZBD" closed-core model, which were based on the design by Gaulard and Gibbs. (Gaulard and Gibbs designed just an open core model). They discovered that all former (coreless or open-core) devices were incapable of regulating voltage, and were therefore impracticable.

Their joint patent described a transformer with no poles and comprised two versions of it, the "closed-core transformer" and the "shell-core transformer. In the closed-core transformer the iron core is a closed ring around which the two coils are arranged uniformly. In the shell type transformer, the copper induction cables are passed through the core. In both designs, the magnetic flux linking the primary and secondary coils travels (almost entirely) in the iron core, with no intentional path through air.

The core consists of iron cables or plates. Based on this invention, it became possible to provide economical and cheap lighting for industry and households." Zipernowsky, Bláthy and Déri discovered the mathematical formula of transformers: $V_s/V_p = N_s/N_p$. With this formula, transformers became calculable and proportionable. Their patent application made the first use of the word "transformer", a word that had been coined by Ottó Bláthy. George Westinghouse had bought both Gaulard and Gibbs' and the "ZBD" patents in 1885. He entrusted William Stanley with the building of a ZBD-type transformer for commercial use. Stanley built the core from interlocking E-shaped iron plates. This design was first used commercially in 1886.

Early Developments and Applications

Russian engineer Mikhail Dolivo-Dobrovolsky developed the first three-phase transformer in 1889. In 1891 Nikola Tesla invented the Tesla coil, an air-cored, dual-tuned resonant transformer for generating very high voltages at high frequency. Audio frequency transformers (at the time called repeating coils) were used by the earliest experimenters in the development of the telephone.

Basic Principles

The transformer is based on two principles: firstly, that an electric current can produce a magnetic field (electromagnetism) and secondly that a changing magnetic field within a coil of wire induces a voltage across the ends of the coil (electromagnetic induction). Changing the current in the primary coil changes the magnitude of the applied magnetic field. The changing magnetic flux extends to the secondary coil where a voltage is induced across its ends.

A simplified transformer design is shown to the left. A current passing through the primary coil creates a magnetic field. The primary and secondary coils are wrapped around a core of very high magnetic permeability, such as iron; this ensures that most of the magnetic field lines produced by the primary current are within the iron and pass through the secondary coil as well as the primary coil.

Induction law

The voltage induced across the secondary coil may be calculated from Faraday's law of induction, which states that:

$$V_S = N_S \frac{d\Phi}{dt}$$

where V_S is the instantaneous voltage, N_S is the number of turns in the secondary coil and Ö equals the magnetic flux through one turn of the coil. If the turns of the coil are oriented perpendicular to the magnetic field lines, the flux is the product of the magnetic field strength B and the area A through which it cuts. The area is constant, being equal to the cross-sectional area of the transformer core, whereas the magnetic field varies with time according to the excitation of the primary. Since the same magnetic flux passes through both the primary and secondary coils in an ideal transformer, the instantaneous voltage across the primary winding equals

$$V_P = N_P \frac{d\Phi}{dt}$$

Taking the ratio of the two equations for V_S and V_P gives the basic equation for stepping up or stepping down the voltage.

Ideal Power Equation

If the secondary coil is attached to a load that allows current to flow, electrical power is transmitted from the primary circuit to the secondary circuit. Ideally, the transformer is perfectly efficient; all the incoming energy is transformed from the primary circuit to the magnetic field and into the secondary circuit. If this condition is met, the incoming electric power must equal the outgoing power.

$$P_{\text{incoming}} = I_P V_P = P_{\text{outgoing}} = I_S V_S$$

giving the ideal transformer equation

If the voltage is increased (stepped up) ($V_S > V_P$), then the current is decreased (stepped down) ($I_S < I_P$) by the same factor. Transformers are efficient so this formula is a reasonable approximation.

The impedance in one circuit is transformed by the *square* of the turns ratio. For example, if an impedance Z_S is attached across the terminals of the secondary coil, it appears to the primary circuit to have an impedance of. This relationship is reciprocal, so that the impedance Z_P of the primary circuit appears to the secondary to be.

Detailed Operation

The simplified description above neglects several practical factors, in particular the primary current required to establish a magnetic field in the core, and the contribution to the field due to current in the secondary circuit.

Models of an ideal transformer typically assume a core of negligible reluctance with two windings of zero resistance. When a voltage is applied to the primary winding, a small current flows, driving flux around the magnetic circuit of the core. The current required to create the flux is termed the *magnetizing current*; since the ideal core has been assumed to have near-zero reluctance, the magnetizing current is negligible, although still required to create the magnetic field.

The changing magnetic field induces an electromotive force (EMF) across each winding. Since the ideal windings have no impedance, they have no associated voltage drop, and so the voltages V_P and V_S measured at the terminals of the transformer, are equal to the corresponding EMFs. The primary EMF, acting as it does in opposition to the primary voltage, is sometimes termed the "back EMF". This is due to Lenz's law which states that the induction of EMF would always be such that it will oppose development of any such change in magnetic field.

Practical Considerations

Leakage Flux

The ideal transformer model assumes that all flux generated by the primary winding links all the turns of every winding, including itself. In practice, some flux traverses paths that take it outside the windings. Such flux is termed *leakage flux*, and results in leakage inductance in series with the mutually coupled transformer windings.

Leakage results in energy being alternately stored in and discharged from the magnetic fields with each cycle of the power supply. It is not directly a power loss, but results in inferior voltage regulation, causing the secondary voltage to fail to be directly proportional to the primary, particularly under heavy load. Transformers are therefore normally designed to have very low leakage inductance. However, in some applications, leakage can be a desirable property, and long magnetic paths, air gaps, or magnetic bypass shunts may be deliberately introduced to a transformer's design to limit the short-circuit current it will supply. Leaky transformers may be used to supply loads that exhibit negative resistance, such as electric arcs, mercury vapour lamps, and neon signs; or for safely handling loads that become periodically short-circuited such as electric arc welders. Air gaps are also used to keep a transformer from saturating, especially audio-frequency transformers in circuits that have a direct current flowing through the windings.

Effect of Frequency

The time-derivative term in Faraday's Law shows that the flux in the core is the integral of the applied voltage. Hypothetically an ideal transformer would work with direct-current excitation, with the core flux increasing linearly with time. In practice, the flux would rise to the point where magnetic saturation of the core occurs, causing a huge increase in the magnetizing current and overheating the transformer. All practical transformers must therefore operate with alternating (or pulsed) current.

Transformer Universal EMF Equation

The EMF of a transformer at a given flux density increases with frequency. By operating at higher frequencies, transformers can be physically more compact because a given core is able to transfer more power without reaching saturation, and fewer turns are needed to achieve the same impedance. However properties such as core loss and conductor skin effect also increase with frequency. Aircraft and military equipment employ 400 Hz power supplies which reduce core and winding weight. Operation of a transformer at its designed voltage but at a higher frequency than intended will lead to reduced magnetizing current; at lower frequency, the magnetizing current will increase. Operation of a transformer at other than its design frequency may require assessment of voltages, losses, and cooling to establish if safe operation is practical. For example, transformers may need to be equipped with "volts per hertz" over-excitation relays to protect the transformer from overvoltage at higher than rated frequency. Knowledge of natural frequencies of transformer windings is of importance for the determination of the transient response of the windings to impulse and switching surge voltages.

Energy Losses

An ideal transformer would have no energy losses, and would be 100 per cent efficient. In practical transformers energy is dissipated in the windings, core, and surrounding structures. Larger transformers are generally more efficient, and those rated for electricity distribution usually perform better than 98 per cent.

Experimental transformers using superconducting windings achieve efficiencies of 99.85 per cent, While the increase in efficiency is small, when applied to large heavily-loaded transformers the annual savings in energy losses are significant. A small transformer, such as a plug-in "wall-wart" or power adapter type used for low-power consumer electronics, may be no more than 85 per cent efficient, with considerable loss even when not supplying any load. Though individual power loss is small, the aggregate losses from the very large number of such devices are coming under increased scrutiny.

The losses vary with load current, and may be expressed as "no-load" or "full-load" loss. Winding resistance dominates load losses, whereas hysteresis and eddy currents losses contribute to over 99 per cent of the no-load loss. The no-load loss can be significant, meaning that even an idle transformer constitutes a drain on an electrical supply, which encourages development of low-loss transformers. Transformer losses are divided into losses in the windings, termed copper loss, and those in the magnetic circuit, termed iron loss. Losses in the transformer arise from:

Winding Resistance

Current flowing through the windings causes resistive heating of the conductors. At higher frequencies, skin effect and proximity effect create additional winding resistance and losses.

Hysteresis Losses

Each time the magnetic field is reversed, a small amount of energy is lost due to hysteresis within the core. For a given core material, the loss is proportional to the frequency, and is a function of the peak flux density to which it is subjected.

Eddy Currents

Ferromagnetic materials are also good conductors, and a solid core made from such a material also constitutes a single short-circuited turn throughout its entire length. Eddy currents therefore circulate within the core in a plane normal to the flux, and are responsible for resistive heating of the core material. The eddy current loss is a complex function of the square of supply frequency and inverse square of the material thickness.

Magnetostriction

Magnetic flux in a ferromagnetic material, such as the core, causes it to physically expand and contract slightly with each cycle of the magnetic field, an effect known as magnetostriction. This produces the buzzing sound commonly associated with transformers, and in turn causes losses due to frictional heating in susceptible cores.

Mechanical Losses

In addition to magnetostriction, the alternating magnetic field causes fluctuating electromagnetic forces between the primary and secondary windings. These incite vibrations within nearby metalwork, adding to the buzzing noise, and consuming a small amount of power.

Stray Losses

Leakage inductance is by itself lossless, since energy supplied to its magnetic fields is returned to the supply with the next half-cycle. However, any leakage flux that intercepts nearby conductive materials such as the transformer's support structure will give rise to eddy currents and be converted to heat.

Equivalent Circuit

The physical limitations of the practical transformer may be brought together as an equivalent circuit model built around an ideal lossless transformer. Power loss in the windings is current-dependent and is represented as in-series resistances R_P and R_S. Flux leakage results in a fraction of the applied voltage dropped without contributing to the mutual coupling, and thus can be modeled as reactances of each leakage inductance X_P and X_S in series with the perfectly-coupled region.

Iron losses are caused mostly by hysteresis and eddy current effects in the core, and are proportional to the square of the core flux for operation at a given frequency. Since the core flux is proportional to the applied voltage, the iron loss can be represented by a resistance R_C in parallel with the ideal transformer.

A core with finite permeability requires a magnetizing current I_M to maintain the mutual flux in the core. The magnetizing current is in phase with the flux; saturation effects cause the relationship between the two to be non-linear, but for simplicity this effect tends to be ignored in most circuit equivalents. With a sinusoidal supply, the core flux lags the induced EMF by 90° and this effect can be modeled as a magnetizing reactance (reactance of an effective inductance) X_M in parallel with the core loss component. R_C and X_M are sometimes together termed the *magnetizing branch* of the model. If the secondary winding is made open-circuit, the current I_0 taken by the magnetizing branch represents the transformer's no-load current.

The secondary impedance R_S and X_S is frequently moved (or "referred") to the primary side after multiplying the components by the impedance scaling factor. The resulting model is sometimes termed the "exact equivalent circuit", though it retains a number of approximations, such as an assumption of linearity.

Analysis may be simplified by moving the magnetizing branch to the left of the primary impedance, an implicit assumption that the magnetizing current is low, and then summing primary and referred secondary impedances, resulting in so-called equivalent impedance. The parameters of equivalent circuit of a transformer can be calculated from the results of two transformer tests: open-circuit test and short-circuit test.

Types

A wide variety of transformer designs are used for different applications, though they share several common features. Important common transformer types include:

Autotransformer

An autotransformer has only a single winding with two end terminals, plus a third at an intermediate tap point. The primary voltage is applied across two of the terminals, and the secondary voltage taken from one of these and the third terminal. The primary and secondary circuits therefore have a number of windings turns in common. Since the volts-per-turn is the same in both windings, each develops a voltage in proportion to its number of turns. An adjustable autotransformer is made by exposing part of the winding coils and making the secondary connection through a sliding brush, giving a variable turns ratio.

Polyphase Transformers

For three-phase supplies, a bank of three individual single-phase transformers can be used, or all three phases can be incorporated as a single three-phase transformer. In this case, the magnetic circuits are connected together, the core thus containing a three-phase flow of flux.

A number of winding configurations are possible, giving rise to different attributes and phase shifts. One particular polyphase configuration is the zigzag transformer, used for grounding and in the suppression of harmonic currents.

Leakage Transformers

A leakage transformer, also called a stray-field transformer, has a significantly higher leakage inductance than other transformers, sometimes increased by a magnetic bypass or shunt in its core between primary and secondary, which is sometimes adjustable with a set screw.

This provides a transformer with an inherent current limitation due to the loose coupling between its primary and the secondary windings. The output and input currents are low enough to prevent thermal overload under all load conditions – even if the secondary is shorted. Leakage transformers are used for arc welding and high voltage discharge lamps (neon lamps and cold cathode fluorescent lamps, which are series-connected up to 7.5 kV AC). It acts then both as a voltage transformer and as a magnetic ballast. Other applications are short-circuit-proof extra-low voltage transformers for toys or doorbell installations.

Resonant Transformers

A resonant transformer is a kind of the leakage transformer. It uses the leakage inductance of its secondary windings in combination with external capacitors, to create one or more resonant circuits. Resonant transformers such as the Tesla coil can generate very high voltages, and are able to provide much higher current than electrostatic high-voltage generation machines such as the Van de Graaff generator.

One of the applications of the resonant transformer is for the CCFL inverter. Another application of the resonant transformer is to couple between stages of a superheterodyne receiver, where the selectivity of the receiver is provided by tuned transformers in the intermediate-frequency amplifiers.

Audio Transformers

Audio transformers are those specifically designed for use in audio circuits. They can be used to block radio frequency interference or the DC component of an audio signal, to split or combine audio signals, or to provide impedance matching between high and low impedance circuits, such as between a high impedance tube (valve) amplifier output and a low impedance loudspeaker, or between a high impedance instrument output and the low impedance input of a mixing console. Such transformers were originally designed to connect different telephone systems to one another while keeping their respective power supplies isolated, and are still commonly used to interconnect professional audio systems or system components.

Being magnetic devices, audio transformers are susceptible to external magnetic fields such as those generated by AC current-carrying conductors. "Hum" is a term commonly used to describe unwanted signals originating from the "mains" power supply (typically 50 or 60 Hz). Audio transformers used for low-level signals, such as those from microphones, often included shielding to protect against extraneous magnetically-coupled signals.

Instrument Transformers

Instrument transformers are used for measuring voltge, current, power and energy in electrical systems, and for protection and control. Where a voltage or current is too large to be conveniently measured by an instrument, it can be scaled down to a standardized low value. Instrument transformers isolate measurement and control circuitry from the high currents or voltages present on the circuits being measured or controlled. A current transformer is a transformer designed to provide a current in its secondary coil proportional to the current flowing in its primary coil. Voltage transformers (VTs), also referred to as "potential transformers" (PTs), are used in high-voltage circuits. They are designed to present a negligible load to the supply being measured, to allow protective relay equipment to be operated at lower voltages, and to have a precise winding ratio for accurate metering.

Classification

Transformers can be classified in different ways:

- *By power capacity*: from a fraction of a volt-ampere (VA) to over a thousand MVA;
- *By frequency range*: power-, audio-, or radio frequency;
- *By voltage class*: from a few volts to hundreds of kilovolts;
- *By cooling type*: air cooled, oil filled, fan cooled, or water cooled;
- *By application*: such as power supply, impedance matching, output voltage and current stabilizer, or circuit isolation;
- *By end purpose*: distribution, rectifier, arc furnace, amplifier output;
- *By winding turns ratio*: step-up, step-down, isolating (equal or near-equal ratio), variable.

Construction

Cores

Laminated Steel Cores

Transformers for use at power or audio frequencies typically have cores made of high permeability silicon steel. The steel has a permeability many times that of free space, and the core thus serves to greatly reduce the magnetizing current, and confine the flux to a path which closely couples the windings.

Early transformer developers soon realized that cores constructed from solid iron resulted in prohibitive eddy-current losses, and their designs mitigated this effect with cores consisting of bundles of insulated iron wires. Later designs constructed the core by stacking layers of thin steel laminations, a principle that has remained in use. Each lamination is insulated from its neighbours by a thin non-conducting layer of insulation. The universal transformer equation indicates a minimum cross-sectional area for the core to avoid saturation.

The effect of laminations is to confine eddy currents to highly elliptical paths that enclose little flux, and so reduce their magnitude. Thinner laminations reduce losses, but are more laborious and expensive to construct. Thin laminations are generally used on high frequency transformers, with some types of very thin steel laminations able to operate up to 10 kHz.

One common design of laminated core is made from interleaved stacks of E-shaped steel sheets capped with I-shaped pieces, leading to its name of "E-I transformer". Such a design tends to exhibit more losses, but is very economical to manufacture. The cut-core or C-core type is made by winding a steel strip around a rectangular form and then bonding the layers together. It is then cut in two, forming two C shapes, and the core assembled by binding the two C halves together with a steel strap. They have the advantage that the flux is always oriented parallel to the metal grains, reducing reluctance.

A steel core's remanence means that it retains a static magnetic field when power is removed. When power is then reapplied, the residual field will cause a high inrush current until the effect of the remaining magnetism is reduced, usually after a few cycles of the applied alternating current. Overcurrent protection devices such as fuses must be selected to allow this harmless inrush to pass. On transformers connected to long, overhead power transmission lines, induced currents due to geomagnetic disturbances during solar storms can cause saturation of the core and operation of transformer protection devices.

Distribution transformers can achieve low no-load losses by using cores made with low-loss high-permeability silicon steel or amorphous (non-crystalline) metal alloy. The higher initial cost of the core material is offset over the life of the transformer by its lower losses at light load.

Solid Cores

Powdered iron cores are used in circuits (such as switch-mode power supplies) that operate above main frequencies and up to a few tens of kilohertz. These materials combine high magnetic permeability with high bulk electrical resistivity. For frequencies extending beyond the VHF band, cores made from non-conductive magnetic ceramic materials called ferrites are common. Some radio-frequency transformers also have movable cores (sometimes called 'slugs') which allow adjustment of the coupling coefficient (and bandwidth) of tuned radio-frequency circuits.

Toroidal Cores

Toroidal transformers are built around a ring-shaped core, which, depending on operating frequency, is made from a long strip of silicon steel or permalloy wound into a coil, powdered iron, or ferrite. A strip construction ensures that the grain boundaries are optimally aligned, improving the transformer's efficiency by reducing the core's reluctance.

The closed ring shape eliminates air gaps inherent in the construction of an E-I core. The cross-section of the ring is usually square or rectangular, but more expensive cores with circular cross-sections are also available. The primary and secondary coils are often wound concentrically to cover the entire surface of the core. This minimizes the length of wire needed, and also provides screening to minimize the core's magnetic field from generating electromagnetic interference. Toroidal transformers are more efficient than the cheaper laminated E-I types for a similar power level. Other advantages compared to E-I types, include smaller size (about half), lower weight (about half), less mechanical hum (making them superior in audio amplifiers), lower exterior magnetic field (about one tenth), low off-load losses (making them more efficient in standby circuits), single-bolt mounting, and greater choice of shapes.

The main disadvantages are higher cost and limited power capacity. Ferrite toroidal cores are used at higher frequencies, typically between a few tens of kilohertz to a megahertz, to reduce losses, physical size, and weight of switch-mode power supplies. A drawback of toroidal transformer construction is the higher cost of windings. As a consequence, toroidal transformers are uncommon above ratings of a few kVA. Small distribution transformers may achieve some of the benefits of a toroidal core by splitting it and forcing it open, then inserting a bobbin containing primary and secondary windings.

Air Cores

A physical core is not an absolute requisite and a functioning transformer can be produced simply by placing the windings in close proximity to each other, an arrangement termed an "air-core" transformer. The air which comprises the magnetic circuit is essentially lossless, and so an air-core transformer eliminates loss due to hysteresis in the core material.

The leakage inductance is inevitably high, resulting in very poor regulation, and so such designs are unsuitable for use in power distribution. They have however very high bandwidth, and are frequently employed in radio-frequency applications, for which a satisfactory coupling coefficient is maintained by carefully overlapping the primary and secondary windings.

Windings

The conducting material used for the windings depends upon the application, but in all cases the individual turns must be electrically insulated from each other to ensure that the current travels throughout every turn. For small power and signal transformers, in which currents are low and the potential difference between adjacent turns is small, the coils are often wound from enameled magnet wire, such as Formvar wire. Larger power transformers operating at high voltages may be wound with copper rectangular strip conductors insulated by oil-impregnated paper and blocks of pressboard.

High-frequency transformers operating in the tens to hundreds of kilohertz often have windings made of braided Litz wire to minimize the skin-effect and proximity effect losses. Large power transformers use multiple-stranded conductors as well, since even at low power frequencies non-uniform distribution of current would otherwise exist in high-current windings. Each strand is individually insulated, and the strands are arranged so that at certain points in the winding, or throughout the whole winding, each portion occupies different relative positions in the complete conductor. The transposition equalizes the current flowing in each strand of the conductor, and reduces eddy current losses in the winding itself. The stranded conductor is also more flexible than a solid conductor of similar size, aiding manufacture.

For signal transformers, the windings may be arranged in a way to minimize leakage inductance and stray capacitance to improve high-frequency response. This can be done by splitting up each coil into sections, and those sections placed in layers between the sections of the other winding. This is known as a stacked type or interleaved winding.

Both the primary and secondary windings on power transformers may have external connections, called taps, to intermediate points on the winding to allow selection of the voltage ratio. The taps may be connected to an automatic on-load tap changer for voltage regulation of distribution circuits. Audio-frequency transformers, used for the distribution of audio to public address loudspeakers, have taps to allow adjustment of impedance to each speaker.

A centre-tapped transformer is often used in the output stage of an audio power amplifier in a push-pull circuit. Modulation transformers in AM transmitters are very similar. Certain transformers have the windings protected by epoxy resin. By impregnating the transformer with epoxy under a vacuum, one can replace air spaces within the windings with epoxy, thus sealing the windings and helping to prevent the possible formation of corona and absorption of dirt or water. This produces transformers more suited to damp or dirty environments, but at increased manufacturing cost.

Coolant

High temperatures will damage the winding insulation. Small transformers do not generate significant heat and are cooled by air circulation and radiation of heat. Power transformers rated up to several hundred kVA can be adequately cooled by natural convective air-cooling, sometimes assisted by fans.

In larger transformers, part of the design problem is removal of heat. Some power transformers are immersed in transformer oil that both cools and insulates the windings. The oil is a highly refined mineral oil that remains stable at transformer operating temperature. Indoor liquid-filled transformers must use a non-flammable liquid, or must be located in fire resistant rooms. Air-cooled dry transformers are preferred for indoor applications even at capacity ratings where oil-cooled construction would be more economical, because their cost is offset by the reduced building construction cost.

The oil-filled tank often has radiators through which the oil circulates by natural convection; some large transformers employ forced circulation of the oil by electric pumps, aided by external fans or water-cooled heat exchangers.

Oil-filled transformers undergo prolonged drying processes to ensure that the transformer is completely free of water vapour before the cooling oil is introduced. This helps prevent electrical breakdown under load. Oil-filled transformers may be equipped with Buchholz relays, which detect gas evolved during internal arcing and rapidly de-energize the transformer to avert catastrophic failure.

Polychlorinated biphenyls have properties that once favoured their use as a coolant, though concerns over their environmental persistence led to a widespread ban on their use. Today, non-toxic, stable silicone-based oils, or fluorinated hydrocarbons may be used where the expense of a fire-resistant liquid offsets additional building cost for a transformer vault. Before 1977, even transformers that were nominally filled only with mineral oils may also have been contaminated with polychlorinated biphenyls at 10-20 ppm. Since mineral oil and PCB fluid mix, maintenance equipment used for both PCB and oil-filled transformers could carry over small amounts of PCB, contaminating oil-filled transformers. Some "dry" transformers (containing no liquid) are enclosed in sealed, pressurized tanks and cooled by nitrogen or sulfur hexafluoride gas. Experimental power transformers in the 2 MVA range have been built with superconducting windings which eliminates the copper losses, but not the core steel loss. These are cooled by liquid nitrogen or helium.

Terminals

Very small transformers will have wire leads connected directly to the ends of the coils, and brought out to the base of the unit for circuit connections. Larger transformers may have heavy bolted terminals, bus bars or high-voltage insulated bushings made of polymers or porcelain. A large bushing can be a complex structure since it must provide careful control of the electric field gradient without letting the transformer leak oil.

Applications

A major application of transformers is to increase voltage before transmitting electrical energy over long distances through wires. Wires have resistance and so dissipate electrical energy at a rate proportional to the square of the current through the wire. By transforming electrical power to a high-voltage (and therefore low-current) form for transmission and back again afterwards, transformers enable economic transmission of power over long distances. Consequently, transformers have shaped the electricity supply industry, permitting generation to be located remotely from points of demand. All but a tiny fraction of the world's electrical power has passed through a series of transformers by the time it reaches the consumer.

Transformers are also used extensively in electronic products to step down the supply voltage to a level suitable for the low voltage circuits they contain.

The transformer also electrically isolates the end user from contact with the supply voltage. Signal and audio transformers are used to couple stages of amplifiers and to match devices such as microphones and record player s to the input of amplifiers. Audio transformers allowed telephone circuits to carry on a two-way conversation over a single pair of wires. Transformers are also used when it is necessary to couple a differential-mode signal to a ground-referenced signal, and for between external cables and internal circuits.

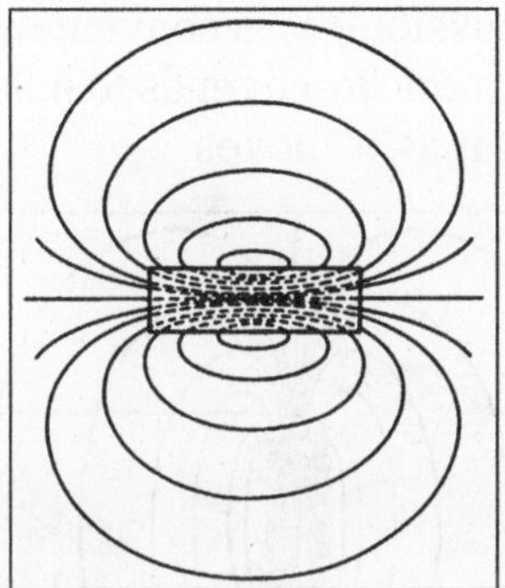

Fig. Schematic Representation of a Bar Magnet and its Magnetic Field. All the Lines are Actually Loops, but some of the Loops are too Large to be Represented on the Diagram.

For instance, if the current should suddenly be doubled, the field would double in strength, as represented in Fig. After doubling the current, there will be twice as many lines of force spread out over a larger volume than before. If one of these hypothetical lines of force were visible while the current were in the process of doubling, it would be seen to move outward.

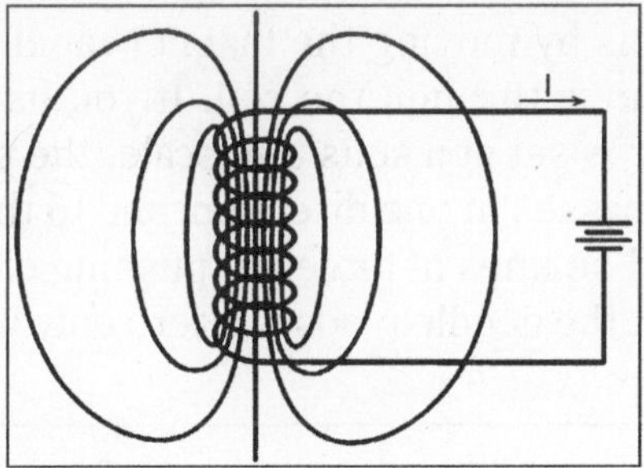

Fig. Schematic Representation of the Magnetic Field Generated by the Passage of Current I Through a Coil of Wire.

In other words, increasing the current through the coil not only adds more lines to the field but also causes the field to expand. Conversely, if the current were reduced, the field would collapse. Now store that concept for a moment and consider the opposite sort of phenomenon, namely, the effect that a magnetic field has on a coil of wire. If you have a permanent magnet at hand, you can demonstrate certain phenomena very readily. Connect your meter directly across the two ends of the coil on your sensitive relay, and set the meter to its most sensitive scale (e.g., 100 microamperes, 1 volt, etc.).

Then put the magnet down next to the relay coils. Some of the lines of force from the magnet will now be passing through the relay coil but you will notice that nothing happens to the meter. In general, a stationary magnetic field has no effect on a coil of wire. But a moving field does affect the coil. Any time that lines of force move with respect to a coil of wire, they induce a voltage into it. (There is one particular direction of motion of a field with respect to a coil which will not induce a voltage into the coil, but that special case will be ignored for the present discussion.) It is convenient to think of this process as one in which each line of magnetic force tends to push some of the free electrons in the wire along in front of it as it moves.

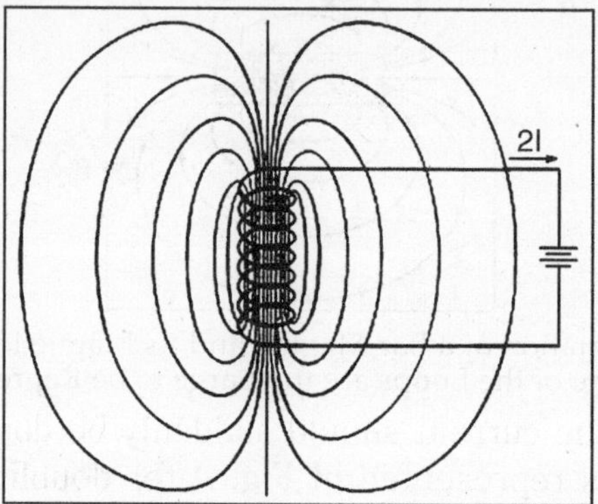

Fig. The Magnetic Field of the Coil when the Current Through it is Doubled. The Field is Represented as having Twice as many Lines, Distributed Over a Larger Area (Volume).

You can observe this by moving the magnet rapidly past the relay coil so that its lines of force move through the coil. If you have a reasonably strong magnet and if the meter is set at a sensitive scale, the pointer will jump to the right when the magnet passes in one direction, and to the left when it moves in the opposite direction. The lines of force are pushing electrons one way or the other. Note, though, that the needle reads nonzero only when the field is actually moving.

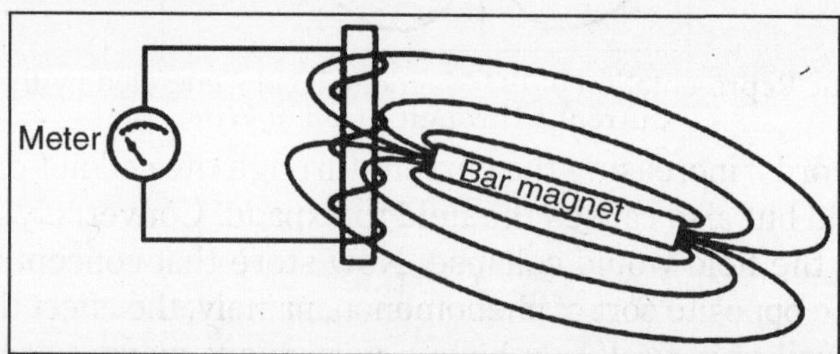

Fig. Representation of a Bar Magnet Close to a Relay Coil. Some of the Lines of Force Intersect the Coil. So Long as the Magnet is Stationary with Respect to the Coil, no Current will Flow Around the Loop Containing the Meter.

These two phenomena, the generation of a magnetic field by current flowing in a coil and the induction of current in a coil by a moving magnetic field, are combined in an extensively used device called the transformer. Its principle of operation is fairly simple. The switch is open and no current flows in either circuit.

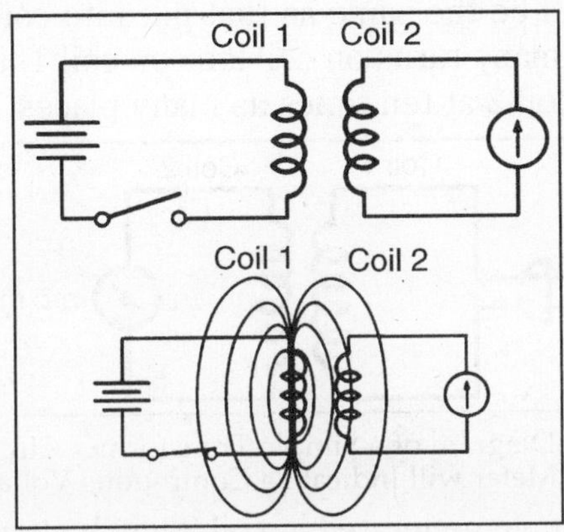

Fig. (*a*) A Circuit Containing two Coils of Wire, Near Each Other but not Connected Together Electrically. (*b*) Representation of the Circuit after the Switch has been Closed. Some of the Lines of Force from Coil I Intersect Coil 2, but so Long as the Current Through Coil I is Constant, no Current will Flow Through Coil 2.

Figure *b* shows the state of affairs when the switch has been closed and current is flowing steadily through coil 1. Although the lines of force from coil 1 are passing through coil 2, no voltage is generated in coil 2 because the lines are not moving. However, between the time when no current flows through coil 1 and when the full current is flowing, that is the lines of force must have multiplied and expanded through coil 2 from zero lines and volume to their positions. During the time when the field was being built up, a voltage must have been induced into coil 2. Similarly, if the switch were opened again, the field would collapse and, during that short time, a voltage would be induced into coil 2 as well.

The voltage in coil 2 would have the opposite polarity in the two cases because the lines of force would be traveling in opposite directions during the buildup and collapse of the field. The buildup and the collapse do not occur instantaneously but take finite times which depend upon the characteristics of the circuit. Now suppose the battery and switch were replaced by an a-c source, for instance, the 115-volt a-c wall socket supply. In this case, the current through coil 1 will build up from zero to a maximum, then back through zero again, etc., repeating this complete cycle 60 times a second.

In turn, the magnetic field will also build up and collapse over and over again, continuously generating a changing and reversing voltage in coil 2. This kind of combination of coils is called a transformer. For more efficient transfer of power from coil I to coil 2, both coils are typically wound on a single iron core, because such a core tends to channel the magnetic fields, keeping the lines of force in the close vicinity of the coils. The schematic diagram for a transformer, together with photographs of actual transformers.

If coil 1 consists of the same number of turns of wire as coil 2, the voltage generated in coil 2 will be the same as that put into coil 1. However, if there are, say, ten times as many turns on coil 2 as on coil 1, the lines of force from coil 1 will cut across coil 2 at ten times as many places.

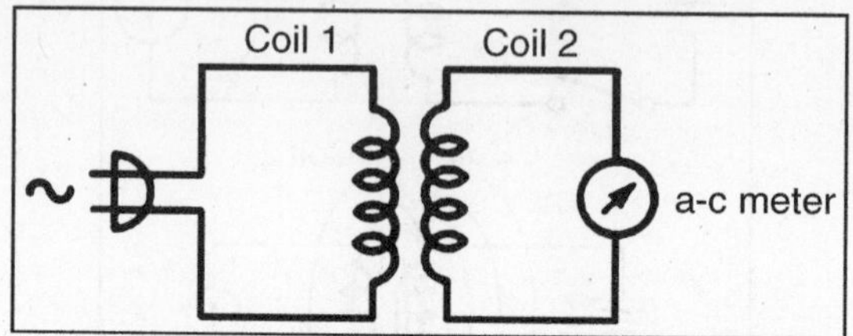

Fig. Diagram of a Simple Transformer Circuit. The Meter will Indicate a Continuous Voltage.

Therefore the voltage generated in coil 2 will be ten times as great as the input voltage to coil 1. In general:

$$\frac{V_i}{V_o} = \frac{N_i}{N_o}$$

where *Vi* = the input voltage *Vo* = the output voltage *Ni* = the number of turns on the input coil *No* = the number of turns on the output coil

It is this phenomenon that makes the transformer so useful. For example, suppose that you were designing a circuit to give a rat a strong constant-current shock. For the shock to have a relatively constant current, a large resistance must be in series with the rat. In fact, the resistance must be large enough that almost all of the voltage from the source is lost across the resistor, and only about one-tenth of it actually affects the rat.

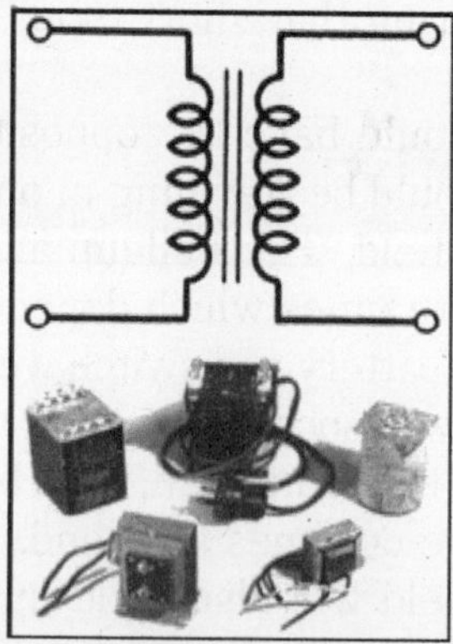

Fig. (*a*) Circuit Symbol for a Transformer, and (*b*) a Photograph of Various Transformers.

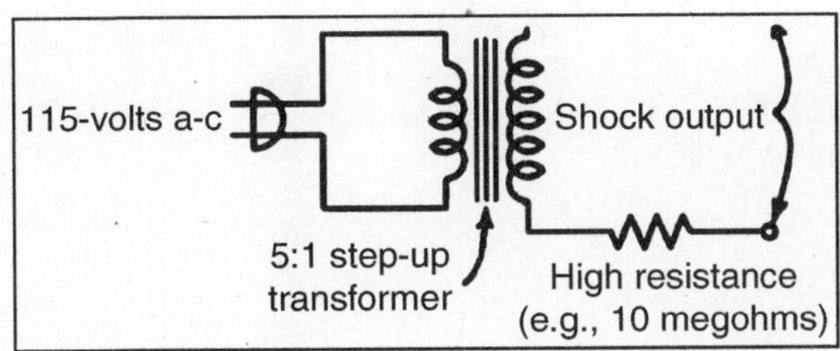

Fig. Constant-current a-c Shock-Delivery Circuit.

To give the rat a good strong shock, the source voltage must be higher than the 115 volts from the wall. A 50-volt shock, for example, requires a 500-volt source. The solution is to put between the wall socket and the shock box a transformer which has about five times as many turns on the output (or secondary) coil as it has on the input (or primary) coil. The transformer is called a step-up transformer. Electric-eye systems which detect the presence of a rat in an alley require a small source of light such as a flashlight bulb or an automobile taillight bulb. These bulbs are typically rated at 6 volts.

Flashlight batteries could be used to light the bulbs, but they burn out quickly, and storage batteries are inconvenient. However, a small *step-down* transformer which takes the 115 volts from a wall socket and transforms it to 6 volts is cheap, available, and never wears out. Such a transformer has approximately 20 times as many turns on its input coil as on its output coil.

Note that a transformer is so named because it transforms one alternating voltage into another alternating voltage. It does not transform direct current into alternating current (that is accomplished by a device called a converter) nor alternating current into direct current (that device is called a rectifier). A transformer is strictly an a-c device, and as such can be quickly and irreversibly damaged if it is connected across a d-c source.

8

Diodes and Rectifiers

INTRODUCTION

A vacuum diode is an electron valve that supports current flow through it in only one direction. It is composed of two electrical "parts" or elements and the evacuated glass envelope. Additionally, there is usually a heater, a coil of special alloy wire that can be driven to red heat by a modest application of voltage. (A vacuum diode without a heater would be a cold cathode diode.)

The heater is coiled inside a cylindrical element called the cathode. The cathode is composed of a special alloy, or, more properly is coated with an alloy, that has high thermal emission characteristics.

That means that if it's heated sufficiently, electrons actually leave the metallic crystal structure of the metal (boil off) and hang around outside in the space nearby forming what is called a space charge. When we built the tube, we pumped all the air out, sealed the glass, and then fired a "getter" inside the tube to chemically bind the remaining bit of gas that was there. The tube really is highly evacuated.

That's important because we don't want all those light weight electrons bumping into gas atoms and molecules when they are hanging around in the space charge, and especially when transiting the gap between the cathode and the other element inside, the anode or plate.

If a voltage is applied across the cathode and plate, and the plate is negative, no current will flow in the tube. Electrons flow from negative to positive, but, darn it, they can't leave the plate to go to the cathode. That's called reverse biased. It's hooked up "backwards" like that. But if the cathode is connected to negative and the positive connected to the plate, electrons that were in the space charge will be driven away from the cathode and attracted to the plate. Current will flow in this direction. That's forward bias.

The hot cathode emits electrons in droves. Many more than are needed. They build up in the volume around the cathode and form a large pool of electrons just waiting for something positive to happen. This pool of electrons is called the space charge. When the plate (anode) is made positive some of the electrons are attracted to it. They impact on it, are absorbed into the metal and electrons flow out of the plate connection into the battery or what ever provided the positive voltage. The other end of the voltage source must be connected back to the cathode in some way.

If the polarity is reversed which makes the plate negative with respect to the cathode the electrons in the space charge are repelled away from the plate and no conduction takes place. The plate is cold and is made of a metal that is a very poor emitter of electrons. The voltages are not nearly as high as those used by Crooks so there is no cold emission from the plate. This makes the diode conduct current in only one direction.

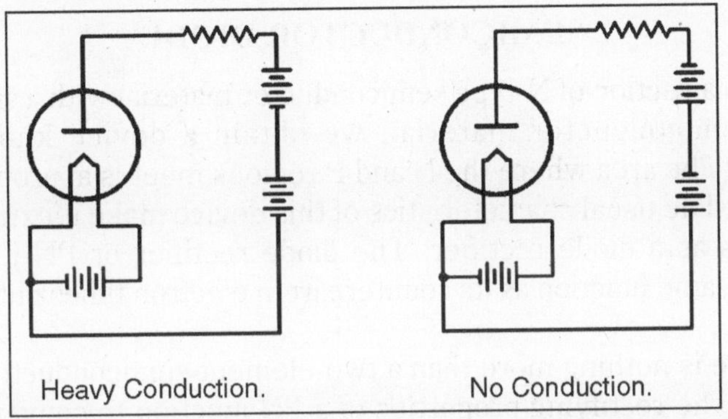

The cathode of a tube is coated with a very special material that emits electrons very easily. It can readily be damaged by operating above or below the proper temperature or by being "poisoned" by the residual air in the tube. The space charge serves another vital function. There is no such thing as a perfect vacuum. In fact the space in near earth orbit is a better vacuum than can be made on the surface of earth. Farther out in space the vacuum is even better. What this means is there are lots of air molecules running around inside of a vacuum tube.

The space charge protects the cathode from these air molecules. Some of these molecules lose an electron and become positive ions. They are attracted to the negative cathode and if it weren't for the space charge they would impact on the surface of the cathode doing considerable damage.

Over time the ability of the cathode to emit electrons would be seriously impaired rendering the tube useless. As these ions run through the space charge they gain electrons from the many collisions and become negative ions. Their direction is reversed turning them away from the delicate cathode.

INDIRECTLY HEATED CATHODES

Most vacuum tubes have a basically cylindrical configuration. Figure two shows how a filament (directly heated) cathode type diode is made along with a heater type (indirectly heated) cathode type diode.

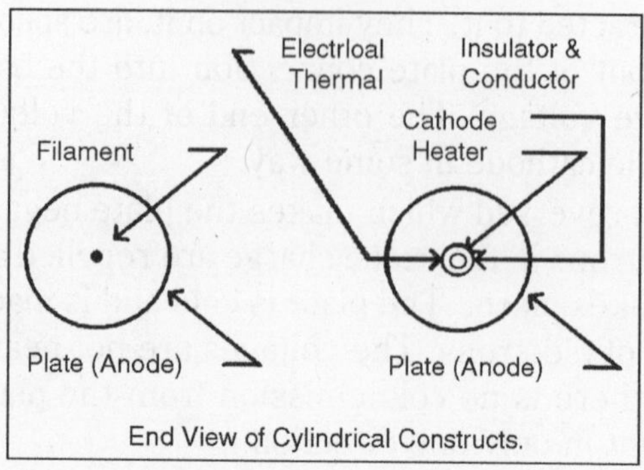

SEMICONDUCTOR DIODE

If we join a section of N-type semiconductor material with a similar section of P-type semiconductor material, we obtain a device known as a PN JUNCTION. (The area where the N and P regions meet is appropriately called the junction.) The usual characteristics of this device make it extremely useful in electronics as a diode rectifier. The diode rectifier or PN junction diode performs the same function as its counterpart in electron tubes but in a different way.

The diode is nothing more than a two-element semiconductor device that makes use of the rectifying properties of a PN junction to convert alternating current into direct current by permitting current flow in only one direction. The schematic symbol of a PN junction diode is shown in figure below. The vertical bar represents thecathode (N-type material) since it is the source of electrons and the arrow represents the anode. (P-type material) since it is the destination of the electrons.

The label "CR1" is an alphanumerical code used to identify the diode. In this figure, we have only one diode so it is labeled CR1 (crystal rectifier number one). If there were four diodes shown in the diagram, the last diode would be labeled CR4. The heavy dark line shows electron flow. Notice it is against the arrow. For further clarification, a pictorial diagram of a PN junction and an actual semiconductor (one of many types) are also illustrated.

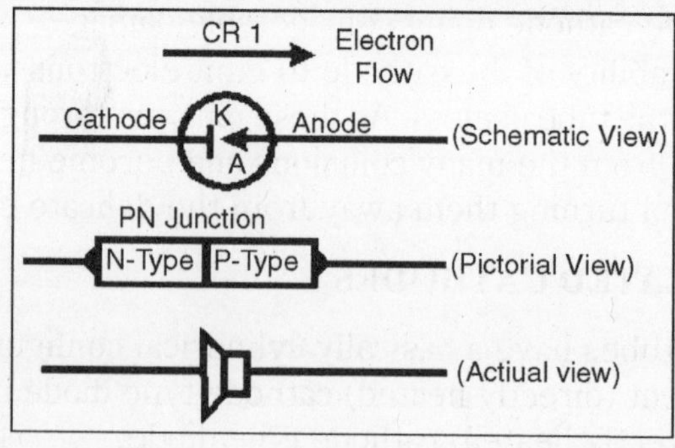

CONSTRUCTION

Merely pressing together a section of P material and a section of N material, however, is not sufficient to produce a rectifying junction. The semiconductor should be in one piece to form a proper PN junction, but divided into a P-type impurity region and an N-type impurity region. This can be done in various ways. One way is to mix P-type and N-type impurities into a single crystal during the manufacturing process. By so doing, a P-region is grown over part of a semiconductor's length and N-region is grown over the other part. This is called a GROWN junction and is illustrated in view A of figure below.

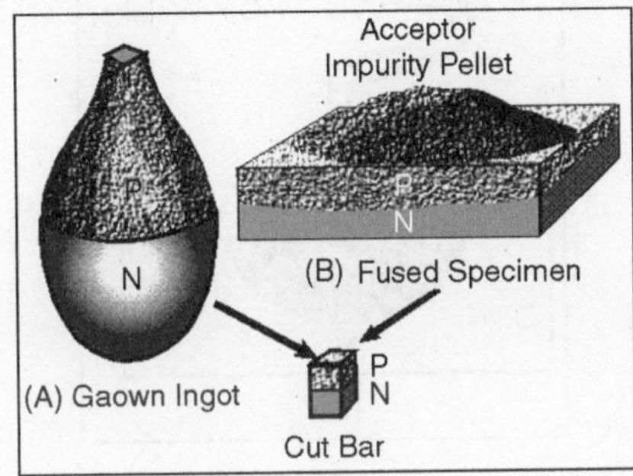

Another way to produce a PN junction is to melt one type of impurity into a semiconductor of the opposite type impurity. For example, a pellet of acceptor impurity is placed on a wafer of N-type germanium and heated. Under controlled temperature conditions, the acceptor impurity fuses into the wafer to form a P-region within it, as shown in view B of figure above.

This type of junction is known as an ALLOY or FUSED-ALLOY junction, and is one of the most commonly used junctions. In figure shown below, a POINT-CONTACT type of construction is shown.

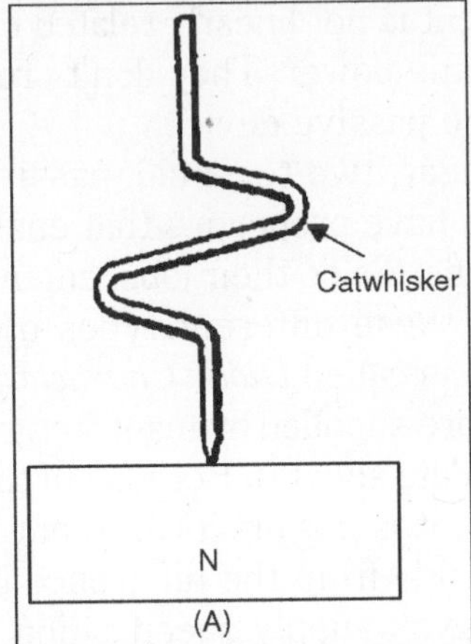

(A)

It consists of a fine metal wire, called a cat whisker, that makes contact with a small area on the surface of an N-type semiconductor as shown in view

A of the figure. The PN union is formed in this process by momentarily applying a high-surge current to the wire and the N-type semiconductor. The heat generated by this current converts the material nearest the point of contact to a P-type material (view B).

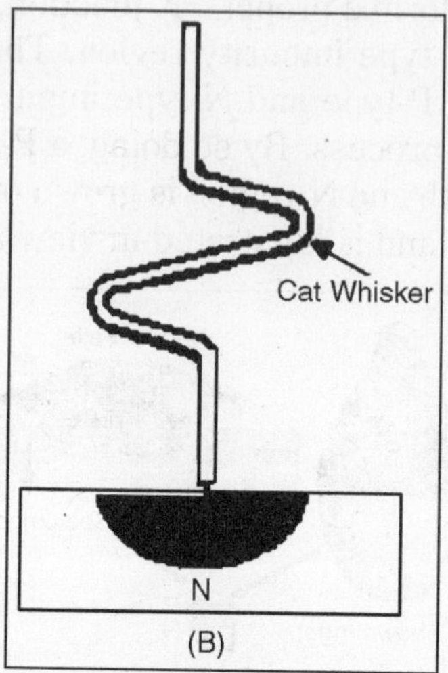

CHARACTERISTICS OF DIODE

Diodes have the following characteristics:

- Diodes are two terminal devices like resistors and capacitors. They don't have many terminals like transistors or integrated circuits.
- In diodes current is directly related to voltage, like in a resistor. They're not like capacitors where current is related to the time derivative of voltage or inductors where the derivative of current is related to voltage.
- In diodes the current is not linearly related to voltage, like in a resistor.
- Diodes only consume power. They don't produce power like a battery. They are said to be passive devices.
- Diodes are nonlinear, two terminal, passive electrical devices.

Semiconductor diodes have properties that enable them to perform many different electronic functions. To do their jobs, engineers and technicians must be supplied with data on these different types of diodes. The information presented for this purpose is called *Diode Characteristics*.

These characteristics are supplied by manufacturers either in their manuals or on specification sheets (data sheets). Because of the scores of manufacturers and numerous diode types, it is not practical to put before you a specification sheet and call it typical. Aside from the difference between manufacturers, a single manufacturer may even supply specification sheets that differ both in format and content. Despite these differences, certain performance and design information is normally required. We will discuss this information in the next few paragraphs. A standard specification sheet usually has a brief description of the diode. Included in this description is the type of diode, the major area of application, and any special features. Of particular interest is the specific application for which the diode is suited. The manufacturer also provides a drawing of the diode which gives dimension, weight, and, if appropriate, any identification marks. In addition to the above data, the following information is also provided: a static operating table (giving spot values of parameters under fixed conditions), sometimes a characteristic curve similar to the one in figure 1–20 (showing how parameters vary over the full operating range), and diode ratings (which are the limiting values of operating conditions outside which could cause diode damage). In general, diodes tend to permit current flow in one direction, but tend to inhibit current flow in the opposite direction. The graph below shows how current can depend upon voltage for a diode.

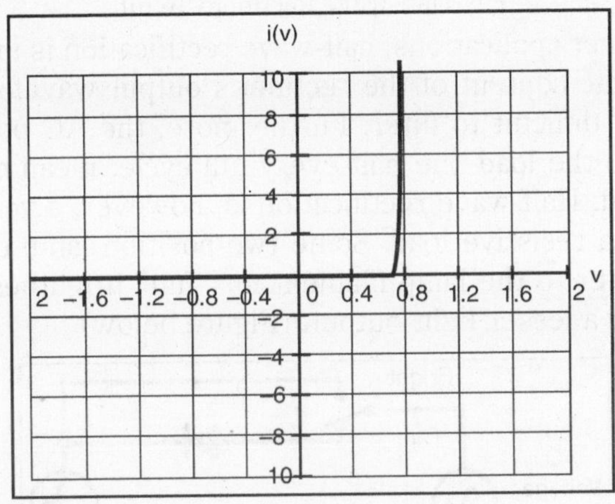

Note the following:

- When the voltage across the diode is positive, a lot of current can flow once the voltage becomes large enough.
- When the voltage across the diode is negative, virtually no current flows.

The circuit symbol for a diode is designed to remind you that current flows easily through a diode in one direction. The circuit symbol for a diode is shown below together with common conventions for current through the diode and voltage across the diode.

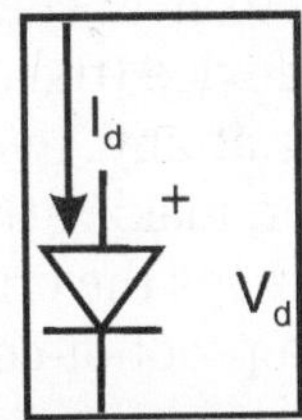

HALF WAVE RECTIFIER

Now we come to the most popular application of the diode: *rectification*. Simply defined, rectification is the conversion of alternating current (AC) to direct current (DC). This involves a device that only allows one-way flow of electrons. As we have seen, this is exactly what a semiconductor diode does. The simplest kind of rectifier circuit is the *half-wave* rectifier. It only allows one half of an AC waveform to pass through to the load. (Figure below)

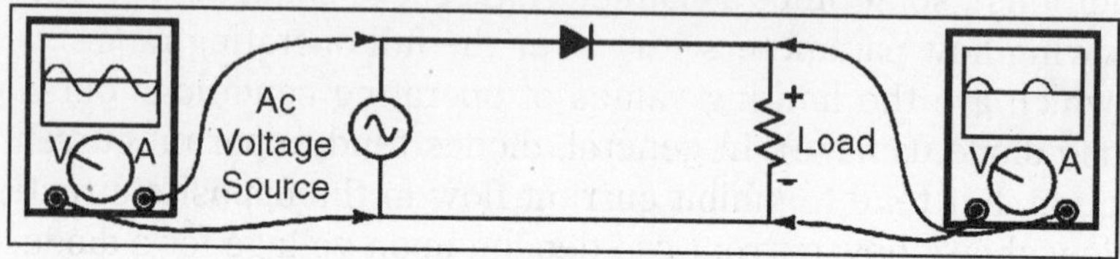

Fig. Half-wave Rectifier Circuit.

For most power applications, half-wave rectification is insufficient for the task. The harmonic content of the rectifier's output waveform is very large and consequently difficult to filter. Furthermore, the AC power source only supplies power to the load one half every full cycle, meaning that half of its capacity is unused. Half-wave rectification is, however, a very simple way to reduce power to a resistive load. Some two-position lamp dimmer switches apply full AC power to the lamp filament for "full" brightness and then half-wave rectify it for a lesser light output. (Figure below)

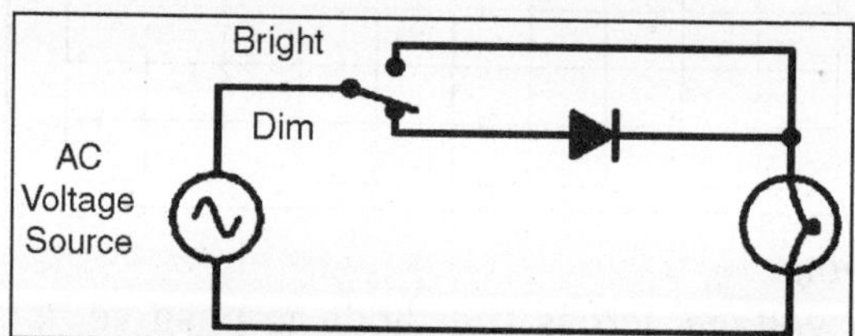

Fig. Half-wave Rectifier Application: Two Level Lamp Dimmer.

In the "Dim" switch position, the incandescent lamp receives approximately one-half the power it would normally receive operating on full-wave AC. Because the half-wave rectified power pulses far more rapidly than the filament has time to heat up and cool down, the lamp does not blink. Instead, its filament merely operates at a lesser temperature than normal, providing less light output. This principle of "pulsing" power rapidly to a slow-responding load device to control the electrical power sent to it is common in the world of industrial electronics. Since the controlling device (the diode, in this case) is either fully conducting or fully nonconducting at any given time, it dissipates little heat energy while controlling load power, making this method of power control very energy-efficient. This circuit is perhaps the crudest possible method of pulsing power to a load, but it suffices as a proof-of-concept application.

VOLTAGE REGULATOR

A Zener Diode is an electronic component which can be used to make a very simple voltage regulator circuit. This circuit enables a fixed stable voltage to be taken from an unstable voltage source such as thebattery bank of a renewable energy system which will fluctuate depending on the state of charge of the bank.

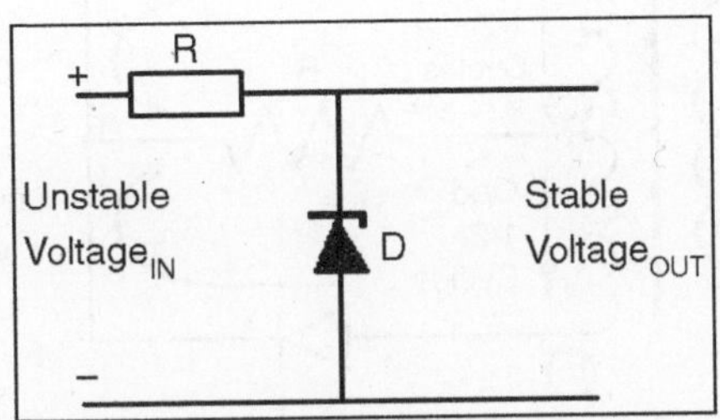

Fig. Zener Diode Voltage Regulator Circuit

Pictured above is a very simple voltage regulator circuit requiring just one zener diode (available from the REUK Shop) and one resistor. As long as the input voltage is a few volts more than the desired output voltage, the voltage across the zener diode will be stable.

As the input voltage increases the current through the Zener diode increases but the voltage drop remains constant - a feature of zener diodes. Therefore since the current in the circuit has increased the voltage drop across the resistor increases by an amount equal to the difference between the input voltage and the zener voltage of the diode.

FULL WAVE RECTIFIER

In a Full-wave rectifier circuit two diodes are now used, together with a transformer whose secondary winding is split equally into two and has a common centre tapped connection, (C). Now each diode conducts in turn when its Anode terminal is positive with respect to the centre point C as shown below.

FULL-WAVE RECTIFIER CIRCUIT

The circuit consists of two *Half-wave* rectifiers connected to a single load resistance with each diode taking it in turn to supply current to the load. When point A is positive with respect to point B, diode D_1conducts in the forward direction as indicated by the arrows. When point B is positive (in the negative half of the cycle) with respect to point A, diode D_2 conducts in the forward direction and the current flowing through resistor R is in the same direction for both circuits.

As the output voltage across the resistor R is the sum of the two waveforms, this type of circuit is also known as a "bi-phase" circuit. As the spaces between each half-wave developed by each diode is now being filled in by the other diode the average DC output voltage across the load resistor is now double that of the single half-wave rectifier circuit and is about $0.637V_{max}$ of the peak voltage, assuming no losses.

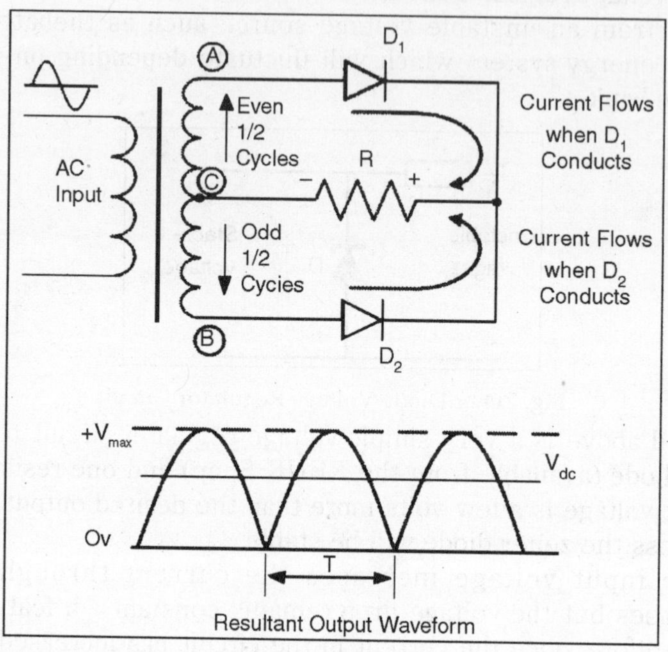

$$V_{d.c.} = \frac{V_{max}}{\pi}$$

$$= 0.637V_{max}$$

$$= 0.9V_S$$

The peak voltage of the output waveform is the same as before for the half-wave rectifier provided each half of the transformer windings have the same rms voltage value. To obtain a different d.c. voltage output different transformer ratios can be used, but one main disadvantage of this type of rectifier is that having a larger transformer for a given power output with two separate windings makes this type of circuit costly compared to a "Bridge Rectifier" circuit equivalent.

BRIDGE RECTIFIER

Another type of circuit that produces the same output as a full-wave rectifier is that of the Bridge Rectifier. This type of single phase rectifier uses 4 individual rectifying diodes connected in a "bridged" configuration to produce the desired output but does not require a special centre tapped transformer, thereby reducing its size and cost. The single secondary winding is connected to one side of the diode bridge network and the load to the other side as shown below.

THE DIODE BRIDGE RECTIFIER

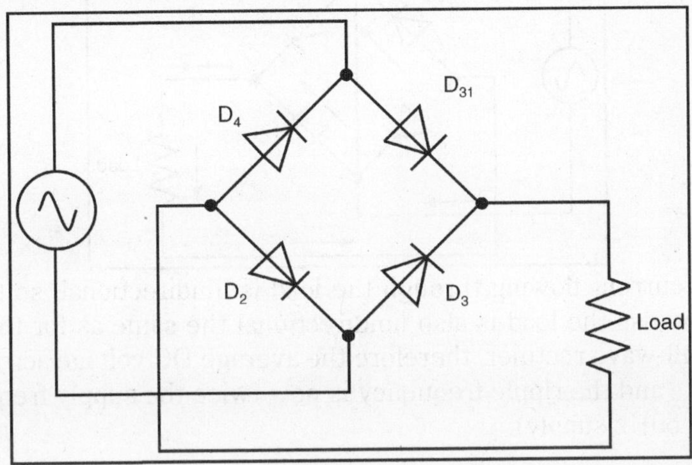

The 4 diodes labelled D_1 to D_4 are arranged in "series pairs" with only two diodes conducting current during each half cycle. During the positive half cycle of the supply, diodes D_1 and D_2 conduct in series while diodes D_3 and D_4 are reverse biased and the current flows through the load as shown below.

The Positive Half-cycle

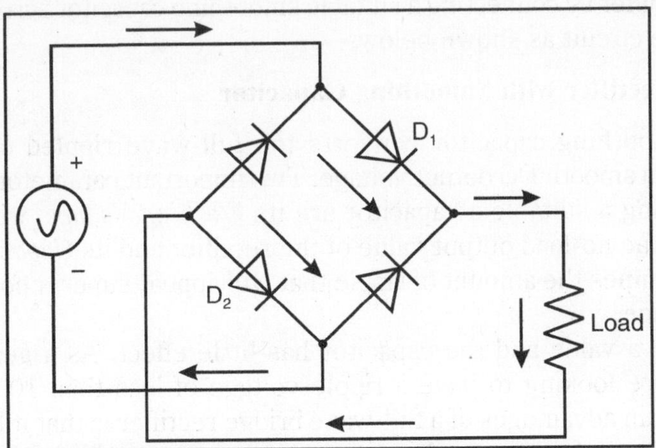

During the negative half cycle of the supply, diodes D_3 and D_4 conduct in series, but diodes D_1 and D_2 switch of as they are now reverse biased. The current flowing through the load is the same direction as before.

The Negative Half-cycle

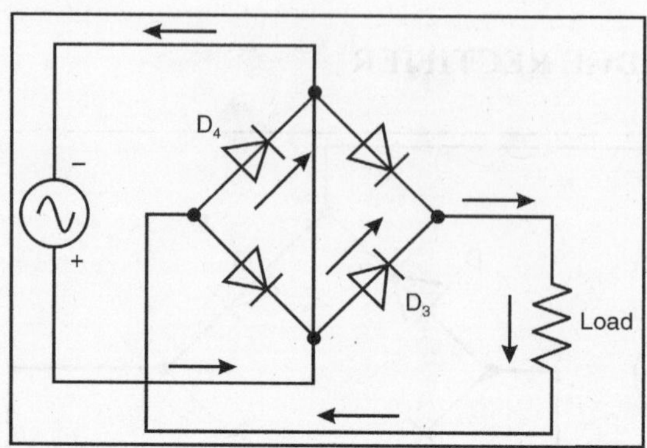

As the current flowing through the load is unidirectional, so the voltage developed across the load is also unidirectional the same as for the previous two diode full-wave rectifier, therefore the average DC voltage across the load is $0.637V_{max}$ and the ripple frequency is now twice the supply frequency (*e.g.* 100Hz for a 50Hz supply).

SMOOTHING CAPACITOR

We saw in the previous section that the single phase half-wave rectifier produces an output wave every half cycle and that it was not practical to use this type of circuit to produce a steady DC supply. The full-wave bridge rectifier however, gives us a greater mean DC value (0.637 Vmax) with less superimposed ripple while the output waveform is twice that of the frequency of the input supply frequency. We can therefore increase its average DC output level even higher by connecting a suitable smoothing capacitor across the output of the bridge circuit as shown below.

Full-wave Rectifier with Smoothing Capacitor

The smoothing capacitor converts the full-wave rippled output of the rectifier into a smooth DC output voltage. Two important parameters to consider when choosing a suitable a capacitor are its *Working Voltage*, which must be higher than the no-load output value of the rectifier and its *Capacitance Value*, which determines the amount of ripple that will appear superimposed on top of the DC voltage.

Too low a value and the capacitor has little effect. As a general rule of thumb, we are looking to have a ripple voltage of less than 100mV peak to peak. The main advantages of a full-wave bridge rectifier is that it has a smaller AC ripple value for a given load and a smaller reservoir or smoothing capacitor than an equivalent half-wave rectifier. Therefore, the fundamental frequency of the ripple voltage is twice that of the AC supply frequency (100Hz) where for the half-wave rectifier it is exactly equal to the supply frequency (50Hz).

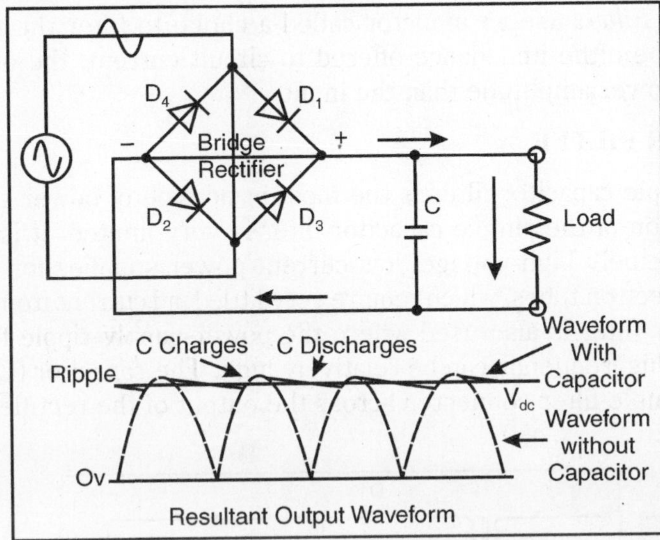

The amount of ripple voltage that is superimposed on top of the DC supply voltage by the diodes can be virtually eliminated by adding a much improved p-filter (pi-filter) to the output terminals of the bridge rectifier. This type of low-pass filter consists of two smoothing capacitors, usually of the same value and a choke or inductance across them to introduce a high impedance path to the alternating ripple component. Another more practical and cheaper alternative is to use a 3-terminal voltage regulator IC, such as a LM7805 which can reduce the ripple by more than 70dB (Datasheet) while delivering over 1amp of output current.

RIPPLE FACTOR

Ripple factor can be defined as the variation of the amplitude of DC (Direct current) due to improper filtering of AC power supply.

it can be measured by,

$$RF = v_{rms} / v_{dc}$$

INDUCTOR FILTER

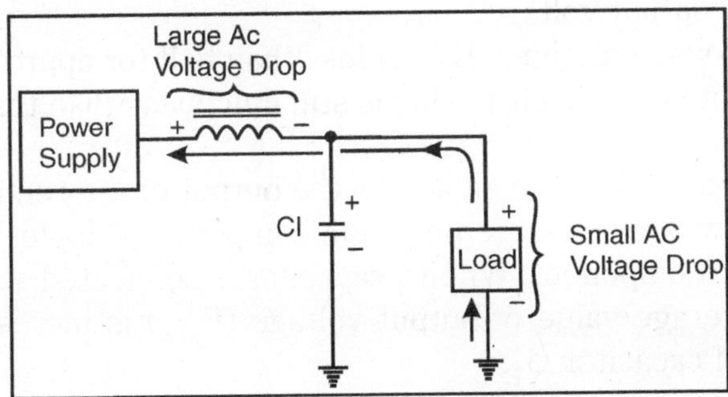

Inductor Filters use an inductor called a choke to filter the pulsating dc input. Because ofthe impedance offered to circuit current, the output of the filter is at a lower amplitude than the input

CAPACITOR FILTER

The simple capacitor filter is the most basic type of power supply filter. The application of the simple capacitor filter is very limited. It is sometimes used on extremely high-voltage, low-current power supplies for cathode-ray and similar electron tubes, which require very little load current from the supply. The capacitor filter is also used where the power-supply ripple frequency is not critical; this frequency can be relatively high. The capacitor (C_1) shown in figure is a simple filter connected across the output of the rectifier in parallel with the load.

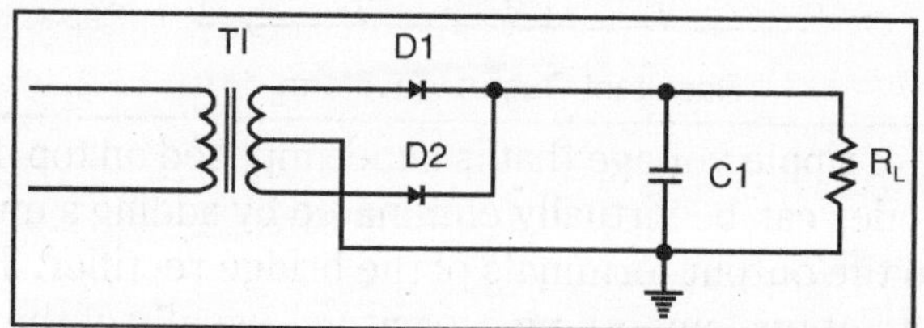

Fig. Full-wave Rectifier with a Capacitor Filter.

When this filter is used, the RC charge time of the filter capacitor (C_1) must be short and the RC discharge time must be long to eliminate ripple action. In other words, the capacitor must charge up fast, preferably with no discharge at all. Better filtering also results when the input frequency is high; therefore, the full-wave rectifier output is easier to filter than that of the half-wave rectifier because of its higher frequency.

For you to have a better understanding of the effect that filtering has on E_{avg}, a comparison of a rectifier circuit with a filter and one without a filter is illustrated in views A and B of figure above.

The output waveforms in figure represent the unfiltered and filtered outputs of the half-wave rectifier circuit. Current pulses flow through the load resistance (R_L) each time a diode conducts. The dashed line indicates the average value of output voltage.

For the half-wave rectifier, E_{avg} is less than half (or approximately 0.318) of the peak output voltage. This value is still much less than that of the applied voltage.

With no capacitor connected across the output of the rectifier circuit, the waveform in view A has a large pulsating component (ripple) compared with the average or dc component. When a capacitor is connected across the output (view B), the average value of output voltage (E_{avg}) is increased due to the filtering action of capacitor C_1.

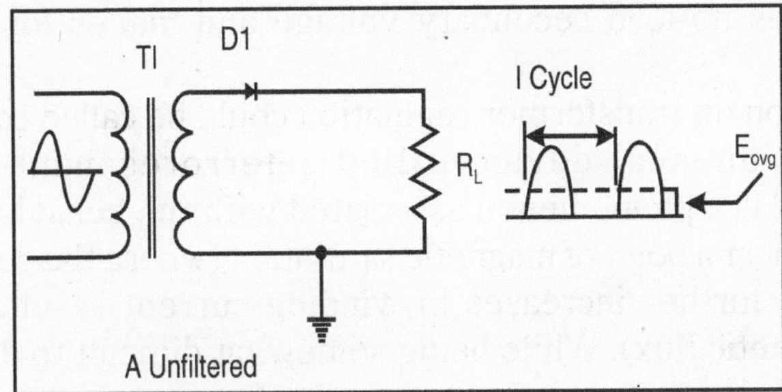

Fig. A. Half-wave Rectifier with and without Filtering.

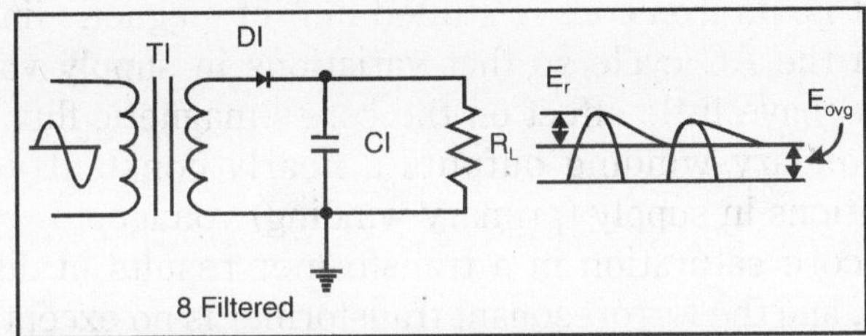

Fig. B. Half-wave Rectifier with and without Filtering.

The value of the capacitor is fairly large (several microfarads), thus it presents a relatively low reactance to the pulsating current and it stores a substantial charge.

LC FILTER

There are some applications, however, where poor regulation is actually desired. One such case is in discharge lighting, where a step-up transformer is required to initially generate a high voltage (necessary to"ignite" the lamps), then the voltage is expected to drop off once the lamp begins to draw current. This is because discharge lamps' voltage requirements tend to be much lower after a current has been established through the arc path. In this case, a step-up transformer with poor voltage regulation suffices nicely for the task of conditioning power to the lamp.

Another application is in current control for AC arc welders, which are nothing more than step-down transformers supplying low-voltage, high-current power for the welding process. A high voltage is desired to assist in"striking" the arc (getting it started), but like the discharge lamp, an arc doesn't require as much voltage to sustain itself once the air has been heated to the point of ionization.

Thus, a decrease of secondary voltage under high load current would be a good thing. Some arc welder designs provide arc current adjustment by means of a movable iron core in the transformer, cranked in or out of the winding assembly by the operator. Moving the iron slug away from the windings reduces the strength of magnetic coupling between the windings, which diminishes no-load secondary voltage and makes for poorer voltage regulation.

No exposition on transformer regulation could be called complete without mention of an unusual device called a ferroresonant transformer." Ferroresonance" is a phenomenon associated with the behaviour of iron cores while operating near a point of magnetic saturation (where the core is so strongly magnetized that further increases in winding current results in little or no increase in magnetic flux). While being somewhat difficult to describe without going deep into electromagnetic theory, the ferroresonant transformer is a power transformer engineered to operate in a condition of persistent core saturation. That is, its iron core is"stuffed full" of magnetic lines of flux for a large portion of the AC cycle so that variations in supply voltage (primary winding current) have little effect on the core's magnetic flux density, which means the secondary winding outputs a nearly constant voltage despite significant variations in supply (primary winding) voltage.

Normally, core saturation in a transformer results in distortion of the sinewave shape, and the ferroresonant transformer is no exception. To combat this side effect, ferroresonant transformers have an auxiliary secondary winding paralleled with one or more capacitors, forming a resonant circuit tuned to the power supply frequency. This"tank circuit" serves as a filter to reject harmonics created by the core saturation, and provides the added benefit of storing energy in the form of AC oscillations, which is available for sustaining output winding voltage for brief periods of input voltage loss (milliseconds' worth of time, but certainly better than nothing).

In addition to blocking harmonics created by the saturated core, this resonant circuit also"filters out" harmonic frequencies generated by nonlinear (switching) loads in the secondary winding circuit and any harmonics present in the source voltage, providing"clean" power to the load.

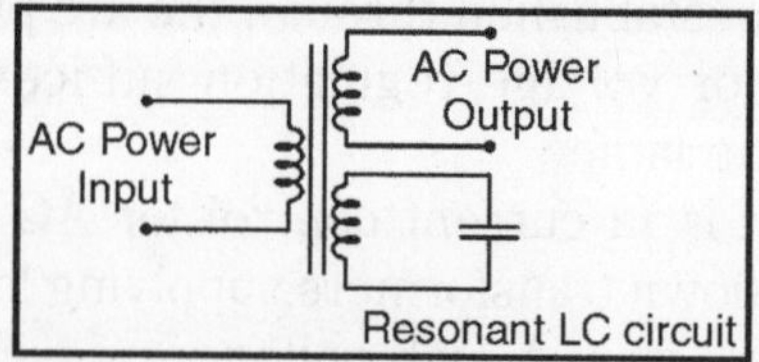

Fig. Ferroresonant Transformer Provides Voltage Regulation of the Output.

Ferroresonant transformers offer several features useful in AC power conditioning: constant output voltage given substantial variations in input voltage, harmonic filtering between the power source and the load, and the ability to"ride through" brief losses in power by keeping a reserve of energy in its resonant tank circuit. These transformers are also highly tolerant of excessive loading and transient (momentary) voltage surges. They are so tolerant, in fact, that some may be briefly paralleled with unsynchronized AC power sources, allowing a load to be switched from one source of power to another in a"make-before-break" fashion with no interruption of power on the secondary side!

Unfortunately, these devices have equally noteworthy disadvantages: they waste a lot of energy (due to hysteresis losses in the saturated core), generating significant heat in the process, and are intolerant of frequency variations, which means they don't work very well when powered by small engine-driven generators having poor speed regulation.

Voltages produced in the resonant winding/capacitor circuit tend to be very high, necessitating expensive capacitors and presenting the service technician with very dangerous working voltages. Some applications, though, may prioritize the ferroresonant transformer's advantages over its disadvantages. Semiconductor circuits exist to"condition" AC power as an alternative to ferroresonant devices, but none can compete with this transformer in terms of sheer simplicity.

CLC FILTER

A simple resistor provides a good example of both homogenous and non-homogeneous systems. If, the input to the system is the voltage across the resistor, *v(t)*, and the output from the system is the current through the resistor, *i(t)*, the system is homogeneous.

Ohm's law guarantees this; if the voltage is increased or decreased, there will be a corresponding increase or decrease in the current.

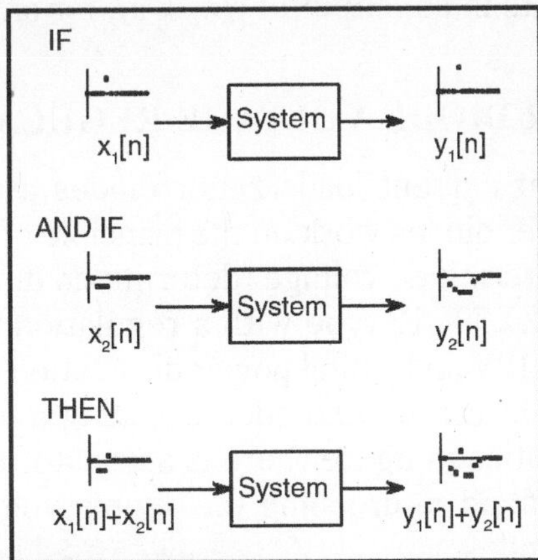

Now, consider another system where the input signal is the voltage across the resistor, *v(t)* but the output signal is the power being dissipated in the resistor, *p(t)*. Since power is proportional to the square of the voltage, if, the input signal is increased by a factor of two, the output signal is increase by a factor of four. This system is not homogeneous and therefore cannot be linear.

The property of additivity is illustrated in Figure; Consider a system where an input of $x_1[n]$ produces an output of $y_1[n]$. Further suppose that a different input, $x_2[n]$, produces another output, $y_2[n]$. The system is said to be additive, if an input of $x_1[n] + x_2[n]$ results in an output of $y_1[n] + y_2[n]$, for all possible input signals. In words, signals added at the input produce signals that are added at the output.

The important point is that added signals pass through the system without interacting. As an example, think about a telephone conversation with your Aunt Edna and Uncle Bernie. Aunt Edna begins a rather lengthy story about how well her radishes are doing this year. In the background, Uncle Bernie is yelling at the dog for having an accident in his favourite chair.

The two voice signals are added and electronically transmitted through the telephone network. Since this system is additive, the sound you hear is the sum of the two voices as they would sound if transmitted individually. You hear Edna and Bernie, not the creature, Ednabernie.

A good example of a nonadditive circuit is the mixer stage in a radio transmitter. Two signals are present: an audio signal that contains the voice or music, and a carrier wave that can propagate through space when applied to an antenna. The two signals are added and applied to a nonlinearity, such as a pn junction diode. This results in the signals merging to form a third signal, a modulated radio wave capable of carrying the information over great distances.

As shown in Figure; shift invariance means that a shift in the input signal will result in nothing more than an identical shift in the output signal. In more formal terms, if an input signal of $x[n]$ results in an output of $y[n]$, an input signal of $x[n + s]$ results in an output of $y[n + s]$, for any input signal and any constant's

ZENER DIODE VOLTAGE REGULATION

For relatively light current loads zener diodes are a cheap solution to voltage regulation. Zener diodes work on the principle of essentially a constant voltage drop at a predetermined voltage (determined during manufacture). An example is a Philips BZX79C12 type with a regulation range between 11.4V and 12.7V but typically 12V and a total power dissipation of 500 mW in a DO-35 package. The dissipation can be extended by using a series pass transistor, see power supplies. Notice in figure there is a resistor to miminmise current drawn but mainly as an aid to dropping the supply voltage and reducing the burden on the zener diodes.

In the second schematic of figure above we have three zener diodes in series providing voltages of 5V, 10V, 12V, 22V and 27V all from a 36V supply. This configuration is not necessarily recommended especially when the current being drawn is seriously mismatched between voltages. It is presented purely out of interest.

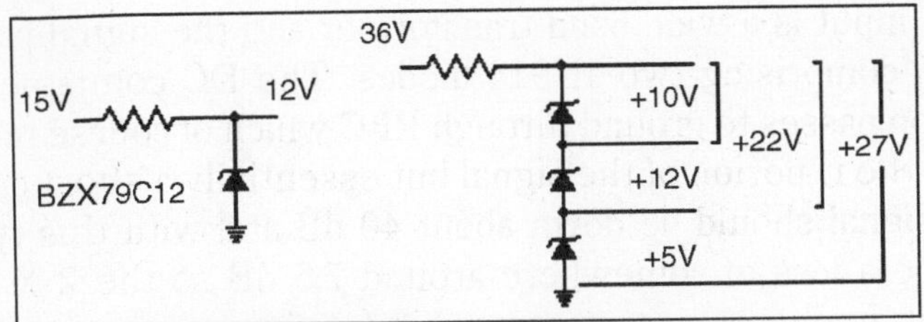

Fig. Zener Voltage Regulation Diodes

VARACTOR OR TUNING DIODES

These types of diodes work on the principle that all diodes exhibit some capacitance. Indeed the zener diode BZX79C12 quoted above has, according to the data book, a capacitance of 65–85 pF at 0V and measured at 1 Mhz. For AM Radio band applications a specific diode has been devised. The Philips BB212 in a TO-92 case is one such type. Each of the diodes has a capacitance of 500 - 620 pF at a reverse bias of 0.5V and <22 pF at 8V. This diode's capacitance ratio is quoted at 22.5:1 which could not be achieved easily if at all with an air variable capacitor. This type of diode is depicted in figure 1 above.

Several obvious advantages come immediately to mind, a small transistor type package, very low cost, ease of construction on a circuit board, can be mounted away from heat generating devices, frequency determining circuitry entirely dependent upon resistor values and ratios, DC voltage control can be either from frequency synthesiser circuits or perhaps a multi-turn potentiometer. Such a potentiometer aids band spreading and fine tuning if two potentiometers are used. The only real limitation is your imagination and the calculations involved.

DIODES AS FREQUENCY MULTIPLIERS

Just one more example of the versatility of diodes is the frquency doubling circuit depicted in Figure. Now if that looks a lot like the full wave rectifier from figure above you would be correct. That is why the ripple frequency for 50/60 Hz always comes out at 100/120 Hz.

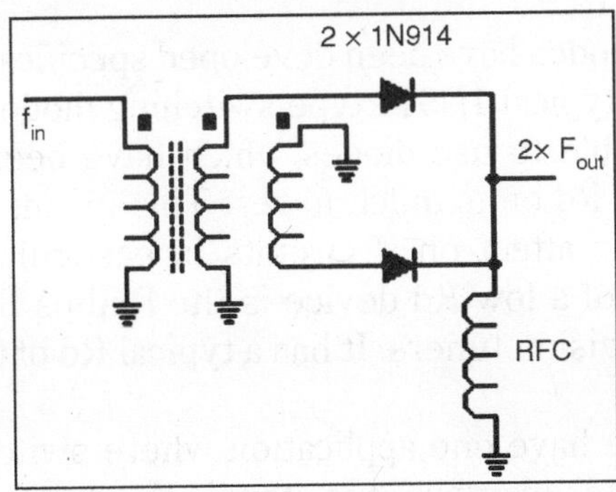

Fig. Diodes as Frequency Multipliers

Here the input is a wide band transformer and the signal passes to a full wave rectifier comprising two 1N914 diodes. The DC component caused by the rectification passes to ground through RFC which of course presents a high impedance to the rf porion of the signal but essentially a short circuit for DC. The original signal should be down about 40 dB and with this type of circuit there would be a loss of somewhere around 7.5 dB so the 2 X signal would require further amplification to restore that loss.

DIODES AS MIXERS

With some subtle re-arrangement to figure we can get the circuit to function as a two diode frequency mixer. Note that there are other diode arrangements as well in this application.

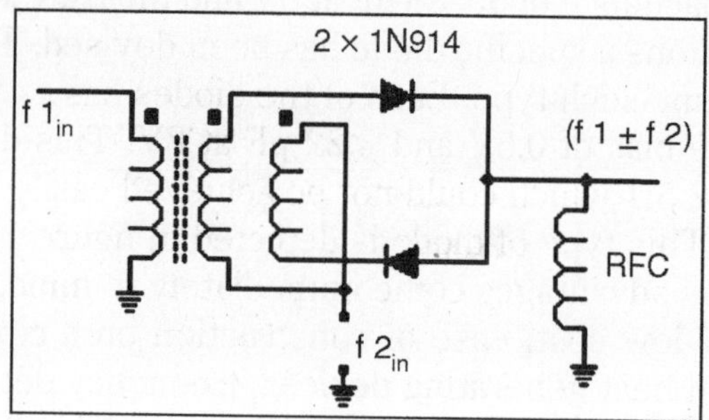

Fig. Diodes as Frequency Mixers

The diodes here act as switches and it can be mathematically shown that only the sum and difference signals will result. For example, if F1 was 5 Mhz and F2 was 3 Mhz then the sum and difference signals from the diodes would be 8 Mhz and 2 Mhz. None of the original signals appear at the output and this is a most important property of using diodes as mixers.

It should be noted that although 1N914 diodes are depicted you would normally use hot carrier diodes in any serious application and the diodes need to be well matched.

Applications of Switching Diodes

Similar types of diodes have been developed specifically for band switching purposes. Although a typical 1N914 type switching diode can be used for such purposes it is preferable to use diodes which have been optimised for such purposes because the Rd on is much lower. This means the diode resistance Rd can have a serious affect on rf circuits in particular the "Q" of a tuned circuit. One example of a low Rd device is the Philips BA482 diode used for band switching in television tuners. It has a typical Rd of 0.4 ohms at a forward current of 10 mA.

In figureabove we have one application where switching diodes operate. All diodes serve to switch in or out capacitors in the diagram which is presented here just to illustrate one single application of switching diodes, many, many more applications exist. Again the limit is your imagination.

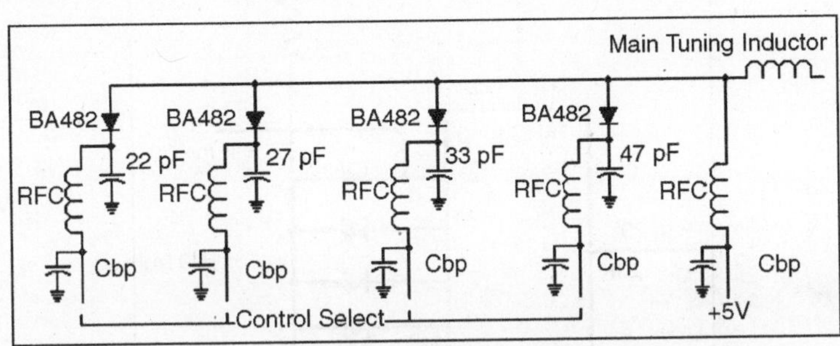

Fig. Applications of Switching Diodes

The switching diodes in figure switch in or out successively higher values of capacitors as each control select line is "grounded". The voltage from the +5V feed line proceeds through the diode at DC thus opening the diode and making it appear "transparent" for rf purposes.

The capacitor with the value attached is then "switched" into circuit. Other components marked RFC and Cbp are chokes and bypass capacitors for "clean" switching. The bypass capacitors and choke values would be determined by the frequency of operation.

We could just have easily have switched inductors instead of capacitors. Note why Rd is quite important on overall circuit performance. If we were using inductors the diode resistance Rd would have a significant affect on inductor "Q" which in turn would affect filter performance, if it was in fact an LC filter application.

SWITCHING DIODES IN LOGIC CIRCUITS

If you you completed the tutorial on digital basics you should be aware of binary numbers. There are a whole range of digital building blocks available and just by way of one illustration of using diodes we have presented the 74HC4040 twelve stage binary ripple counter (there are others with varying number of stages). In the schematic of figure we have this counter which divides by successive division of two for twelve stages. Initially because there is no voltage drop across the resistor a high appears on all anodes as well as on pin 4 the master reset causing the counter to reset forcing all outputs low and in turn a voltage drop across each diode and across the resistor and a low on reset.

Progressively each of the outputs change from low to high for a certain period of time and without unduly complicating matters when all outputs as selected by our diode combination (in this particular case 1 + 2 + 32 + 64 = 99) are simultaneously high the voltage drop across the resistor will cease and cause pin 11 (reset which was formerly low) to go high and reset all the internal ripple counters.

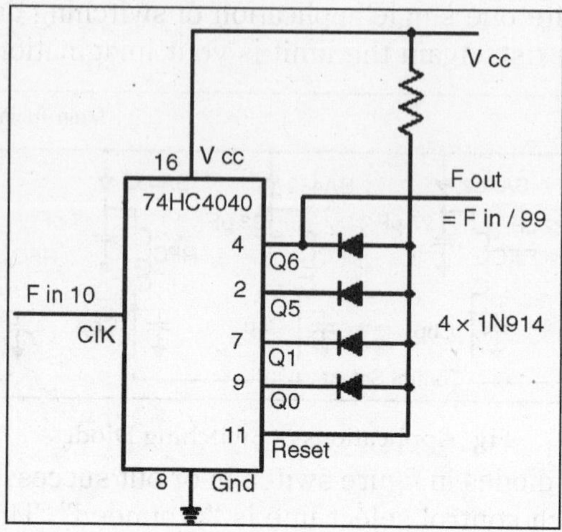

Fig. Applications of Switching Diodes in Digital Logic Circuits

At the same time pin 4 changes state also with reset. It can been shown this happens once every 99 periods. Simply by placing diodes on the right outputs we can select to divide by any number up to 4095 using this particular counter.

LIGHT-EMITTING-DIODES OR LED'S

Many circuits use a led as a visual indicator of some sort even if only as an indicator of power supply being turned on. A sample calculation of the dropping resistor is included in figure.

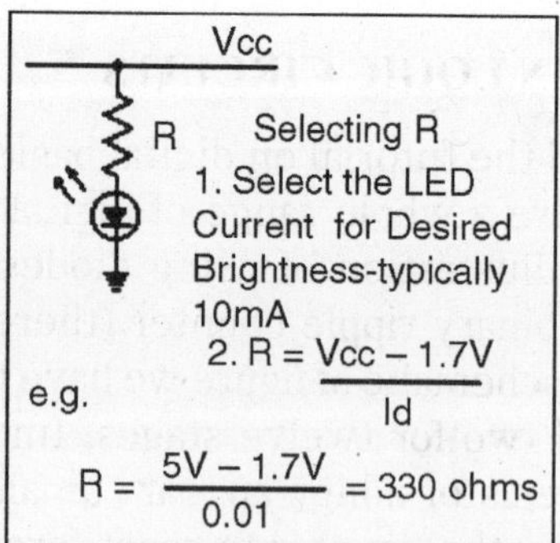

Fig. Connecting Light Emitting Diodes (LED's) to Supply

Most leds operate at 1.7V although this is not always the case and it is wise to check. The dropping resistor is simply the net of supply voltage minus the 1.7V led voltage then divided by the led brightness current expressed as "amps" (ohms law). Note the orientation of both cathode and anode with respect to the ground end and the supply end. Usually with a led the longer lead is the anode.

RECTIFIERS

The first diode in figure below is a semiconductor diode which could be a small signal diode of the 1N914 type commonly used in switching applications, a rectifying diode of the 1N4004 (400V 1A) type or even one of the high power, high current stud mounting types.

You will notice the straight bar end has the letter "k", this denotes the "cathode" while the "a" denotes anode. Current can only flow from anode to cathode and not in the reverse direction, hence the "arrow" appearance. This is one very important property of diodes.

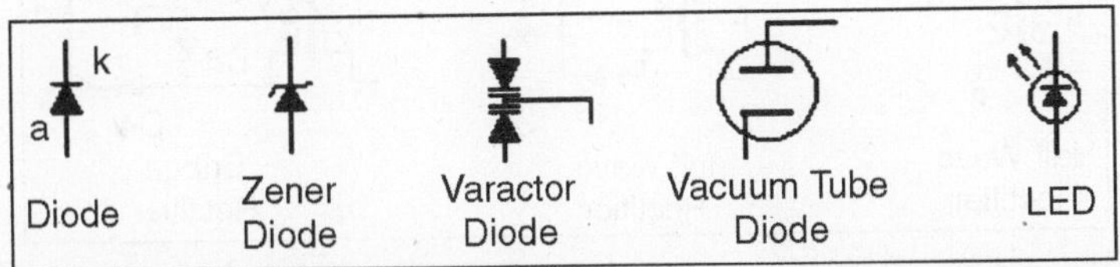

The second of the diodes is a zener diode which are fairly popular for the voltage regulation of low current power supplies. Whilst it is possible to obtain high current zener diodes, most regulation today is done electronically with the use of dedicated integrated circuits and pass transistors.

The next of the diodes in the schematic is a varactor or tuning diode. Depicted here is actually two varactor diodes mounted back to back with the DC control voltage applied at the common junction of the cathodes. These cathodes have the double bar appearance of capacitors to indicate a varactor diode. When a DC control voltage is applied to the common junction of the cathodes, the capacitance exhibited by the diodes (all diodes and transistors exhibit some degree of capacitance) will vary in accordance with the applied voltage.

A typical example of a varactor diode would be the Philips BB204G tuning diodes of which there are two enscapsulated in a TO-92 transistor package. At a reverse voltage Vr (cathode to anode) of 20V each diode has a capacitance of about 16 pF and at Vr of 3V this capacitance has altered to about 36 pF. Being low cost diodes, tuning diodes have virtually replaced air variable capacitors in radio applications today.

The next diode is the simplest form of vacuum tube or valve. It simply has the old cathode and anode. These terms were passed on to modern solid state devices. Vacuum tube diodes are mainly only of interest to restorers and tube enthusiasts. The last diode depicted is of course a light emitting diode or LED. A led actually doesn't emit as much light as it first appears, a single LED has a plastic lens installed over it and this concentrates the amount of light. Seven LED's can be arranged in a bar fashion called a seven segment LED display and when decoded properly can display the numbers 0–9 as well as the letters A to F.

RECTIFYING DIODES

The principal early application of diodes was in rectifying 50/60 Hz AC mains to raw DC which was later smoothed by choke transformers and/or capacitors. This procedure is still carried out today and a number of rectifying schemes for diodes have evolved, half wave, full wave and bridge rectifiers.

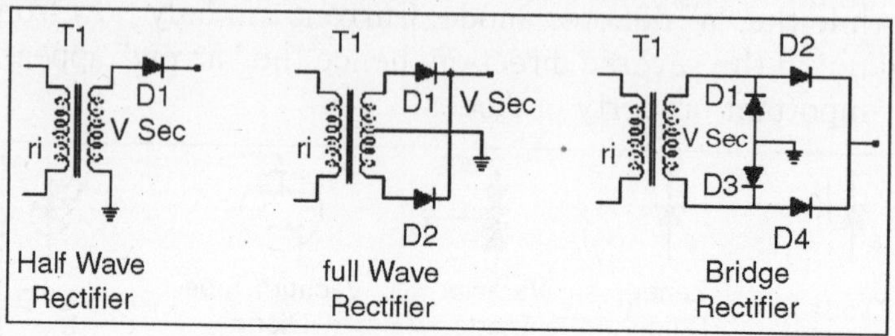

Fig. Rectifying Diodes

As examples in these applications the half wave rectifier passes only the positive half of successive cycles to the output filter through D1. During the negative part of the cycle D1 does not conduct and no current flows to the load.

In the full wave application it essentially is two half wave rectifiers combined and because the transformer secondary is centre tapped, D1 conducts on the positive half of the cycle while D2 conducts on the negative part of the cycle. Both add together. This is more efficient. The full wave bridge rectifier operates essentially the same as the full wave rectifier but does not require a cetre tapped transformer. Further discussion may be seen on the topic power supplies. A further application of rectifying diodes is in the conversion or detection of rf modulated signals to audio frequencies. Typical examples are am modulated signals being detected and early detection schemes for fm also used diodes for detecting modulation.

DIODE AS A RECTIFIER

THE POWER DIODE

We saw that a semiconductor signal diode will only conduct current in one direction from its anode to its cathode (forward direction), but not in the reverse direction acting a bit like an electrical one way valve. A widely used application of this feature is in the conversion of an alternating voltage (AC) into a continuous voltage (DC). In other words, Rectification.

But small signal diodes can also be used as Rectifiers in low-power, low current (less than 1-amp) rectifiers or applications, but were larger forward bias currents or higher reverse bias blocking voltages are involved the PN junction of a small signal diode would eventually overheat and melt so larger more robust Power Diodes are used instead. The power semiconductor diode, known simply as the Power Diode, has a much larger PN junction area compared to its smaller signal diode cousin, resulting in a high forward current capability of up to several hundred amps (KA) and a reverse blocking voltage of up to several thousand volts (KV).

Since, the power diode has a large PN junction, it is not suitable for high frequency applications above 1MHz, but special and expensive high frequency, high current diodes are available. For high frequency rectifier applications Schottky Diodes are generally used because of their short reverse recovery time and low voltage drop in their forward bias condition.

Power diodes provide uncontrolled rectification of power and are used in applications such as battery charging and DC power supplies as well as AC rectifiers and inverters. Due to their high current and voltage characteristics they can also be used as free-wheeling diodes and snubber networks.

Power diodes are designed to have a forward "ON" resistance of fractions of an Ohm while their reverse blocking resistance is in the mega-Ohms range. Some of the larger value power diodes are designed to be "stud mounted" onto heatsinks reducing their thermal resistance to between 0.1 to 1° C/Watt.

If an alternating voltage is applied across a power diode, during the positive half cycle the diode will conduct passing current and during the negative half cycle the diode will not conduct blocking the flow of current. Then conduction through the power diode only occurs during the positive half cycle and is therefore unidirectional *i.e.* DC as shown.

Power Diode Rectifier

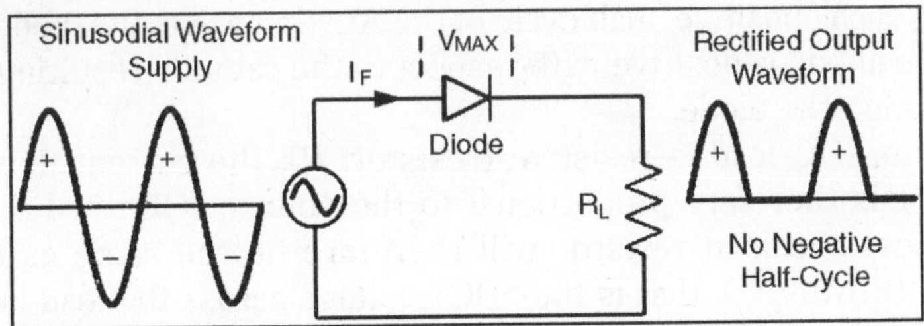

Power diodes can be used individually as above or connected together to produce a variety of rectifier circuits such as "Half-Wave", "Full-Wave" or as "Bridge Rectifiers". Each type of rectifier circuit can be classed as either uncontrolled, half-controlled or fully controlled were an uncontrolled rectifier uses only power diodes, a fully controlled rectifier uses thyristors (SCRs) and a half controlled rectifier is a mixture of both diodes and thyristors.

The most commonly used individual power diode for basic electronics applications is the general purpose 1N400x Series Glass Passivated type rectifying diode with standard ratings of continuous forward rectified current of 1.0 amp and reverse blocking voltage ratings from 50v for the 1N4001 up to

1000v for the 1N4007, with the small 1N4007GP being the most popular for general purpose mains voltage rectification.

Half Wave Rectification

A rectifier is a circuit which converts the *Alternating Current* (AC) input power into a *Direct Current*(DC) output power. The input power supply may be either a single-phase or a multi-phase supply with the simplest of all the rectifier circuits being that of the Half Wave Rectifier.

The power diode in a half wave rectifier circuit passes just one half of each complete sine wave of the AC supply in order to convert it into a DC supply. Then this type of circuit is called a "half-wave" rectifier because it passes only half of the incoming AC power supply as shown below.

Half Wave Rectifier Circuit

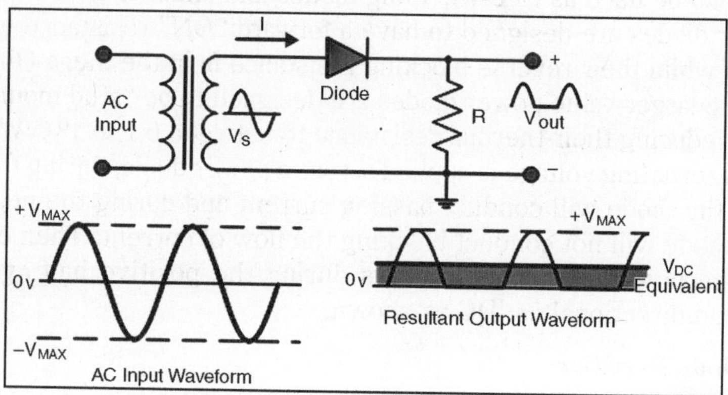

During each "positive" half cycle of the AC sine wave, the diode is *forward biased* as the anode is positive with respect to the cathode resulting in current flowing through the diode.

Since, the DC load is resistive (resistor, *R*), the current flowing in the load resistor is therefore proportional to the voltage (Ohm´s Law), and the voltage across the load resistor will therefore be the same as the supply voltage, V_s (minus V_f), that is the "DC" voltage across the load is sinusoidal for the first half cycle only so $V_{out} = V_s$.

During each "negative" half cycle of the AC sinusoidal input waveform, the diode is *reverse biased* as the anode is negative with respect to the cathode.

Therefore, no current flows through the diode or circuit. Then in the negative half cycle of the supply, no current flows in the load resistor as no voltage appears across it so therefore, $V_{out} = 0$.

The current on the DC side of the circuit flows in one direction only making the circuit Unidirectional. As the load resistor receives from the diode a positive half of the waveform, zero volts, a positive half of the waveform, zero volts, etc, the value of this irregular voltage would be equal in value to an equivalent

DC voltage of $0.318 \times V_{max}$ of the input sinusoidal waveform or $0.45 \times V_{rms}$ of the input sinusoidal waveform.

Then the equivalent DC voltage, V_{DC} across the load resistor is calculated as follows.

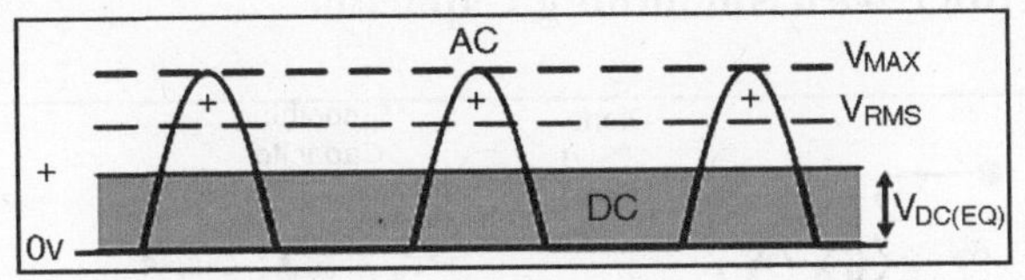

$$V_{d.c.} = \frac{V_{max}}{\pi} = 0.318V_{max} = 0.45V_s$$

Where V_{max} is the maximum or peak voltage value of the AC sinusoidal supply, and V_S is the RMS (Root Mean Squared) value of the supply.

Example of Power Diode

Calculate the voltage across V_{DC} and the current I_{DC}, flowing through a 100Ω resistor connected to a 240 V_{rms} single phase half-wave rectifier as shown above. Also calculate the DC power consumed by the load.

$$V_{max} = V_{rms} \times 1.414,$$

or,

$$V_{rms} = V_{max} \times 0.7071$$

$$V_{DC} = 0.45\ V_{rms} = 0.45 \times 240 = 108\text{Volts}$$

or,

$$V_{DC} = 0.318\ V_{max} = 0.318 \times (240 \times 1.414) = 108\text{Volts}$$

$$I_{DC} = \frac{V_{DC}}{R} = \frac{108V}{100\Omega} = 1.08\text{ Amps}$$

$$\text{Power} = I^2R = 1.08^2 \times 100 = 116\text{watts}$$

During the rectification process the resultant output DC voltage and current are therefore both "on" and "off" during every cycle. As the voltage across the load resistor is only present during the positive half of the cycle (50 per ecnt of the input waveform), this results in a low average DC value being supplied to the load.

The variation of the rectified output waveform between this "on" and "off" condition produces a waveform which has large amounts of "ripple" which is an undesirable feature. The resultant DC ripple has a frequency that is equal to that of the AC supply frequency.

Very often when rectifying an alternating voltage we wish to produce a "steady" and continuous DC voltage free from any voltage variations or ripple.

One way of doing this is to connect a large value Capacitor across the output voltage terminals in parallel with the load resistor as shown below. This type of capacitor is known commonly as a "Reservoir" or *Smoothing Capacitor*.

Half-wave Rectifier with Smoothing Capacitor

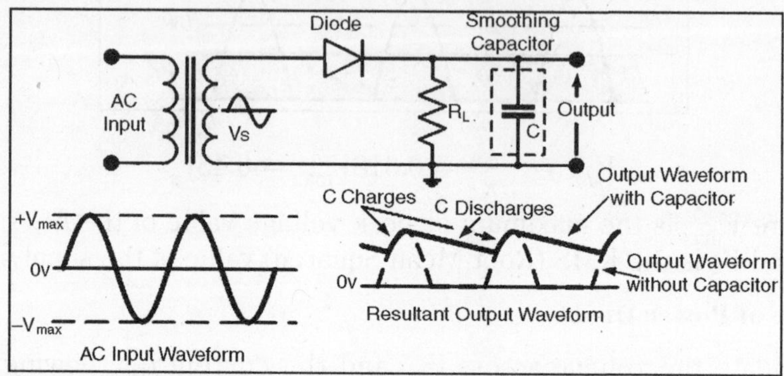

When rectification is used to provide a direct voltage (DC) power supply from an alternating (AC) source, the amount of ripple voltage can be further reduced by using larger value capacitors but there are limits both on cost and size to the types of smoothing capacitors used.

For a given capacitor value, a greater load current (smaller load resistance) will discharge the capacitor more quickly (RC Time Constant) and so increases the ripple obtained. Then for single phase, half-wave rectifier circuit using a power diode it is not very practical to try and reduce the ripple voltage by capacitor smoothing alone. In this instance it would be more practical to use "Full-wave Rectification" instead.

In practice, the half-wave rectifier is used most often in low-power applications because of their major disadvantages being. The output amplitude is less than the input amplitude, there is no output during the negative half cycle so half the power is wasted and the output is pulsed DC resulting in excessive ripple.

DIFFERENT TYPES OF DIODE

There are many different types of diodes that are available for use in electronics design.

Different semiconductor diode types can be used to perform different functions as a result of the properties of these different diode types.

Semiconductor diodes can be used for many applications. The basic application is obviously to rectify waveforms. This can be used within power supplies or within radio detectors. Signal diodes can also be used for many other functions within circuits where the "one way" effect of a diode may be required.

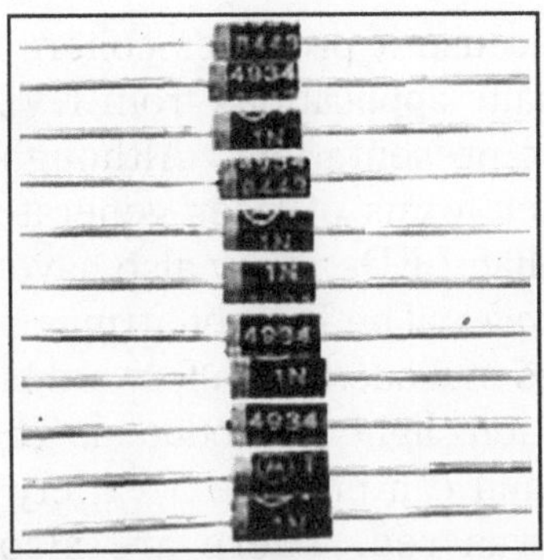

Fig. Leaded Semiconductor Diodes

Diodes are not just used as rectifiers, as various other types of diode can be used in many other applications. Some other different types of diodes include: light emitting diodes, photo-diodes, laser diodes.

Many of the different types of diodes mentioned below have further pages providing in-depth information about them including their structures, method of operation, how they may be used in circuits, and precautions and tips for using them in electronics design.

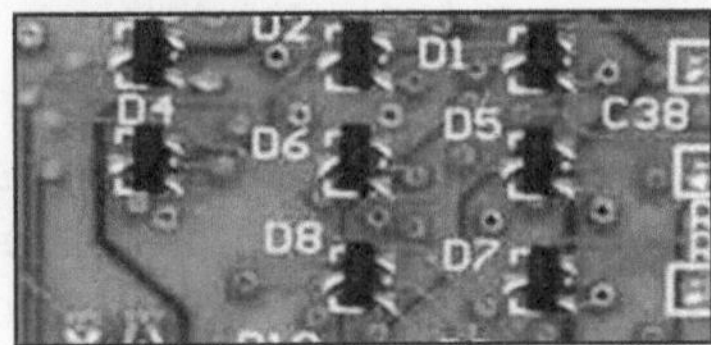

Fig. SMT Diodes on a PCB

TYPES OF DIODES

It is sometimes useful to summarise the different types of diode that are available. Some of the categories may overlap, but the various definitions may help to narrow the field down and provide an overview of the different diode types that are available.

- *Backward Diode:* This type of diode is sometimes also called the back diode. Although not widely used, it is a form of PN junction diode that is very similar to the tunnel diode in its operation. It finds a few specialist applications where its particular properties can be used.
- *BARITT Diode:* This form of diode gains its name from the words Barrier Injection Transit Time diode. It is used in microwave applications and bears many similarities to the more widely used IMPATT diode.
- *Gunn Diode:* Although not a diode in the form of a PN junction, this type of diode is a semiconductor device that has two terminals. It is generally used for generating microwave signals.
- *Laser Diode:* This type of diode is not the same as the ordinary light

emitting diode because it produces coherent light. Laser diodes are widely used in many applications from DVD and CD drives to laser light pointers for presentations. Although laser diodes are much cheaper than other forms of laser generator, they are considerably more expensive than LEDs. They also have a limited life.

- *Light Emitting Diodes:* The light emitting diode or LED is one of the most popular types of diode. When forward biased with current flowing through the junction, light is produced. The diodes use component semiconductors, and can produce a variety of colours, although the original colour was red. There are also very many new LED developments that are changing the way displays can be used and manufactured. High output LEDs and OLEDs are two examples.
- *Photodiode:* The photo-diode is used for detecting light. It is found that when light strikes a PN junction it can create electrons and holes. Typically photo-diodes are operated under reverse bias conditions where even small amounts of current flow resulting from the light can be easily detected. Photo-diodes can also be used to generate electricity. For some applications, PIN diodes work very well as photodetectors.
- *PIN Diode:* This type of diode is typified by its construction. It has the standard P-type and N-type areas, but between them there is an area of Intrinsic semiconductor which has no doping. The area of the intrinsic semiconductor has the effect of increasing the area of the depletion region which can be useful for switching applications as well as for use in photodiodes, etc.
- *PN Junction:* The standard PN junction may be thought of as the normal or standard type of diode in use today. These diodes can come as small signal types for use in radio frequency, or other low current applications which may be termed as signal diodes. Other types may be intended for high current and high voltage applications and are normally termed rectifier diodes.

Semiconductor diodes are widely used throughout all areas of the electronics industry from electronics design through to production and repair. The semiconductor diode is very versatile, and there are very many variants and different types of diode that enable all the variety of different applications to be met. The different diode types of types of diodes include those for small signal applications, high current and voltage as well as different types of diodes for light emission and detection as well as types for low forward voltage drops, and types to give variable capacitance. In addition to this there are a number of diode types that are used for microwave applications.

Bibliography

K.D. Abhyankar and A.W. Joshi An Overview of Basic Theoretical Physics, Universities Press, 2009.

Mrinal Chakraborty: Encyclopaedia of Theoretical Physics, Anmol Pub, 2011.

A.K. Jha : *A Textbook of Applied Physics*, I.K. International Publication, Delhi, 2009.

A.S. Vasudeva : *A Textbook of Engineering Physics* , S. Chand Publisher, Delhi, 2008.

Ajit Kumar Sharma : *Textbook of Modern Physics* : , Discovery Publishing, Delhi, 2011.

Becchi : *Introduction to the Basic Concepts of Modern Physics*, Springer, Paperback, 1998

D R Brown : *Advance Dictionary of Physics*, Ivy Publication, 2007.

G K Bose : *A-Z Molecular Physics*, Centrum Press, Delhi, 2009.

G. Chatwal : *A Textbook of Objective Modern Physics* , Wisdom Press, Delhil, 2012.

Gaylon S. Campbell and John M. Norman : *An Introduction to Environmental Biophysics*, Springer Publisher, 2009.

Gnadig : *Physics Problems : with Hints and Solutions*, Cambridge University Press, Paperback, 1998.

Harish Parthasarathy : *Advanced Engineering Physics*, Ane Books Pvt. Ltd., 2009.

Harish Parthasarathy : *Advanced Signal Analysis and its Applications to Mathematical Physics*, I.K. International, Delhi, 2009.

J.G. Arapura : *Approaches to Personhood in Indian Thought : Essays in Descriptive Metaphysics*, Satguru Publisher, 1998.

J.K. Mathews : *Applied Physics*, Pacific Publisher, Delhi, 2011.

Jasprit Singh : *Modern Physics for Engineers*, Wiley India, 2011.

John D Miller : *A Handbook on Nanophysics*, Dominant Publication, Delhi, 2008.

John F. James : *A Students Guide to Fourier Transforms : With Applications in Physics and Engineering*, Cambridge University Press, 1998.

Jyoti Kumar Roshan : *Advanced Biophysics*, Anmol Publication, Delhi, 2008.

K.P. Das : *An Introduction to Biophysics*, Cyber Tech Publisher, Delhi, 2012.

I

J

K

L

M

N

O

P

Q

R

S

T

U

V

W

X

Z